Genetic Conditions Affecting the Skeleton: Congenital, Idiopathic Scoliosis and Arthrogryposis

Genetic Conditions Affecting the Skeleton: Congenital, Idiopathic Scoliosis and Arthrogryposis

Editors

Philip Giampietro
Nancy Hadley-Miller
Cathy L. Raggio

MDPI • Basel • Beijing • Wuhan • Barcelona • Belgrade • Manchester • Tokyo • Cluj • Tianjin

Editors
Philip Giampietro
University of Illinois-Chicago
USA

Nancy Hadley-Miller
Children's Hospital Colorado
Anschutz Medical Campus
USA

Cathy L. Raggio
Hospital for Special Surgery
USA

Editorial Office
MDPI
St. Alban-Anlage 66
4052 Basel, Switzerland

This is a reprint of articles from the Special Issue published online in the open access journal *Genes* (ISSN 2073-4425) (available at: https://www.mdpi.com/journal/genes/special_issues/Genetic_Skeleton).

For citation purposes, cite each article independently as indicated on the article page online and as indicated below:

LastName, A.A.; LastName, B.B.; LastName, C.C. Article Title. *Journal Name* **Year**, *Volume Number*, Page Range.

ISBN 978-3-0365-5975-9 (Hbk)
ISBN 978-3-0365-5976-6 (PDF)

Contents

Editorial

Overview of Gene Special Issue "Genetic Conditions Affecting the Skeleton: Congenital, Idiopathic Scoliosis and Arthrogryposis"

Philip F. Giampietro [1,*], **Nancy Hadley-Miller** [2] and **Cathy L. Raggio** [3]

[1]   Department of Pediatrics, University of Illinois-Chicago, Chicago, IL 60612, USA
[2]   Musculoskeletal Research Center, Children's Hospital Colorado, Aurora, CO 80045, USA; nancy.miller@childrenscolorado.org
[3]   Hospital for Special Surgery, New York, NY 10021, USA; raggioc@hss.edu
*   Correspondence: philipg@uic.edu

Citation: Giampietro, P.F.; Hadley-Miller, N.; Raggio, C.L. Overview of Gene Special Issue "Genetic Conditions Affecting the Skeleton: Congenital, Idiopathic Scoliosis and Arthrogryposis". *Genes* **2022**, *13*, 1194. https://doi.org/10.3390/genes13071194

Received: 7 May 2022
Accepted: 17 May 2022
Published: 4 July 2022

Publisher's Note: MDPI stays neutral with regard to jurisdictional claims in published maps and institutional affiliations.

Introduction

In this Special Issue of *Genes* entitled "Genetic Conditions Affecting the Skeleton: Congenital, Idiopathic Scoliosis and Arthrogryposis", evidence is presented which suggests that congenital, idiopathic scoliosis, and arthrogryposis share similar overlapping, but also distinct etiopathogenic mechanisms, including connective tissue and neuromuscular mechanisms. Congenital scoliosis (CS) is defined by the presence of an abnormal spinal curvature due to an underlying vertebral bony malformation (VM). Idiopathic scoliosis (IS) is defined by the presence of an abnormal structural spinal curvature of ≥10 degrees in the sagittal plane in the absence of an underlying VM. Arthrogryposis is defined by the presence of congenital contractures in two or more joints of the appendicular skeleton. All three conditions have complex genetic causes. This Special Issue highlights the complex nature of these conditions and current concepts in our approach to better understand their genetics.

In 2007, Yves Cotrel postulated genetic, biomechanic, neurological, oto-rhino-laryngology, molecular biology, endocrinology, neurophysiology, biochemistry, sensory physiology, and anatomical pathological mechanisms for the occurrence of IS [1]. GWAS studies have confirmed the complexity of IS and the existence of low-penetrance genes associated with IS. Genetic loci in muscle developmental genes (*LBX1* and *BNC2*) and extracellular matrix genes (*FBN1*) have been identified as potential predisposition genes for IS. Additional evidence for cilia-related genes in the development of AIS has been presented in this Special Issue by Mathieu et al. [2] with autosomal dominant transmitted *POC5* coding gene variants occurring at a higher frequency in patients with AIS as compared with the general population. *POC5* plays an important role in centriole assembly.

Terhune et al. [3] hypothesize that the development of IS may be due to damaging variants within a specific set of pathways or molecular classes, rather than being driven by just a few select 'AIS genes'. Performing whole-exome sequence analysis on 23 multigeneration families, the authors identified an enrichment of variants in cytoskeletal- and extracellular-matrix-related processes. One hundred and thirty-two genes were shared by two or more families. Ten genes were shared by >4 families, although no genes were shared by all, supporting the polygenic nature of IS. The combination of the inability to relate specific genetic variants to IS development suggests a potential role of environmental and/or epigenetic factors in the etiology of IS.

Epigenetic causes need to be considered as modulating factors for curve progression in IS. Using discordant monozygotic twin pairs for IS, Carry et al. [4] identified 57 differentially methylated regions (DMRs) where hyper- or hypo-methylation was consistent across the region and 28 DMRs had a consistent association with curve severity. Twenty-one DMRs were correlated with bone methylation, including WNT10A (WNT signaling) and NPY (regulator of bone and energy homeostasis).

In this Special Issue, Janusz et al. [5] studied gender differences associated with curve progression in IS. Genetic differences in *ESR1* and *ESR2* have been hypothesized to be a contributing factor in some cases for AIS. Using muscle tissue obtained from the paravertebral muscles of girls with AIS, differences in DNA methylation between patients with a Cobb angle $\leq 70°$ and $>70°$ in the T-DMR2 region at the concave side of the curvature were identified, suggesting that *ESR1* and *ESR2* DNA methylation may be associated with AIS severity.

The relationship between CS and IS remains an interesting area for hypothesis generation and future research. One hypothesis is that the two conditions may have similar genetic mechanisms. In zebrafish, the depletion of ptk7 results in both congenital and idiopathic scoliosis. In this edition, Su and colleagues [6] provide evidence that mutations in *PTK7*, which functions in canonical and non-canonical Wnt signaling, are associated with the development of both CS and IS. Preliminary evidence is presented that gradations in expression may correlate with CS and IS, with more severe expression being associated with CS and diminished expression being associated with IS. Apart from the 16p11.2 microdeletion, which contains the *TBX6* gene associated with CS, Lai and colleagues studied copy number variants in a cohort of 67 patients with CS [7]. Some of the CNV contained genes such as *DHX40, NBPF20, RASA2*, and *MYSM1*, which have been found to be associated with syndromes characterized by scoliosis or thought to play a role in bone/spine development. They hypothesize that CNVs in consort with single nucleotide variants, in addition to somatic mutation and environmentally mediated effects, contribute to the occurrence of CS. Wang et al. [8]. identified several hypomorphic sequence variants in *FGFR1*, a transmembrane cytokine receptor involved in gastrulation, organ specification, and the patterning of various tissues in patients with mild spine and heart defects with CS.

The search for genes, pathways, and modifier genes for all three conditions is ongoing and may lead to potential therapies. From a total of 908 genes linked with scoliosis and 444 genes linked with AMC identified in the literature, 227 genes were associated with AMC-SC by Latypova and colleagues [9]. This group of genes is associated with a wide range of cellular functions, including transcription regulation, transmembrane receptor, growth factor, and ion channels. Further research using genomics and animal models to identify prognostic factors and therapeutic targets for AMC needs to be implemented. Dahan et al. [10] described the clinical and genetic heterogeneity in patients with multiple pterygium syndrome and scoliosis, harboring mutations in *CHNRG* and *MYH3*.

Due to the overlap in pathophysiology for arthrogryposis and IS, overlapping treatment mechanisms for both conditions can be conceptualized. *MYH3*-associated distal arthrogryposis in humans may be modeled in smyhc1 zebrafish which develop scoliosis as discussed by Whittle and colleagues [11]. Para-aminoblebbistatin inhibits myosin heavy chain ATPase activity, which chemically relaxes the skeletal muscle and prevents the curved phenotype of treated smyhc1 mutant fish. It is anticipated that our continued developing understanding of the genetic causes of these conditions will facilitate the development of targeted therapeutic interventions.

Author Contributions: Conceptualization: P.F.G., N.H.-M. and C.L.R.; Writing-draft preparation: P.F.G.; Review of manuscript: N.H.-M. and C.L.R. All authors have read and agreed to the published version of the manuscript.

Funding: This research received no external funding.

Institutional Review Board Statement: Not applicable.

Informed Consent Statement: Not applicable.

Data Availability Statement: Not applicable.

Conflicts of Interest: The authors have no conflict of interest.

References

1. Tang, N.L.S.; Dobbs, M.B.; Gurnett, C.A.; Qiu, Y.; Lam, T.P.; Cheng, J.C.Y.; Hadley-Miller, N. A Decade in Review after Idiopathic Scoliosis Was First Called a Complex Trait—A Tribute to the Late Dr. Yves Cotrel for His Support in Studies of Etiology of Scoliosis. *Genes* **2021**, *12*, 1033. [CrossRef] [PubMed]
2. Mathieu, H.; Spataru, A.; Aragon-Martin, J.A.; Child, A.; Barchi, S.; Fortin, C.; Parent, S.; Moldovan, F. Prevalence of POC5 Coding Variants in French-Canadian and British AIS Cohort. *Genes* **2021**, *12*, 1032. [CrossRef] [PubMed]
3. Terhune, E.A.; Wethey, C.I.; Cuevas, M.T.; Monley, A.M.; Baschal, E.E.; Bland, M.R.; Baschal, R.; Trahan, G.D.; Taylor, M.R.G.; Jones, K.L.; et al. Whole Exome Sequencing of 23 Multigeneration Idiopathic Scoliosis Families Reveals Enrichments in Cytoskeletal Variants, Suggests Highly Polygenic Disease. *Genes* **2021**, *12*, 922. [CrossRef] [PubMed]
4. Carry, P.M.; Terhune, E.A.; Trahan, G.D.; Vanderlinden, L.A.; Wethey, C.I.; Ebrahimi, P.; McGuigan, F.; Åkesson, K.; Hadley-Miller, N. Severity of Idiopathic Scoliosis Is Associated with Differential Methylation: An Epigenome-Wide Association Study of Monozygotic Twins with Idiopathic Scoliosis. *Genes* **2021**, *12*, 1191. [CrossRef] [PubMed]
5. Janusz, P.; Chmielewska, M.; Andrusiewicz, M.; Kotwicka, M.; Kotwicki, T. Methylation of Estrogen Receptor 1 Gene in the Paraspinal Muscles of Girls with Idiopathic Scoliosis and Its Association with Disease Severity. *Genes* **2021**, *12*, 790. [CrossRef] [PubMed]
6. Su, Z.; Yang, Y.; Wang, S.; Zhao, S.; Zhao, H.; Li, X.; Niu, Y.; Deciphering Disorders Involving Scoliosis and COmorbidities (DISCO) Study Group; Qiu, G.; Wu, Z.; et al. The Mutational Landscape of PTK7 in Congenital Scoliosis and Adolescent Idiopathic Scoliosis. *Genes* **2021**, *12*, 1791. [CrossRef] [PubMed]
7. Lai, W.; Feng, X.; Yue, M.; Cheung, P.W.H.; Choi, V.N.T.; Song, Y.-Q.; Luk, K.D.K.; Cheung, J.P.Y.; Gao, B. Identification of Copy Number Variants in a Southern Chinese Cohort of Patients with Congenital Scoliosis. *Genes* **2021**, *12*, 1213. [CrossRef] [PubMed]
8. Wang, S.; Chai, X.; Yan, Z.; Zhao, S.; Yang, Y.; Li, X.; Niu, Y.; Lin, G.; Su, Z.; Wu, Z.; et al. Novel FGFR1 Variants Are Associated with Congenital Scoliosis. *Genes* **2021**, *12*, 1126. [CrossRef] [PubMed]
9. Latypova, X.; Creadore, S.G.; Dahan-Oliel, N.; Gustafson, A.G.; Wei-Hung Hwang, S.; Bedard, T.; Shazand, K.; van Bosse, H.J.P.; Giampietro, P.F.; Dieterich, K. A Genomic Approach to Delineating the Occurrence of Scoliosis in Arthrogryposis Multiplex Congenita. *Genes* **2021**, *12*, 1052. [CrossRef] [PubMed]
10. Dahan-Oliel, N.; Dieterich, K.; Rauch, F.; Bardai, G.; Blondell, T.N.; Gustafson, A.G.; Hamdy, R.; Latypova, X.; Shazand, K.; Giampietro, P.F.; et al. The Clinical and Genotypic Spectrum of Scoliosis in Multiple Pterygium Syndrome: A Case Series on 12 Children. *Genes* **2021**, *12*, 1220. [CrossRef] [PubMed]
11. Whittle, J.; Johnson, A.; Dobbs, M.B.; Gurnett, C.A. Models of Distal Arthrogryposis and Lethal Congenital Contracture Syndrome. *Genes* **2021**, *12*, 943. [CrossRef] [PubMed]

Review

A Decade in Review after Idiopathic Scoliosis Was First Called a Complex Trait—A Tribute to the Late Dr. Yves Cotrel for His Support in Studies of Etiology of Scoliosis

Nelson L. S. Tang [1,2,*], Matthew B. Dobbs [3], Christina A. Gurnett [4], Yong Qiu [5], T. P. Lam [6], Jack C. Y. Cheng [6] and Nancy Hadley-Miller [7]

[1] KIZ/CUHK Joint Laboratory of Bioresources and Molecular Research in Common Diseases, Department of Chemical Pathology, Li Ka Shing Institute of Health Sciences, The Chinese University of Hong Kong, Hong Kong SAR, China

[2] Functional Genomics and Biostatistical Computing Laboratory, CUHK Shenzhen Research Institute, Shenzhen 518000, China

[3] Dobbs Clubfoot Center, Paley Orthopedic and Spine Institute, West Palm Beach, FL 33401, USA; mdobbs@paleyinstitute.org

[4] Department of Neurology, Washington University in St Louis, St Louis, MO 63110, USA; gurnettc@wustl.edu

[5] Department of Spine Surgery, The Affiliated Drum Tower Hospital of Nanjing University Medical School, Nanjing 210000, China; scoliosis2002@sina.com

[6] Department of Orthopaedics & Traumatology and SH Ho Scoliosis Research Lab, Joint Scoliosis Research Center of the Chinese University of Hong Kong and Nanjing University, The Chinese University of Hong Kong, Hong Kong SAR, China; tplam@cuhk.edu.hk (T.P.L.); jackcheng@cuhk.edu.hk (J.C.Y.C.)

[7] Department of Orthopedics, University of Colorado Anschutz Medical Campus, Aurora, CO 80012, USA; Nancy.Miller@ChildrensColorado.org

* Correspondence: nelsontang@cuhk.edu.hk

Citation: Tang, N.L.S.; Dobbs, M.B.; Gurnett, C.A.; Qiu, Y.; Lam, T.P.; Cheng, J.C.Y.; Hadley-Miller, N. A Decade in Review after Idiopathic Scoliosis Was First Called a Complex Trait—A Tribute to the Late Dr. Yves Cotrel for His Support in Studies of Etiology of Scoliosis. *Genes* **2021**, *12*, 1033. https://doi.org/10.3390/genes12071033

Academic Editors: Gil Atzmon and Toshifumi Yokota

Received: 30 April 2021
Accepted: 28 June 2021
Published: 1 July 2021

Publisher's Note: MDPI stays neutral with regard to jurisdictional claims in published maps and institutional affiliations.

Abstract: Adolescent Idiopathic Scoliosis (AIS) is a prevalent and important spine disorder in the pediatric age group. An increased family tendency was observed for a long time, but the underlying genetic mechanism was uncertain. In 1999, Dr. Yves Cotrel founded the Cotrel Foundation in the Institut de France, which supported collaboration of international researchers to work together to better understand the etiology of AIS. This new concept of AIS as a complex trait evolved in this setting among researchers who joined the annual Cotrel meetings. It is now over a decade since the first proposal of the complex trait genetic model for AIS. Here, we review in detail the vast information about the genetic and environmental factors in AIS pathogenesis gathered to date. More importantly, new insights into AIS etiology were brought to us through new research data under the perspective of a complex trait. Hopefully, future research directions may lead to better management of AIS, which has a tremendous impact on affected adolescents in terms of both physical growth and psychological development.

Keywords: idiopathic scoliosis; genetic predisposition; complex trait; model animal; genome wide association study; genetic linkage study

In 2005, Tang and Miller et al. received grant support from the Fondation Yves Cotrel in the Institut de France to investigate the genetic etiology of adolescent Idiopathic Scoliosis (AIS) [1]. During an academic tour to Hong Kong and China, we presented to Dr. Cotrel our new genetic concept for scoliosis etiology. Together with the breakthroughs of the time, such as the completion of Human Genome Project and International HapMap Project, Dr. Cotrel was convinced that it would be a "Prime Time" to forward genetic research related to AIS. The prevailing concept of AIS etiology at that time was that the disorder was due to one or two genes with major effects [2,3]. Conversely, we argued that AIS might be caused by an interplay of multiple genes and environmental factors. These ideas and hypotheses were later published in a paper titled "Genetic Association of Complex Traits, using Idiopathic Scoliosis as an Example" [4]. Although similar ideas were suggested, it was the first time that AIS was explicitly called a complex trait. Now, more than a decade has passed, and therefore, we took this opportunity to review how far our original hypothesis was supported and where we are in terms of the "Prime Time" of genetic research in AIS while we remember the contributions and teachings of Dr. Cotrel.

1. Setting the Scene

Adolescent Idiopathic Scoliosis (AIS) is the most common form of spinal deformity [5]. It affects up to 4% of otherwise healthy adolescent girls in the population. However, there is still little understanding about the etiology of AIS. Many hypotheses were proposed which included genetic predisposition, growth, and hormonal disturbances, and developmental neuromuscular dysfunction (as illustrated in Figure 1 by Dr. Cotrel). Based on his categorization, The Cotrel Fondation gathered basic scientists and clinical researchers in Paris every year to discuss and propose primary etiologic mechanisms for AIS based on our area of expertise (as illustrated in Figure 2). These annual gatherings snowballed into a series of genetic studies into the etiology of AIS (as illustrated in Figure 3).

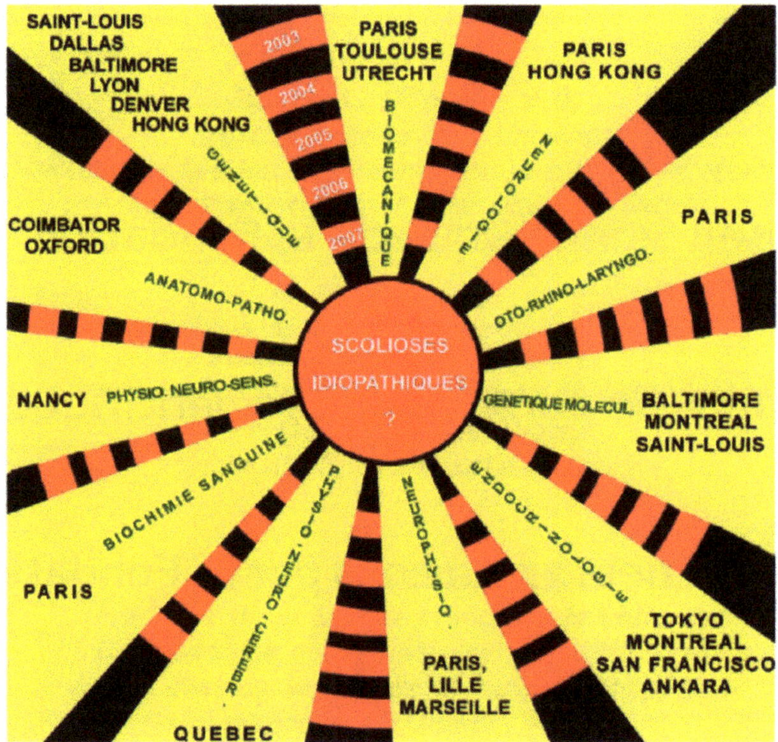

Figure 1. In 2007, Dr. Cotrel and team listed 10 potential categories of underlying causes of scoliosis. It was a key mission of the Cotrel Foundation to find out the primary etiology of adolescent idiopathic scoliosis (AIS). They include genetics, biomechanics, neurology, Oto-Rhino-laryngology, molecular biology, endocrinology, neurophysiology, biochemistry, sensory physiology, and anatomical pathology (courtesy of the Cotrel Foundation).

Figure 2. A group photo of international research colleagues and Dr. Cotrel (right on the first row) taken during 2007 Cotrel Foundation Annual Scientific meeting (courtesy of the Cotrel Foundation).

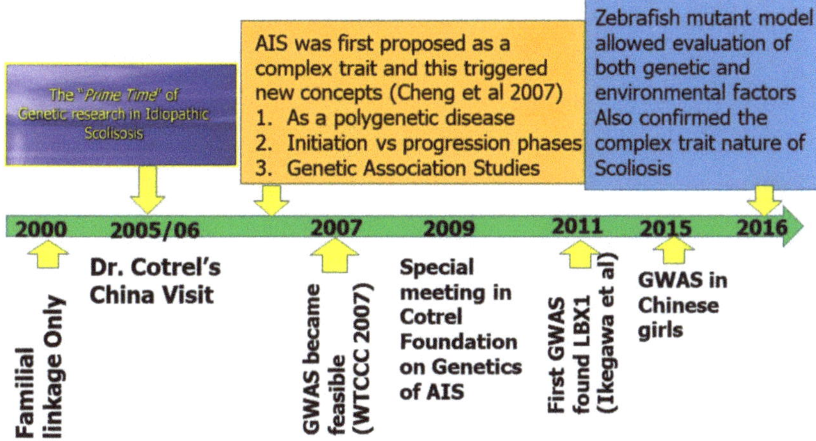

Figure 3. Timeline of genetic research milestones of AIS showing the essential role played by Cotrel Foundation in cultivating complex trait model of AIS.

Although the tendency of AIS to run in families was described for a long time, the mechanism and mode of inheritance were unknown. Linkage analyses were carried out by different research groups to locate the causative genetic loci. Both parametric and nonparametric linkage analysis were applied to study familial forms of AIS and resulted in different modes of inheritance, including autosomal dominant and X-linked while multiple loci in chromosomes 9, 16, 17, 19, and X and others were implicated as the loci of major AIS disease genes [6–9]. Although some progress was made in understanding the candidate genetic loci, the familial aggregation pattern is not fully understood, and few highly confident causative genes were identified.

2. AIS Was First Considered as a Complex Trait

Genetic susceptibility for AIS is confusing and its inheritance mechanism is uncertain, however, it is now widely accepted that the complex trait (a polygenic inheritance working together with environmental factors) fits well with AIS. When we put forward this bold new theory in 2007 [4], we borrowed the concept from the case of breast cancer in which high penetrance genes (BRCA1 and BRCA2) were recognized as causing the autosomal dominantly inherited familial breast cancer. Excluding the familial cases, weak but significant genetic predisposition was established for the majority of sporadic breast cancer patients, as evident in the Claus model, and subsequently developed models [10,11]. With reference to these concepts, we hypothesized that genetic susceptibility for AIS potentially also operates in two forms: (a) rare familial scoliosis due to single gene defects, and (b) low penetrance genetic variants acting collectively accounting for genetic predisposition in the majority of "sporadic AIS".

Figure 4 is an updated complex trait model that we put forward in the 2007 paper [4] after incorporating information from the latest research efforts in AIS. A number of AIS predisposition genes were discovered during this decade, and they are represented in the model. Implications of individual genes will be further discussed in this review article.

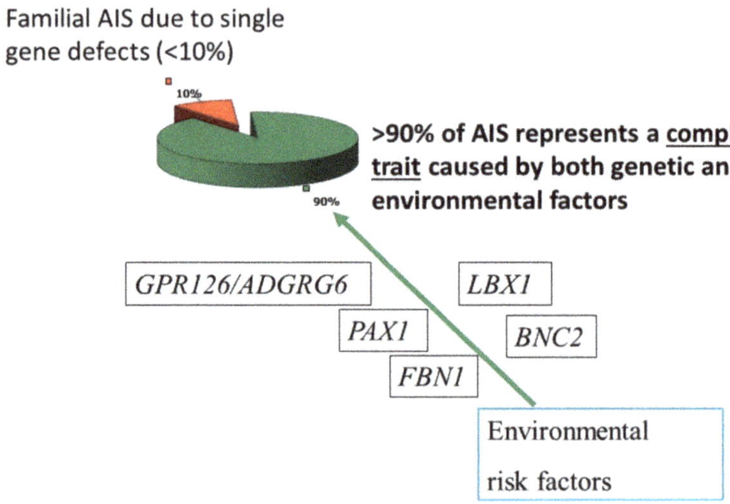

Figure 4. Genetic predisposition for AIS under perspective of a complex trait model. Plausible genes contributing to scoliosis are also shown of which many were identified by GWAS.

3. A High Heritability of Liability to AIS

Unlike other common diseases, there was a concern that a specific value of heritability was not yet determined for AIS [4,12,13]. Without this figure, the contribution of genetics in AIS etiology could not be compared with that of other common traits. For example, the heritability of body height was estimated to be between 0.89 to 0.93 [14], which indicated that a significant proportion of variation in body height was accounted for by genetic factors. Similarly, genetic liability to disease could be evaluated in terms of heritability.

Tang et al. examined the sibling recurrent risk and heritability of AIS in first-degree relatives of 415 Chinese patients who were first diagnosed by a community screening program [15]. Out of the total 531 sibs, 94 sibs had scoliosis (17.7%). The prevalence of AIS among male and female sibs were 11.5% (95% CI: 7.5–15.5%) and 23.0% (95% CI:

18.1–27.9%), respectively. These recurrent risks were significantly higher than the risk in the general population ($p < 0.0001$) supporting a strong familial tendency of AIS. The average sibling recurrent risk ratio was 13-fold higher than that of the general population. Overall, heritability for AIS was estimated to be 87.5%, which was comparable to that of body height.

These results support the prevailing impression of a strong genetic liability for development of AIS [13]. Due to the polygenic nature of complex traits, inheriting genetic risk alleles does not definitely result in the disease. It just represents one additional risk factor among many others. The final phenotype outcome might represent an integration of both genetic and environmental (including lifestyle) factors. This was the first report of the standard genetic aggregation parameters for sibling recurrent risk and heritability of AIS in the Chinese population. The findings confirm that AIS has a surprisingly strong genetic predisposition which is comparable to that of other common diseases or complex traits. Such a high level of heritability was also later confirmed in Caucasian AIS patients who were recruited among those with more severe disease (higher Cobb angle or received surgery) [16,17]. The results of sibling recurrent risk were also replicated and validated in other populations as well [18].

Therefore, a high heritability of AIS liability is (1) universal across populations and (2) applied to both mild and severe disease [16]. However, the figures for heritability may be overestimated due to study design and methodology; for example, there are shared environmental factors among siblings in addition to shared inheritance in the family. Although a high level of heritability was found by different methodology [13,15], the exact contribution of the shared environmental factors in relation to the heritability of AIS liability is difficult to determine.

4. AIS Delineated into 2 Phases: Initiation and Progression

Heterogeneity both in terms of phenotype and genotype are enormous in AIS. It was widely reported that discordant profiles of spine curvature are present among siblings in a family [17,19] and even between genetically identical twins [20–22]. On the other hand, these studies confirm a very high concordance rate and degree of familial segregation, thus supporting the conclusion of the high heritability of AIS. At first glance, these two observations may seem contradictory, as there is a high heritability on one hand and low reproducibility of specific phenotype of spine deformity on the other.

To reconcile these observations, we proposed to understand AIS in two phases. Firstly, an initiation (onset) phase when the spine curvature starts to form, and a second progression phase, which determines the primary direction, severity, and the outcome of the curvature. With this 2-phase concept, the contradictory observations mentioned above could be reasoned if different genetic and environmental factors operate at different phases. Firstly, some genetic and environmental risk factors trigger the initiation of curvature. Then, it is followed by a separate battery of factors (i.e., genetic and/or environmental), which determine how far the curve would advance during the progression phase. In 2007, Cheng et al. elaborated this concept [4]. Figure 5 shows an updated version of the concept. Subsequently, this scheme and nomenclature were cited by other researchers in their publications [5,16] and implemented into orthopedic textbooks.

The implication of the 2-phase model is far-reaching. In the past, the traditional research approach lumped the 2 phases together and results were poorly reproducible due to risk factors of different phases that were confounding each other. Recent research design incorporates the 2-phase model and leads to ground-breaking discoveries of disease predisposition genes in initiation of AIS (to be discussed in the next section).

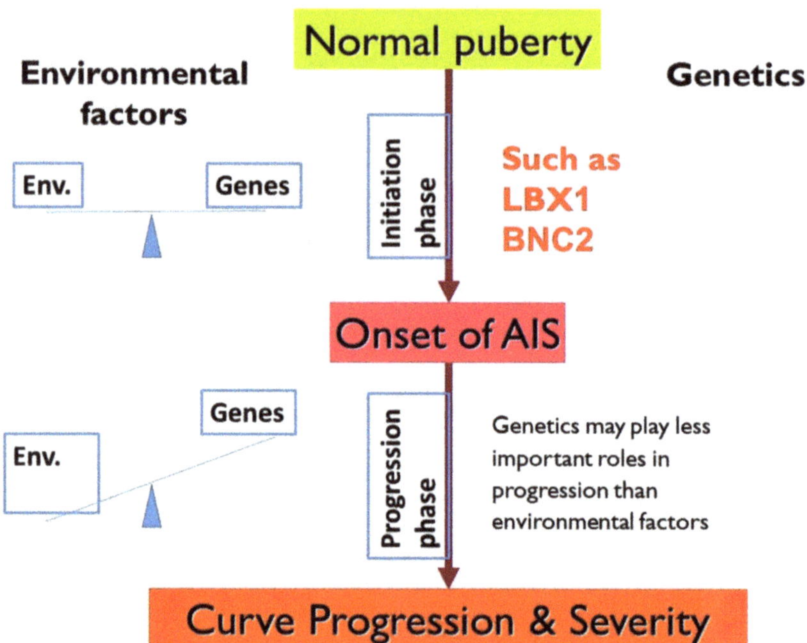

Figure 5. Updated 2-phase concept of AIS. With more genetic studies, more information could be filled in compared to that of first version we proposed more than a decade ago. Genes like LBX1 and BNC2 are AIS predisposition genes that contribute to initiation phase. On the other hand, environmental (Env.) factors may outweigh genetics in determining how far curve would advance.

While genetic studies are very informative, it is important to tease out false-positive associations. That is why replication of studies and confirmation of association signals in other sample sets, or even better in other population samples, are fundamental principles of high-quality genetic studies. While predisposition genes leading to the initiation of spinal curvature were found and validated by different studies, genetic findings were poorly reproducible for prediction of the extent of curve progression (commonly represented by the final Cobb's angle). A biotechnology startup company developed a single nucleotide polymorphism (SNP) based prognostic test to predict curve progression among AIS patients [23]. It is composed of genotype data of 53 SNPs, and the resulting data divided patients into low, medium, and high risk for curve progression. Query was raised early after the first publication as it appeared that the test was only good at the identification of those patients without curve progression [24]. Subsequently, the initial results and potential utility of the test could not be replicated in multiple secondary studies involving both Caucasian and Asian patients [25–28].

Collectively, data are suggestive that the progression phase is potentially under secondary influences including both environmental influence and genetic effects together with potential epigenetic influence. This hypothesis was further supported by the lack of correlation between curve severity and family history of scoliosis [29]. Therefore, we depicted with more detail the 2-phase model in Figure 5, which highlights the differential influence of genetic and environmental factors in each phase. While genetic factors predominantly determine the liability during the initiation of the curve, environmental/together with (epi) genetic factors may govern how far the curve will progress. Many factors are involved in curve progression, such as basic demographics (age of onset, sex), growth velocity, bone mineral density [30], mobility, and morphology of the scoliosis spine on imaging [31,32].

The effect on curve progression due to anthropometric, environmental, and lifestyle factors are reviewed in these publications [33–35].

5. Genetic Association Study in AIS and GWAS

In view of the complex trait nature of AIS, genetic association studies would be the most appropriate methodology in finding the susceptibility genes. Linkage analysis typically used a few families [36] and few of them resulted in identification of a specific gene involved (for example, [37]). In general, conventional linkage analyses were not particularly successful in finding causative genes for AIS. Linkage studies with even a large number of families did not arrive at convincing predisposition genes by using the standard linkage analysis [6,7].

The issue of genetic heterogeneity in AIS is also evident in linkage study results, as no single locus found by linkage analysis was replicated by another research center. This lack of overlap in the reported linkage loci may be a result of several reasons. Firstly, they might be false-positive linkage loci. Secondly, it is also plausible that every family has their own private locus, which implies that many scoliosis causative genes are present in the genome. Furthermore, a genealogical study of IS patients from the Intermountain West [3] also suggested multiple different genes might be involved in the predisposition of Idiopathic Scoliosis in different families. Currently, the exact reason for this lack of replication of linkage results is not known and both possibilities are equally valid.

In the 2007 annual Cotrel foundation research meeting, the genetic association study methodology was proposed. The complex trait paper was written and published before the era of genome-wide association study. The objectives of the earlier research supported by the Cotrel Foundation were to perform genetic association studies and to fine map a previously linkage-supported locus and other candidate genes. On the other hand, genetic association analysis of a dichotomous (discrete) trait may be more suitable to identify genetic etiology of the multifactorial complex trait that could account for a great proportion of AIS patients [4].

While genome-wide association study (GWAS) would provide an unbiased representation of association signals across the whole genome, large funding support was needed. Immediately after the first GWAS WTCCC paper [38], the Cotrel research group and others gathered in Paris to create a proposal to perform a GWAS in AIS. Although the GWAS budget exceeded that of the regular grant funding mechanism, the Cotrel Foundation arranged a special meeting in 2008, invited leading scientists, including Prof. Francois Gros and Prof. Stuart Edelstein to discuss issues related to genetics of AIS. Subsequently, a research team in Riken, led by Prof. Ikegawa, reported the discovery of LBX1 gene in a GWAS of Japanese AIS patients in 2011 [39]. Although this gene was not detected by another GWAS with only 419 AIS families [40], the LBX1 gene turned out to be the first predisposition gene that was then replicated in multiple subsequent studies. Londono et al. also replicated the finding as part of the International Consortium for Scoliosis Genetics [41]. Similar to most other predisposition alleles in other complex traits, the SNPs carried a small increase in the risk of AIS (Odds ratio~1.5 fold for the high risk allele of rs11190870), and thus, accounted for a small fraction of the heritability of AIS. Additional genetic predisposition loci are yet to be found. In 2015, Zhu et al. reported the first GWAS in Chinese AIS patients [42,43] (as illustrated in Figure 6). These and other latest GWAS confirmed the complex trait nature of AIS and collectively they formed the category of low penetrance genes leading to the initiation (onset) of AIS.

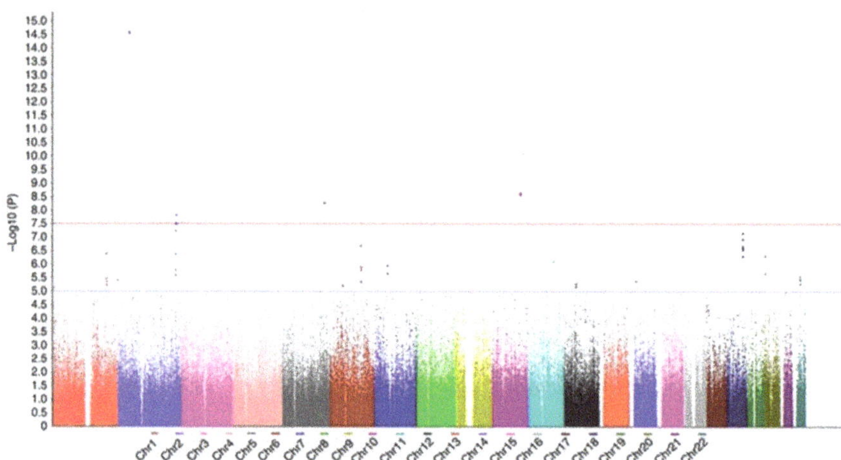

Fig 1 in Zhu et al 2015 with permission to re-use obtained from Nature publisher group.

Figure 6. An example GWAS results in AIS showing putative predisposition SNPs and their statistical significance in a Manhattan plot from Chinese AIS GWAS study [42].

6. LBX1 as the First Confirmed AIS Genetic Predisposition Locus

The research field understands that false-positive findings are common in candidate gene associations due to spurious associations. GWAS is also susceptible to spurious association as it performs so many statistical tests in one study. Although stringent Bonferroni correction is applied to define the cutoff *p*-values, it does not guarantee all GWAS hits are true positives. Replication provides the ultimate validation of any association [44,45]. LBX1 is the most highly replicated AIS predisposition locus to date [39,41]. However, the biological role of LBX1 in AIS is largely unknown. LBX1 is a member of a large family of homeobox transcription factors and regulates upper limb muscle precursor cell migration during embryo mesoderm development in chicken and mice [46–48]. LBX1 has a long evolutionary history in most chordates and is required for the formation of the neuromuscular unit of hypaxial or upper limb musculature [48]. LBX1 also regulates the differentiation and maintenance of satellite cells in postnatal mice muscle [49]. In contrast, the role of LBX1 on bone is less obvious.

Genetic associations were detected near to but outside the protein-coding sequence of LBX1 gene (for example, rs11190870, rs625039, and rs11598564 are intergenic SNPs associated with AIS), together with rs678741 in a nearby antisense transcript called *LBX1-AS1*. The predisposition allele of rs11190870 was the major (prevalent or common in the healthy population) allele in various populations with an allele frequency ranging from 0.49 to 0.58) [41,50]. It is unknown why such a prevalent allele would predispose to a prevalent disease, as the challenge of natural selection would be expected to select out defective (disease predisposition) alleles or for those alleles to become less frequent.

7. BNC2 Is Another Replicated AIS Locus Related to Muscle Development

Rs3904778 located in an intron of BNC2 was first associated with AIS in a GWAS of Japanese patients [51]. The high-risk haplotype was suggested to bind to a muscular cell transcriptional factor, YY1. The results were further replicated in Chinese patients and Caucasian patients [51–53].

Both LBX1 and BNC2 are functionally related to early muscle development. These findings provide potential new insight into the initiation (onset) of scoliosis. AIS was traditionally viewed as a bone disease involving the various levels of vertebrae and their growth. With these highly replicated genetic predisposition loci identified, which do not have an obvious function in bone, the traditional understanding of AIS as a primary bone disease needs to be reconsidered. Instead, a primary functional alteration of muscle with subsequent interplay between muscle and bone tissues may be occurring during the initiation of scoliosis.

8. Extracellular Matrix and Fibers

Patients with Marfan syndrome and Ehlers–Danlos syndrome often develop scoliosis among other inherited types of connective tissue diseases. They are autosomal dominant diseases with mutations in fibrillin (FBN1) or one of the collagen genes [54,55]. Rare mutations in FBN1 and FBN2 were associated with scoliosis, even though these scoliosis patients do not have typical features consistent with the syndromic condition. In addition, it was the load of mutations among these genes rather than any single gene that determined the predisposition to AIS [56,57].

9. Fish Studies Confirmed Scoliosis as a Complex Trait

A major hurdle in the study of AIS is the development of a reliable animal model that is representative of the disease phenotype. Previous attempts of inducing scoliosis with surgical intervention, such as using pinealectomy chicks and bipedal mice, were largely unsatisfactory. A spontaneous scoliosis animal model with spinal curvature developing during a rapid growth phase (reminiscence of the pubertal growth) was long sought. Zygotic ptk7 (Protein tyrosine kinase-7) mutant zebrafish provide important clues to the etiology of scoliosis and represent a breakthrough in animal model research of AIS [58,59]. Ptk7 and planar cell polarity (PCP) signaling are normal during the embryonic period as supported by maternal ptk7 deposited in the eggs which facilitate normal morphogenesis. Only after physiological degradation of maternally derived gene products will a deficiency of ptk7 occur in the larval stage of the transgenic fish [60]. This indicates that a single defect during the postnatal growth phase can cause scoliosis. Being an essential pathway, PCP signaling is involved in various pathologies related to embryonic development, such as neural tube defects [61–63]. While the ptk7 fish model faithfully demonstrated the vertebral curve phenotype during its rapid growth phase, unfortunately, the mutant fish also developed hydrocephalus, which was attributed to defects in ependymal cell cilia function [50]. Despite this, the results have far-reaching implications. Unlike those induced scoliosis in chicken and mice models, it was one of the few models that best resembled the situation in human disease. Another one was recessive kif6 mutant in zebrafish [64]. The ptk7 mutant has an additional advantage that the onset of spinal curvature is later in the life cycle and coincides with the rapid growing juvenile stage.

In zebrafish, due to the small CSF space, cilia may be essential for a normal CSF flow. On the other hand, Chu et al. measured the CSF flow in patients with neural axis anomalies but failed to demonstrate any observable impairment [65]. A study by Zhang et al provided the mechanism linking abnormal CSF flow and scoliosis [66]. They found that cilia action was essential to maintain CSF flow in the zebrafish embryos as well as the transport of adrenergic signals for neuromuscular development. Both CSF flow disruption and urotensin receptor mutations induced scoliosis in zebrafish [67,68]. The results suggest that neuromuscular development of the dorsal somite is indeed the common and critical pathway leading to scoliosis and that the cilia abnormality was potentially a trigger for the spinal curvature instead of being an essential mechanism. The findings in zebrafish reinforce the concept that initiation of AIS may be due to a neuromuscular problem rather than being a primary bone disease. Similar scoliosis-related phenotypes were also induced in other zebrafish mutants (for example, stat3 and myh3) although the mechanisms may be different [69,70].

With these scoliosis of spontaneous-onset fish models, various treatment strategies can now be examined. Traditional scoliosis treatments target primarily mechanical correction by using bracing or surgery. Few attempts of medical treatment other than melatonin were tried. Melatonin treatment was based on the pinealectomy induced scoliosis model in chicken, however, the efficiency in patients was limited [71]. The zygotic mutant zebrafish established that the onset of scoliosis in the zebrafish model could be ameliorated by drug treatment alone. Treatment with aspirin or N-acetylcysteine lowered the prevalence of scoliosis in ptk7 mutant zebrafish [72]. These treatments either delayed the onset of scoliosis or reduced the prevalence of scoliosis. For example, the prevalence of scoliosis reduced from 81% to 17% after treatment with N-acetylcysteine [72]. These early zebrafish findings provide new hope for potential medical intervention in human AIS. If such nonsurgical interventions can be potentially effective, they will represent a major milestone in scoliosis treatment.

A key difference between single-gene disease and the polygenic complex trait is that the role played by environmental factors should have a prominent contribution in the latter. Cheng et al and other researchers showed that systemic low bone mineral density (BMD) was a significant and independent prognostic factor for curve progression. Furthermore, over 30% of AIS had low BMD as compared to 16% in that of the general population [5]. Low BMD results from gene–environmental interaction and can be considered as an environmental factor contributing to liability to AIS. Osteopenia represents a risk factor that could be modified, and it remains to be seen if such interventions could have beneficial therapeutic effects on scoliosis.

10. Discussion and Looking Forward

10.1. What Genetic Studies Inform about the Etiology of Scoliosis?

This is a question frequently raised by colleagues: how can genetic studies benefit patients? It is an excellent question, and it deserves an answer. We hope that after reading this review, many colleagues will appreciate that the knowledge and understanding of AIS were much enhanced. More than 20 years ago, Dr. Cotrel initiated the visions as shown in Figure 1 in which he figured out 10 potential categories of pathogenesis that were equally likely to cause AIS. Now, we have arrived at strong data supporting genetic loci in muscle developmental genes and extracellular matrix as potential predisposition genes.

The next question is whether such understanding could be obtained without utilizing genetic studies. We believe that the advances in the past decade are largely related to genetic research. Genetic studies have a unique advantage that all genetic alleles, mutation, or genetic risk factors must be the primary event in any causative analysis. Identifying a causative chain of events, and determining which are the primary and secondary adaptive events in pathogenesis is always difficult. Furthermore, it was a century-long argument about which tissue (representing the 10 potential categories of pathogenesis in Figure 1) holds the primary defect in AIS (be it bone, cartilage, ligament, spinal cord, muscle, nerve, intervertebral disc, etc.). In addition, other research approaches could only collect samples from patients requiring surgery who, by default, must have an advanced curve and tissues are obtained late during the disease process. The pathology found in these settings is virtually impossible to differentiate between the primary causative change or secondary adaptive response. On the other hand, alleles and mutations in a gene used for a particular function are fixed at the time of conception. Therefore, it predates the onset of any disease so that mutations and alleles must represent a primary event and pinpoint the functional pathway of pathogenesis. When working with genetic association or linkage results, the concern regarding the differentiation between primary cause or secondary adaptation is largely diminished. Our task is merely (1) to confirm that the genetic findings are genuine and not false positive, and (2) to make biological sense from the findings to understand the underlying functional defects.

LBX1 and BNC2 are primarily related to early muscle development. Fibrillin and collagen genes form the extracellular matrix and connective tissue. They stand out among others in GWAS and mutation analysis of AIS and provided strong support that pathogenesis of a proportion of AIS patients is related to soft tissue abnormalities. Complex traits are characterized by patient heterogeneity meaning that patients may have different etiologies. Given the understanding of complexity in AIS, soft tissue anomalies could be an initiation event in a significant proportion of AIS patients, in particular since LBX1 risk alleles are so prevalent in the population. We think that these findings may help to settle the century-long argument of the tissue of primary defect in AIS and provide an answer to Dr. Cotrel's questions about the 10 categories of etiologies. With this new concept, new innovative approaches to treatment and management can be envisioned. It is hoped that the results of research in this field will help a proportion of patients whose scoliosis are due to soft tissue pathogenesis to reduce the morbidity of AIS in the future.

10.2. What Should Be the Future Research Direction?

The identification of soft tissue as a potential primary defect leading to the initiation of spinal curvature will be an important direction for future research. It suggests that the trajectory or disease path after AIS onset is amenable to modification. Two new ideas for future research direction are obvious with these findings: (1) early curve control/correction, and (2) screening for early curve.

Remodeling of soft tissue is more feasible than remodeling hard tissue. For example, soft tissues can be remodeled by various means, e.g., exercise, stretching, bracing, etc. Inspired by the soft tissue remodeling idea, Figure 7 shows the follow-up X-ray of a male AIS patient whose curve was controlled with lifestyle and environment modifications alone without bracing.

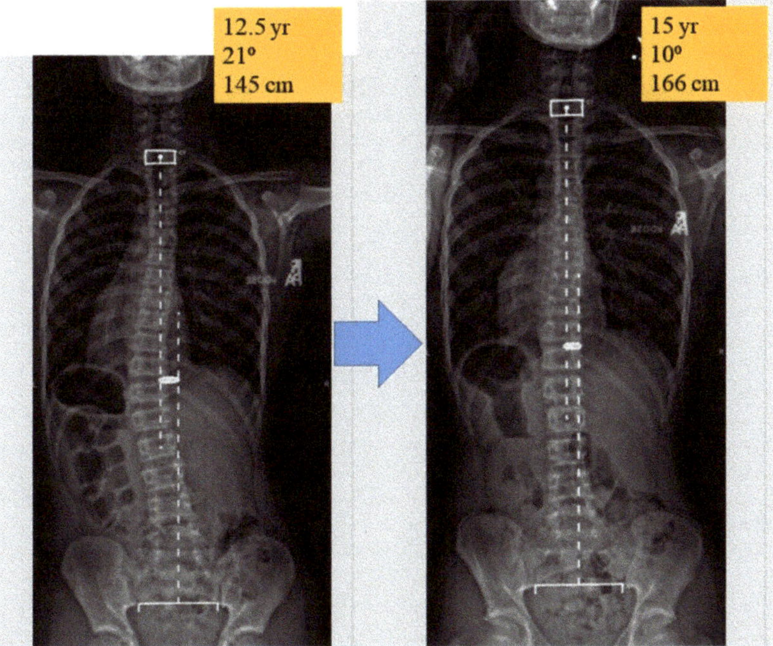

Figure 7. Spine X-rays of a male patient who was first diagnosed at 12.5 years old with a 21-degree curve. Environmental and lifestyle modifications were implemented like upper limb exercise, posture awareness, and change of seating furniture. His curve reverted to 10 degrees at the age of 15 years old and the follow-up x-ray was taken after he had passed through his peak growth spurt. Bracing was not prescribed in this case. This case demonstrates that soft tissue remodeling is feasible in some selected patients.

Bracing provides regional control of the spine curvature and was used for a long time. However, there is no biological ground to determine when bracing should be started. The implication that being a soft tissue disease in some AIS patients from the outset would suggest that we need to start bracing early in the course of disease before secondary adaptive bone changes occur which perpetuate a vicious cycle of maladaptation.

For an early intervention to become available and effective, patients need to be identified early. As a matter of fact, early diagnosis is also required to evaluate the effectiveness of any early intervention. Hong Kong is among the few regions in the World that practice universal scoliosis screening of school children. Screening enables detection of early patients with milder curves, whereas hospital cases are usually biased toward severe cases so that a comprehensive, full-spectrum understanding of the epidemiology of AIS is possible. In addition, patients with mild curves could be studied so that risk factors for curve progression can be analyzed. The recent advance in low-dose radiation imaging instruments allows accurate and close monitoring of curve progression which was not possible before. These new technologies will very likely bring us new understandings of the biomechanics of early stage of disease when targeted intervention could be possible. All such new concepts, of course, need further evidence and clinical trial to support.

In conclusion, genetic studies of AIS yielded breakthroughs in the understanding of its etiology. Genetic predisposition in genes related to muscle development and extracellular matrix components indicates that initiation of AIS in some individuals is potentially due to soft tissue pathology. The new insights clarify the dated arguments in pathogenesis and provide new directions for future research and potential treatment of the disease.

Author Contributions: Conceptualization and first draft: N.L.S.T.; provision of clinical perspectives: M.B.D., Y.Q., T.P.L., J.C.Y.C., N.H.-M.; provision of genetic perspectives: N.L.S.T., M.B.D., C.A.G., N.H.-M. All authors were involved in drafting or substantively revising the manuscript and read and agreed to the published version of the manuscript.

Funding: Authors would like to acknowledge the following grants: Natural Science Foundation of China (No. 81871747, No.81661168013 & No. 8197090027) to Y.Q., J.C.K.C., and Joint Research Scheme sponsored by the National Natural Science Foundation of China and the Research Grants Council of the Hong Kong Special Administrative Region (N_CUHK416/16) to J.C.K.C., Y.Q., T.P.L., N.L.S.T. NIH grant R01 AR 67715-06 to C.A.G. and M.B.D.

Institutional Review Board Statement: Not applicable.

Informed Consent Statement: Not applicable for this review.

Data Availability Statement: Not applicable for this review.

Conflicts of Interest: Authors received research grant support for research or training in clinical setting through the Cotrel Foundation or Dr. Cotrel.

References

1. Génétique—Les projets soutenus. *Fond. Yves Cotrel. Inst. Fr.*. Available online: https://www.fondationcotrel.org/genetique-les-projets-soutenus/ (accessed on 4 May 2021).
2. Axenovich, T.I.; Zaidman, A.M.; Zorkoltseva, I.V.; Tregubova, I.L.; Borodin, P.M. Segregation analysis of idiopathic scoliosis: Demonstration of a major gene effect. *Am. J. Med. Genet.* **1999**, *86*, 389–394. [CrossRef]
3. Ogilvie, J.W.; Braun, J.; Argyle, V.; Nelson, L.; Meade, M.; Ward, K. The search for idiopathic scoliosis genes. *Spine* **2006**, *31*, 679–681. [CrossRef]
4. Cheng, J.C.Y.; Tang, N.L.S.; Yeung, H.-Y.; Miller, N. Genetic association of complex traits: Using idiopathic scoliosis as an example. *Clin. Orthop.* **2007**, *462*, 38–44. [CrossRef]
5. Cheng, J.C.; Castelein, R.M.; Chu, W.C.; Danielsson, A.J.; Dobbs, M.B.; Grivas, T.B.; Gurnett, C.A.; Luk, K.D.; Moreau, A.; Newton, P.O.; et al. Adolescent idiopathic scoliosis. *Nat. Rev. Dis. Primer* **2015**, *1*, 15068. [CrossRef]
6. Miller, N.H.; Justice, C.M.; Marosy, B.; Doheny, K.F.; Pugh, E.; Zhang, J.; Dietz, H.C.; Wilson, A.F. Identification of candidate regions for familial idiopathic scoliosis. *Spine* **2005**, *30*, 1181–1187. [CrossRef]
7. Justice, C.M.; Miller, N.H.; Marosy, B.; Zhang, J.; Wilson, A.F. Familial idiopathic scoliosis: Evidence of an X-linked susceptibility locus. *Spine* **2003**, *28*, 589–594. [CrossRef]

8. Gao, X.; Gordon, D.; Zhang, D.; Browne, R.; Helms, C.; Gillum, J.; Weber, S.; Devroy, S.; Swaney, S.; Dobbs, M.; et al. CHD7 gene polymorphisms are associated with susceptibility to idiopathic scoliosis. *Am. J. Hum. Genet.* **2007**, *80*, 957–965. [CrossRef]

9. Wise, C.A. The Genetic Architecture of Idiopathic Scoliosis. In *Molecular Genetics of Pediatric Orthopaedic Disorders*; Wise, C.A., Rios, J.J., Eds.; Springer: New York, NY, USA, 2015; pp. 71–89. ISBN 978-1-4939-2169-0.

10. Claus, E.B.; Risch, N.; Thompson, W.D. Genetic analysis of breast cancer in the cancer and steroid hormone study. *Am. J. Hum. Genet.* **1991**, *48*, 232–242.

11. Amir, E.; Freedman, O.C.; Seruga, B.; Evans, D.G. Assessing women at high risk of breast cancer: A review of risk assessment models. *J. Natl. Cancer Inst.* **2010**, *102*, 680–691. [CrossRef]

12. Tenesa, A.; Haley, C.S. The heritability of human disease: Estimation, uses and abuses. *Nat. Rev. Genet.* **2013**, *14*, 139–149. [CrossRef]

13. Grauers, A.; Rahman, I.; Gerdhem, P. Heritability of scoliosis. *Eur. Spine J.* **2012**, *21*, 1069–1074. [CrossRef] [PubMed]

14. Silventoinen, K.; Sammalisto, S.; Perola, M.; Boomsma, D.I.; Cornes, B.K.; Davis, C.; Dunkel, L.; De Lange, M.; Harris, J.R.; Hjelmborg, J.V.B.; et al. Heritability of adult body height: A comparative study of twin cohorts in eight countries. *Twin Res. Off. J. Int. Soc. Twin Stud.* **2003**, *6*, 399–408. [CrossRef] [PubMed]

15. Tang, N.L.S.; Yeung, H.-Y.; Hung, V.W.Y.; Di Liao, C.; Lam, T.-P.; Yeung, H.-M.; Lee, K.-M.; Ng, B.K.-W.; Cheng, J.C.-Y. Genetic epidemiology and heritability of AIS: A study of 415 Chinese female patients. *J. Orthop. Res. Off. Publ. Orthop. Res. Soc.* **2012**, *30*, 1464–1469. [CrossRef]

16. Khanshour, A.M.; Wise, C.A. The Genetic Architecture of Adolescent Idiopathic Scoliosis. In *Pathogenesis of Idiopathic Scoliosis*; Machida, M., Weinstein, S.L., Dubousset, J., Eds.; Springer: Tokyo, Japan, 2018; pp. 51–74. ISBN 978-4-431-56541-3.

17. Ward, K.; Ogilvie, J.; Argyle, V.; Nelson, L.; Meade, M.; Braun, J.; Chettier, R. Polygenic inheritance of adolescent idiopathic scoliosis: A study of extended families in Utah. *Am. J. Med. Genet. A.* **2010**, *152A*, 1178–1188. [CrossRef] [PubMed]

18. Aulisa, A.G.; Guzzanti, V.; Galli, M.; Bottaro, G.; Vitelli, O.; Ferrara, P.; Logroscino, G. The familiarity of idiopathic scoliosis: Statistical analysis and clinical considerations. *Eur. J. Orthop. Surg. Traumatol. Orthop. Traumatol.* **2013**, *23*, 781–784. [CrossRef]

19. Grauers, A.; Danielsson, A.; Karlsson, M.; Ohlin, A.; Gerdhem, P. Family history and its association to curve size and treatment in 1463 patients with idiopathic scoliosis. *Eur. Spine J.* **2013**, *22*, 2421–2426. [CrossRef]

20. Inoue, M.; Minami, S.; Kitahara, H.; Otsuka, Y.; Nakata, Y.; Takaso, M.; Moriya, H. Idiopathic scoliosis in twins studied by DNA fingerprinting: The incidence and type of scoliosis. *J. Bone Jt. Surg. Br.* **1998**, *80*, 212–217. [CrossRef]

21. Simony, A.; Carreon, L.Y.; Jmark, K.H.; Kyvik, K.O.; Andersen, M.Ø. Concordance Rates of Adolescent Idiopathic Scoliosis in a Danish Twin Population. *Spine* **2016**, *41*, 1503–1507. [CrossRef]

22. Schlösser, T.P.C.; Simony, A.; Gerdhem, P.; Andersen, M.Ø.; Castelein, R.M.; Kempen, D.H.R. The heritability of coronal and sagittal phenotype in idiopathic scoliosis: A report of 12 monozygotic twin pairs. *Spine Deform.* **2021**, *9*, 51–55. [CrossRef]

23. Ward, K.; Ogilvie, J.W.; Singleton, M.V.; Chettier, R.; Engler, G.; Nelson, L.M. Validation of DNA-based prognostic testing to predict spinal curve progression in adolescent idiopathic scoliosis. *Spine* **2010**, *35*, E1455–E1464. [CrossRef]

24. Dobbs, M.B.; Gurnett, C.A. Re: Ward, K.; Ogilvie, J.W.; Singleton, M.V.; et al. Validation of DNA-based prognostic testing to predict spinal curve progression in adolescent idiopathic scoliosis. Spine 2010; 35:E1455-64. *Spine* **2011**, *36*, 1257, author reply 1257. [CrossRef]

25. Ogura, Y.; Takahashi, Y.; Kou, I.; Nakajima, M.; Kono, K.; Kawakami, N.; Uno, K.; Ito, M.; Minami, S.; Yanagida, H.; et al. A Replication Study for Association of 53 Single Nucleotide Polymorphisms in a Scoliosis Prognostic Test with Progression of Adolescent Idiopathic Scoliosis in Japanese. *Spine* **2013**, *38*, 1375–1379. [CrossRef]

26. Xu, L.; Qin, X.; Sun, W.; Qiao, J.; Qiu, Y.; Zhu, Z. Replication of Association between 53 Single-Nucleotide Polymorphisms in a DNA-Based Diagnostic Test and AIS Progression in Chinese Han Population. *Spine* **2016**, *41*, 306–310. [CrossRef]

27. Tang, Q.L.; Julien, C.; Eveleigh, R.; Bourque, G.; Franco, A.; Labelle, H.; Grimard, G.; Parent, S.; Ouellet, J.; Mac-Thiong, J.-M.; et al. A Replication Study for Association of 53 Single Nucleotide Polymorphisms in ScoliScore Test with Adolescent Idiopathic Scoliosis in French-Canadian Population. *Spine* **2015**, *40*, 537–543. [CrossRef] [PubMed]

28. Roye, B.D.; Wright, M.L.; Matsumoto, H.; Yorgova, P.; McCalla, D.; Hyman, J.E.; Roye, D.P.; Shah, S.A.; Vitale, M.G. An Independent Evaluation of the Validity of a DNA-Based Prognostic Test for Adolescent Idiopathic Scoliosis. *J. Bone Jt. Surg. Am.* **2015**, *97*, 1994–1998. [CrossRef]

29. Rudnick, S.B.; Zabriskie, H.; Ho, J.; Gurnett, C.A.; Dobbs, M.B. Scoliosis severity does not impact the risk of scoliosis in family members. *J. Pediatr. Orthop. Part. B* **2018**, *27*, 147–151. [CrossRef]

30. Yip, B.H.K.; Yu, F.W.P.; Wang, Z.; Hung, V.W.Y.; Lam, T.P.; Ng, B.K.W.; Zhu, F.; Cheng, J.C.Y. Prognostic Value of Bone Mineral Density on Curve Progression: A Longitudinal Cohort Study of 513 Girls with Adolescent Idiopathic Scoliosis. *Sci. Rep.* **2016**, *6*, 39220. [CrossRef]

31. Vergari, C.; Skalli, W.; Abelin-Genevois, K.; Bernard, J.C.; Hu, Z.; Cheng, J.C.Y.; Chu, W.C.W.; Assi, A.; Karam, M.; Ghanem, I.; et al. Effect of curve location on the severity index for adolescent idiopathic scoliosis: A longitudinal cohort study. *Eur. Radiol.* **2021**. [CrossRef]

32. Haller, G.; Zabriskie, H.; Spehar, S.; Kuensting, T.; Bledsoe, X.; Syed, A.; Gurnett, C.A.; Dobbs, M.B. Lack of joint hypermobility increases the risk of surgery in adolescent idiopathic scoliosis. *J. Pediatr. Orthop. Part. B* **2018**, *27*, 152–158. [CrossRef]

33. Noshchenko, A.; Hoffecker, L.; Lindley, E.M.; Burger, E.L.; Cain, C.M.; Patel, V.V.; Bradford, A.P. Predictors of spine deformity progression in adolescent idiopathic scoliosis: A systematic review with meta-analysis. *World J. Orthop.* **2015**, *6*, 537–558. [CrossRef]

34. Lenz, M.; Oikonomidis, S.; Harland, A.; Fürnstahl, P.; Farshad, M.; Bredow, J.; Eysel, P.; Scheyerer, M.J. Scoliosis and Prognosis-a systematic review regarding patient-specific and radiological predictive factors for curve progression. *Eur. Spine J.* **2021**. [CrossRef]

35. Altaf, F.; Gibson, A.; Dannawi, Z.; Noordeen, H. Adolescent idiopathic scoliosis. *BMJ* **2013**, *346*, f2508. [CrossRef] [PubMed]

36. Edery, P.; Margaritte-Jeannin, P.; Biot, B.; Labalme, A.; Bernard, J.-C.; Chastang, J.; Kassai, B.; Plais, M.-H.; Moldovan, F.; Clerget-Darpoux, F. New disease gene location and high genetic heterogeneity in idiopathic scoliosis. *Eur. J. Hum. Genet.* **2011**, *19*, 865–869. [CrossRef]

37. Patten, S.A.; Margaritte-Jeannin, P.; Bernard, J.-C.; Alix, E.; Labalme, A.; Besson, A.; Girard, S.L.; Fendri, K.; Fraisse, N.; Biot, B.; et al. Functional variants of POC5 identified in patients with idiopathic scoliosis. *J. Clin. Invest.* **2015**, *125*, 1124–1128. [CrossRef]

38. The Wellcome Trust Case Control Consortium. Genome-wide association study of 14,000 cases of seven common diseases and 3000 shared controls. *Nature* **2007**, *447*, 661–678. [CrossRef]

39. Takahashi, Y.; Kou, I.; Takahashi, A.; Johnson, T.A.; Kono, K.; Kawakami, N.; Uno, K.; Ito, M.; Minami, S.; Yanagida, H.; et al. A genome-wide association study identifies common variants near LBX1 associated with adolescent idiopathic scoliosis. *Nat. Genet.* **2011**, *43*, 1237–1240. [CrossRef]

40. Sharma, S.; Gao, X.; Londono, D.; Devroy, S.E.; Mauldin, K.N.; Frankel, J.T.; Brandon, J.M.; Zhang, D.; Li, Q.-Z.; Dobbs, M.B.; et al. Genome-wide association studies of adolescent idiopathic scoliosis suggest candidate susceptibility genes. *Hum. Mol. Genet.* **2011**, *20*, 1456–1466. [CrossRef]

41. Londono, D.; Kou, I.; Johnson, T.A.; Sharma, S.; Ogura, Y.; Tsunoda, T.; Takahashi, A.; Matsumoto, M.; Herring, J.A.; Lam, T.-P.; et al. A meta-analysis identifies adolescent idiopathic scoliosis association with LBX1 locus in multiple ethnic groups. *J. Med. Genet.* **2014**, *51*, 401–406. [CrossRef] [PubMed]

42. Zhu, Z.; Tang, N.L.-S.; Xu, L.; Qin, X.; Mao, S.; Song, Y.; Liu, L.; Li, F.; Liu, P.; Yi, L.; et al. Genome-wide association study identifies new susceptibility loci for adolescent idiopathic scoliosis in Chinese girls. *Nat. Commun.* **2015**, *6*, 8355. [CrossRef]

43. Zhu, Z.; Xu, L.; Leung-Sang Tang, N.; Qin, X.; Feng, Z.; Sun, W.; Zhu, W.; Shi, B.; Liu, P.; Mao, S.; et al. Genome-wide association study identifies novel susceptible loci and highlights Wnt/beta-catenin pathway in the development of adolescent idiopathic scoliosis. *Hum. Mol. Genet.* **2017**, *26*, 1577–1583. [CrossRef]

44. Ott, J. Association of genetic loci: Replication or not, that is the question. *Neurology* **2004**, *63*, 955–958. [CrossRef] [PubMed]

45. Huffman, J.E. Examining the current standards for genetic discovery and replication in the era of mega-biobanks. *Nat. Commun.* **2018**, *9*, 1–4. [CrossRef]

46. Gross, M.K.; Moran-Rivard, L.; Velasquez, T.; Nakatsu, M.N.; Jagla, K.; Goulding, M. Lbx1 is required for muscle precursor migration along a lateral pathway into the limb. *Dev. Camb. Engl.* **2000**, *127*, 413–424.

47. Mootoosamy, R.C.; Dietrich, S. Distinct regulatory cascades for head and trunk myogenesis. *Development* **2002**, *129*, 573–583. [CrossRef]

48. Wotton, K.R.; Schubert, F.R.; Dietrich, S. Hypaxial muscle: Controversial classification and controversial data? *Vertebr. Myogenesis Stem Cells Precursors* **2015**, 25–48. [CrossRef]

49. Watanabe, S.; Kondo, S.; Hayasaka, M.; Hanaoka, K. Functional analysis of homeodomain-containing transcription factor Lbx1 in satellite cells of mouse skeletal muscle. *J. Cell Sci.* **2007**, *120*, 4178–4187. [CrossRef] [PubMed]

50. Man, G.C.-W.; Tang, N.L.-S.; Chan, T.F.; Lam, T.P.; Li, J.W.; Ng, B.K.-W.; Zhu, Z.; Qiu, Y.; Cheng, J.C.-Y. Replication Study for the Association of GWAS-associated Loci with Adolescent Idiopathic Scoliosis Susceptibility and Curve Progression in a Chinese Population. *Spine* **2019**, *44*, 464–471. [CrossRef]

51. Ogura, Y.; Kou, I.; Miura, S.; Takahashi, A.; Xu, L.; Takeda, K.; Takahashi, Y.; Kono, K.; Kawakami, N.; Uno, K.; et al. A Functional SNP in BNC2 Is Associated with Adolescent Idiopathic Scoliosis. *Am. J. Hum. Genet.* **2015**, *97*, 337–342. [CrossRef]

52. Xu, L.; Xia, C.; Qin, X.; Sun, W.; Tang, N.L.-S.; Qiu, Y.; Cheng, J.C.-Y.; Zhu, Z. Genetic variant of BNC2 gene is functionally associated with adolescent idiopathic scoliosis in Chinese population. *Mol. Genet. Genom.* **2017**, *292*, 789–794. [CrossRef]

53. Kou, I.; Otomo, N.; Takeda, K.; Momozawa, Y.; Lu, H.-F.; Kubo, M.; Kamatani, Y.; Ogura, Y.; Takahashi, Y.; Nakajima, M.; et al. Genome-wide association study identifies 14 previously unreported susceptibility loci for adolescent idiopathic scoliosis in Japanese. *Nat. Commun.* **2019**, *10*, 1–9. [CrossRef]

54. Dietz, H. Marfan Syndrome. In *GeneReviews®*; Adam, M.P., Ardinger, H.H., Pagon, R.A., Wallace, S.E., Bean, L.J., Mirzaa, G., Amemiya, A., Eds.; University of Washington: Seattle, WA, USA, 1993.

55. Malfait, F.; Wenstrup, R.; De Paepe, A. Classic Ehlers-Danlos Syndrome. In *GeneReviews®*; Adam, M.P., Ardinger, H.H., Pagon, R.A., Wallace, S.E., Bean, L.J., Mirzaa, G., Amemiya, A., Eds.; University of Washington: Seattle, WA, USA, 1993.

56. Buchan, J.G.; Alvarado, D.M.; Haller, G.E.; Cruchaga, C.; Harms, M.B.; Zhang, T.; Willing, M.C.; Grange, D.K.; Braverman, A.C.; Miller, N.H.; et al. Rare variants in FBN1 and FBN2 are associated with severe adolescent idiopathic scoliosis. *Hum. Mol. Genet.* **2014**, *23*, 5271–5282. [CrossRef]

57. Haller, G.; Alvarado, D.; Mccall, K.; Yang, P.; Cruchaga, C.; Harms, M.; Goate, A.; Willing, M.; Morcuende, J.A.; Baschal, E.; et al. A polygenic burden of rare variants across extracellular matrix genes among individuals with adolescent idiopathic scoliosis. *Hum. Mol. Genet.* **2016**, *25*, 202–209. [CrossRef] [PubMed]

58. Grimes, D.T.; Boswell, C.W.; Morante, N.F.C.; Henkelman, R.M.; Burdine, R.D.; Ciruna, B. Zebrafish models of idiopathic scoliosis link cerebrospinal fluid flow defects to spine curvature. *Science* **2016**, *352*, 1341–1344. [CrossRef] [PubMed]

59. Boswell, C.W.; Ciruna, B. Understanding Idiopathic Scoliosis: A New Zebrafish School of Thought. *Trends Genet.* **2017**, *33*, 183–196. [CrossRef]

60. Hayes, M.; Gao, X.; Yu, L.X.; Paria, N.; Henkelman, R.M.; Wise, C.A.; Ciruna, B. ptk7 mutant zebrafish models of congenital and idiopathic scoliosis implicate dysregulated Wnt signalling in disease. *Nat. Commun.* **2014**, *5*, 4777. [CrossRef]

61. Aw, W.Y.; Devenport, D. Planar cell polarity: Global inputs establishing cellular asymmetry. *Curr. Opin. Cell Biol.* **2017**, *44*, 110–116. [CrossRef]

62. Wang, M.; de Marco, P.; Capra, V.; Kibar, Z. Update on the Role of the Non-Canonical Wnt/Planar Cell Polarity Pathway in Neural Tube Defects. *Cells* **2019**, *8*, 1198. [CrossRef]

63. Hirst, C.E.; Marcelle, C. The avian embryo as a model system for skeletal myogenesis. *Results Probl. Cell Differ.* **2015**, *56*, 99–122.

64. Buchan, J.G.; Gray, R.S.; Gansner, J.M.; Alvarado, D.M.; Burgert, L.; Gitlin, J.D.; Gurnett, C.A.; Goldsmith, M.I. Kinesin family member 6 (kif6) is necessary for spine development in zebrafish. *Dev. Dyn. Off. Publ. Am. Assoc. Anat.* **2014**, *243*, 1646–1657. [CrossRef]

65. Chu, W.C.W.; Man, G.C.W.; Lam, W.W.M.; Yeung, B.H.Y.; Chau, W.-W.; Ng, B.K.W.; Lam, T.-P.; Lee, K.-M.; Cheng, J.C.Y. A detailed morphologic and functional magnetic resonance imaging study of the craniocervical junction in adolescent idiopathic scoliosis. *Spine* **2007**, *32*, 1667–1674. [CrossRef]

66. Zhang, X.; Jia, S.; Chen, Z.; Chong, Y.L.; Xie, H.; Feng, D.; Wu, X.; Song, D.Z.; Roy, S.; Zhao, C. Cilia-driven cerebrospinal fluid flow directs expression of urotensin neuropeptides to straighten the vertebrate body axis. *Nat. Genet.* **2018**, *50*, 1666–1673. [CrossRef]

67. Bearce, E.A.; Grimes, D.T. On being the right shape: Roles for motile cilia and cerebrospinal fluid flow in body and spine morphology. *Semin. Cell Dev. Biol.* **2021**, *110*, 104–112. [CrossRef]

68. Ringers, C.; Olstad, E.W.; Jurisch-Yaksi, N. The role of motile cilia in the development and physiology of the nervous system. *Philos. Trans. R. Soc. B Biol. Sci.* **2020**, *375*, 20190156. [CrossRef] [PubMed]

69. Xiong, S.; Wu, J.; Jing, J.; Huang, P.; Li, Z.; Mei, J.; Gui, J.-F. Loss of stat3 function leads to spine malformation and immune disorder in zebrafish. *Sci. Bull.* **2017**, *62*, 185–196. [CrossRef]

70. Whittle, J.; Antunes, L.; Harris, M.; Upshaw, Z.; Sepich, D.S.; Johnson, A.N.; Mokalled, M.; Solnica-Krezel, L.; Dobbs, M.B.; Gurnett, C.A. MYH3-associated distal arthrogryposis zebrafish model is normalized with para-aminoblebbistatin. *EMBO Mol. Med.* **2020**, *12*, e12356. [CrossRef] [PubMed]

71. Sánchez-Barceló, E.J.; Mediavilla, M.D.; Tan, D.X.; Reiter, R.J. Scientific basis for the potential use of melatonin in bone diseases: Osteoporosis and adolescent idiopathic scoliosis. *J. Osteoporos.* **2010**, *2010*, 830231. [CrossRef] [PubMed]

72. Van Gennip, J.L.M.; Boswell, C.W.; Ciruna, B. Neuroinflammatory signals drive spinal curve formation in zebrafish models of idiopathic scoliosis. *Sci. Adv.* **2018**, *4*, eaav1781. [CrossRef] [PubMed]

genes

Article

The Clinical and Genotypic Spectrum of Scoliosis in Multiple Pterygium Syndrome: A Case Series on 12 Children

Noémi Dahan-Oliel [1,2,*], Klaus Dieterich [3], Frank Rauch [1,2], Ghalib Bardai [1,2], Taylor N. Blondell [4], Anxhela Gjyshi Gustafson [5], Reggie Hamdy [1,2], Xenia Latypova [3], Kamran Shazand [5], Philip F. Giampietro [6] and Harold van Bosse [4,*]

[1] Shriners Hospitals for Children, Montreal, QC H4A 0A9, Canada; frauch@shriners.mcgill.ca (F.R.); GBardai@shriners.mcgill.ca (G.B.); rhamdy@shriners.mcgill.ca (R.H.)
[2] Faculty of Medicine and Health Sciences, McGill University, Montreal, QC H3G 2M1, Canada
[3] Inserm, U1216, Grenoble Institut Neurosciences, Génétique médicale, Université Grenoble Alpes, CHU Grenoble Alpes, 38000 Grenoble, France; KDieterich@chu-grenoble.fr (K.D.); XMartin@chu-grenoble.fr (X.L.)
[4] Shriners Hospitals for Children, Philadelphia, PA 19140, USA; tblondell@shrinenet.org
[5] Shriners Hospitals for Children Headquarters, Tampa, FL 33607, USA; agustafson@shrinenet.org (A.G.G.); kshazand@shrinenet.org (K.S.)
[6] Pediatric Genetics, University of Illinois, Chicago, IL 60612, USA; philipg@uic.edu
* Correspondence: ndahan@shrinenet.org (N.D.-O.); hvanbosse@shrinenet.org (H.v.B.)

Abstract: Background: Multiple pterygium syndrome (MPS) is a genetically heterogeneous rare form of arthrogryposis multiplex congenita characterized by joint contractures and webbing or pterygia, as well as distinctive facial features related to diminished fetal movement. It is divided into prenatally lethal (LMPS, MIM253290) and nonlethal (Escobar variant MPS, MIM 265000) types. Developmental spine deformities are common, may present early and progress rapidly, requiring regular fo llow-up and orthopedic management. Methods: Retrospective chart review and prospective data collection were conducted at three hospital centers. Molecular diagnosis was confirmed with whole exome or whole genome sequencing. Results: This case series describes the clinical features and scoliosis treatment on 12 patients from 11 unrelated families. A molecular diagnosis was confirmed in seven; two with *MYH3* variants and five with *CHRNG*. Scoliosis was present in all but our youngest patient. The remaining 11 patients spanned the spectrum between mild (curve $\leq 25°$) and malignant scoliosis ($\geq 50°$ curve before 4 years of age); the two patients with *MYH3* mutations presented with malignant scoliosis. Bracing and serial spine casting appear to be beneficial for a few years; non-fusion spinal instrumentation may be needed to modulate more severe curves during growth and spontaneous spine fusions may occur in those cases. Conclusions: Molecular diagnosis and careful monitoring of the spine is needed in children with MPS.

Keywords: *CHRNG*; distal arthrogryposis type 8; Escobar; multiple pterygium syndrome; *MYH3*; scoliosis

Citation: Dahan-Oliel, N.; Dieterich, K.; Rauch, F.; Bardai, G.; Blondell, T.N.; Gustafson, A.G.; Hamdy, R.; Latypova, X.; Shazand, K.; Giampietro, P.F.; et al. The Clinical and Genotypic Spectrum of Scoliosis in Multiple Pterygium Syndrome: A Case Series on 12 Children. *Genes* **2021**, *12*, 1220. https://doi.org/10.3390/genes12081220

Academic Editor: Stephen Robertson

Received: 18 June 2021
Accepted: 2 August 2021
Published: 6 August 2021

Publisher's Note: MDPI stays neutral with regard to jurisdictional claims in published maps and institutional affiliations.

1. Introduction

The term pterygium is used most commonly to describe an acquired ophthalmologic condition, with a conjunctival "wing" or flap that can cross the cornea. The term also describes joints with congenital webbing or winging of soft tissue which limits the joint motion. Usually, more than one joint and body region is involved, meeting the criteria for arthrogryposis multiplex congenita [1]. While pterygia may be an incidental finding in persons with arthrogryposis, such as the occasional knee pterygium found associated with severe contractures in Amyoplasia, they can also manifest as a more generalized syndrome, such as popliteal pterygium syndrome or multiple pterygium syndrome (MPS).

MPS is a rare form of arthrogryposis multiplex characterized by a constellation of congenital anomalies [2]. The webbing of skin and contractures of the joints that are found in this disorder may restrict movement. Examples of joint involvement in the upper extremities include axillary pterygia (both the anterior and posterior folds), elbow flexion contractures with antecubital webbing, mildly dorsiflexed wrists, and fingers with camptodactyly, interdigital pterygia and thumb-in-palm deformities. In the lower extremities, perineal pterygia can span medially from one thigh to the other, knee flexion contractures with pterygia can be severe, and foot deformities include both clubfoot and congenital vertical talus (rocker bottom foot).

Other characteristic findings of MPS are short stature, webbing of the neck, and distinctive facial features including micrognathia, cleft palate, down-turned corners of the mouth, an elongated philtrum, down-slanting palpebral fissures, epicanthal folds, ptosis, and low-set ears, all of which are related to diminished fetal movement. Developmental spine deformities are common, not only as a coronal plane deformity (scoliosis), but frequently with a substantial associated sagittal plane deformity, making it a kyphoscoliosis. Spontaneous spinal fusion abnormalities occur often in MPS and are congenital.

MPS can be separated into the lethal pterygium syndromes (LMPS, MIM253290) and the non-lethal syndromes; the latter conditions are categorically referred to as Escobar syndrome (or Escobar variant MPS, MIM265000) and most are autosomal recessive. Here we will only be discussing the non-lethal forms of MPS.

Several genes associated with MPS give rise to what has recently been described as a prenatal form of myasthenia, first associated with variants in *CHRNG*. *CHRNG* codes for the γ subunit of the acetylcholine receptor (AChR) in the developing fetus. Mutations that impact the expression of this subunit, its integration in the AChR, or the transport of the AChR to the sarcolemma will have major consequences to the neuromuscular junction. At 33 weeks gestation though, the γ subunit starts to be replaced by the ε subunit of AChR, finally leading to a functional adult neuromuscular junction. However, effects of the fetal akinesia have become manifest before that stage of development and the damage is irreversible. Interestingly, individuals with Escobar syndrome do not have muscle weakness or electrophysiological symptoms associated with myasthenia gravis postnatally, as the AChR essentially functions normally after birth. Other causes of Escobar syndrome recessive gene mutations of the AChR include *CHRNA1* (α1-AChR subunit), *CHRNB1* (β1-subunit), *CHRND* (δ-subunit), and *RAPSN* (AChR binding protein). Escobar syndrome can also be caused by pathogenic variants in *CNTN1* (contactin 1) and *DOCK7* (dedicator of cytokinesis 7) [3–5]. See Figure 1.

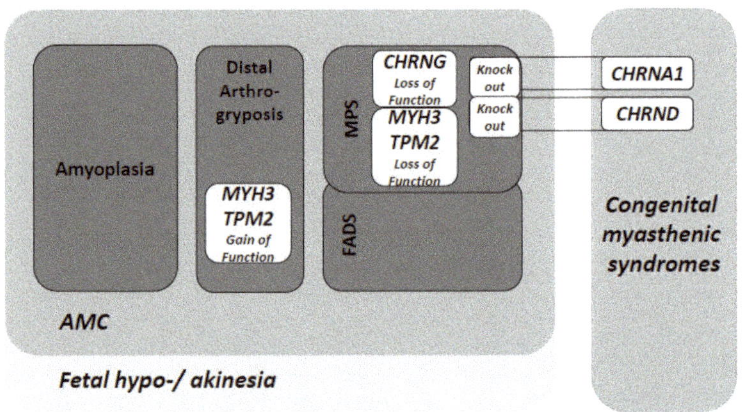

Figure 1. Clinical spectrum and overlap of molecular causes of arthrogryposis multiplex congenita (AMC). The main categories of AMC phenotypes are represented as dark grey boxes. Some of the genetic determinants that are responsible for multiple phenotypes among AMC subtypes are represented in white boxes. Both *MYH3* and *TPM2* biallelic loss of function have been associated to MPS, although dominant heterozygous variants of these genes are linked to several forms of distal arthrogryposis. FADS: fetal akinesia deformation sequence; MPS: multiple pterygium syndrome.

More recently, cases of MPS with an autosomal dominant form, also classified as distal arthrogryposis type 8 (DA8) [6,7], have been associated with mutations in the embryonic myosin heavy chain gene, *MYH3*. This is the same gene that is found underlying other forms of distal arthrogryposis (DA1, DA2A or Freeman Sheldon syndrome, and DA2B or Sheldon Hall syndrome).

Spinal curvatures in Escobar syndrome appear early, often present at birth, and can progress quickly. Treatment options include spine casting, bracing, and expandable implant surgery, to allow as much chest growth and development as possible, and ultimately spinal fusion, with a goal of well-balanced spine [8]. Although advances in the phenotypic spectrum, disease progression and genetic etiology of MPS have been made [9,10], definite phenotype–genotype correlations need yet to be discovered. The objective of this case series was to describe the phenotypic presentations of a small multisite collection of patients with Escobar syndrome, in particular as to how it relates to their spinal deformity. By determining the genetic basis of their MPS, we hope to associate the genotype with their phenotype and natural history of their scoliosis, and describe the clinical interventions aimed at reducing the spinal curvatures associated with these cases.

2. Materials and Methods

2.1. Design

Case series using a retrospective chart review and prospective data collection at three hospital centers; Shriners Hospitals for Children in Montreal, Canada, Shriners Hospitals for Children in Philadelphia, United States, and Grenoble Alpes University Hospital Center in France. The patients at the two Shriners hospitals were enrolled in a multisite pediatric arthrogryposis registry that collects clinical, patient-reported outcomes, and provides whole genome sequencing to map the phenotype to genotype in this population.

2.2. Ethical Approval and Consent

Written informed consent according to local ethical committees in all participating centers was obtained for all patients. Clinical and genetic data were anonymized and entered into a secured database accessible only to the research team.

2.3. Patients

We describe the phenotypic presentation of 12 patients presenting with MPS and scoliosis. Clinical data were analyzed retrospectively. Clinical data (sex, age), scoliosis (i.e., age at onset, congenital/early onset/late onset), and a description of the phenotype was extracted from the electronic medical record and evaluated in clinic (contractures, limb anomalies, webbing/pterygia, other system involvement). A description of interventions, including type (observation, bracing, and surgery) and age at treatment was collected. Spine X-rays were reviewed and classified in order to better understand the curves in children with Escobar syndrome using the following classification system of curve behavior. A mild curve was defined as a curve $\leq 25°$ at any age. A moderate curve was a curve $> 25°$ but $\leq 50°$ at any age. Severe curves exceeded $50°$, but not until after 4 years of age, whereas malignant curves were $\geq 50°$ before 4 years of age.

2.4. Genotyping and Data Analysis

Genomic DNA was isolated from saliva collected in Oragene OGR-600 tubes (DNA Genotek, Kanata, ON, Canada) according to manufacturer's instructions and subsequently extracted using the Chemagic360 platform (PerkinElmer, Waltham, MA, USA). DNA was quantified using Qubit. Average gDNA obtained from extractions was 40 ng/uL. Libraries were prepared in the Zephyr G3 NGS Automated Workstation (PerkinElmer) using the Illumina DNA PCR-Free Prep, Tagmentation (Illumina, San Diego, CA, USA) with 1000 ng of input gDNA. Resulting libraries were quantified prior to pooling using the Kapa Library Quantification Kit (Roche, Basel, Switzerland). Libraries were pooled at a concentration of 1.4 nM and were sequenced on the NovaSeq 6000 sequencing platform (Illumina). Paired-end sequencing (2 × 150 bp) was performed at >30× coverage per sample. GRCh38 reference genome was used. Genotyping at the Grenoble arthrogryposis reference center was conducted using DNA extracted from whole blood samples collected on EDTA tubes. Libraries were prepared with the Nextera DNA Flex Library Prep Kit (Illumina). Exome sequencing was performed on a NextSeq 500 (Illumina).

Data analysis, including variant calling, was performed using the Illumina TruSight™ Software Suite platform. This platform performs alignment, variant calling, variant annotation and filtering. For the Grenoble hospital center, bioinformatic analysis was performed with an in-house pipeline using an NGS platform, using the Burrows-Wheeler Aligner, Picard Tools 2.18.23-0 and GATK v4.0.12. Variants were interpreted with ANNOVAR and prioritized through python annotation scripts. The strategy for genome data interpretation was primarily based on disease and phenotype gene target definition. Data were descriptively analyzed.

3. Results

3.1. Clinical Features

The 12 patients (7 females) with MPS and scoliosis were from 11 unrelated non-consanguineous families. They were born in Algeria ($n = 1$), Canada ($n = 1$), France ($n = 4$) and the United States ($n = 6$) and were between 1 and 20 years old at the time of last follow up (Tables 1 and 2). Family history was positive in one family (F6), with presence of clinical features of Escobar and scoliosis in the mother and maternal aunt of the index patient (P7). In the family of two affected siblings (F2), one pregnancy had been terminated due to severe fetal deformities; muscle histology in that fetus was in accordance with a diagnosis of Escobar syndrome.

Table 1. Phenotype: pterygia and joints affected.

Patient	Family	Sex	Age (Years)	Multiple Pterygia (HP: 0001059)	Neck Pterygia (HP: 0009759)	Axillary Pterygia (HP: 0001040)	Pterygia—Other Locations	Scoliosis (HP: 0002650)	Micrognathia (HP: 0000347) and/or Retrognathia (HP: 0000278)	Talipes Equinovarus (HP: 0001762)	Congenital Vertical Talus (HP: 0001838)
P1	F1	M	7	+	+	+	Elbows, knees, digits	+	+	-	-
P2	F2	M	20	+	+	-	Knees, elbows, digits	+	-	-	+
P3	F2	F	13	+	-	-	Knees	+	-	-	-
P4	F3	F	10	+	+	+	-	+	-	-	+
P5	F4	F	18	+	-	-	Elbows, knees	+	+	-	+
P6	F5	F	2	+	+	-	Elbows, knees	+	+	-	+
P7	F6	F	8	+	+	+	-	+	+	+	-
P8	F7	M	11	+	+	+	Hips, knees	+	+	-	-
P9	F8	F	1	+	+	+	Elbows	-	+	-	+
P10	F9	M	13	+	-	+	Hips, knees	+	+	-	-
P11	F10	F	8	+	+	+	Digits	+	+	-	+
P12	F11	M	6	+	+	+	Digits	-	+	-	-

The children were born after 34–41 weeks of pregnancy (Table 3). Six of the children were born vaginally, while the other six were born via c-section due to breech position and complicated pregnancy with oligohydramnios or polyhydramnios. Eight children were admitted to the neonatal intensive care unit after birth due to feeding difficulties and/or respiratory issues.

Characteristic phenotypic features of Escobar were observed in all 12 patients, including webbing over the neck, axillary, elbow, knee and/or fingers (Table 1). Downslanting palpebral fissures were observed in 10 patients (Table 2). Three patients had cleft palate and another five had high palate. Four patients had both posterior rotated ears and low-set ears, whereas another three patients either had posterior rotated ears or low-set ears. Figures 2 and 3 showcase pterygia and characteristic features in MPS among two patients.

Table 2. Phenotype: facial features and other characteristics.

Patient	Downslanting Palpebral Fissures (HP: 0000494)	Posteriorly Rotated Ears (HP: 0000358)	Low Set Ears (HP: 0000369)	Cleft Palate (HP: 0000175)	High Palate (HP: 0000218)	Other
P1	+	+	+	-	-	High nasal bridge (HP:0000426); hydrocele; inguinal hernia; tongue tie which was retracted and cauterized to divide frenulum
P2	+	-	-	-	+	Trismus, clenched hands (HP:0001188); small mouth; fusion of posterior elements of C2 and C3
P3	-	-	-	-	-	Extra cervical rib (HP:000089)
P4	+	+	+	-	+	Laryngomalacia (HP:00000060); Sprengel deformity (HP:0003745)
P5	+	-	+	-	+	Laryngomalacia (HP:00000060); corkscrew esophagus; tethering of spinal cord at age 1
P6	+	+	-	-	-	Band around right distal thigh; hypoglycemia at birth; hypoplasia of occipital lobes with prominent subarachnoid spaces
P7	-	-	-	+	-	-
P8	+	-	-	-	-	Ptosis (HP:0000508); convergent strabismus
P9	+	+	+	-	+	Facial hemangioma (HP:0000329); postnatal growth restriction
P10	+	-	+	-	+	Tongue atrophy (HP:0012473); ptosis (HP:0000508); postnatal growth restriction
P11	+	+	+	+	-	Tongue atrophy (HP:0012473)
P12	+	-	-	+	-	Right pneumothorax at birth

Seven of the 12 patients had foot deformities, six with congenital vertical tali and one with clubfeet (Table 1). Contractures were widespread across several joints (e.g., shoulders, elbows, wrists, fingers, hips, knees, ankles) and varied in severity among patients. Regarding functional mobility (data available in 11 patients), five patients were independent walkers, one used a stroller as she did not crawl or walk yet at age 22 months, two used a manual wheelchair for outdoor mobility, and three used a motorized chair for outdoor mobility. Of the five children who ambulated without a mobility aid, four wore knee ankle foot orthoses and two walked with a crouched gait.

Table 3. Pre- and postnatal information.

Patient	Prenatal Detection				Delivery		Postnatal		
	Oligohydramnios	Polyhydramnios	Lack of Fetal Move-ment/Contractures	Other Findings	GA (Weeks)	Type	Feeding Difficulty	Intubation	NICU
P1	-	-	+	Initial hydrops fetalis which resolved	38	c-section	+	-	+
P2	-	+	+	Fluid in right lung which resolved, right diaphragmatic plication at 2 weeks of age	41.5	Vaginal	+	-	+
P3	-	-	-		40	Vaginal	-	-	+
P4	+	-	+	Initial clinical impression of trisomy 18	34.5	c-section	+	+	+
P5	-	+	+		34	c-section	+	+	+
P6	-	+	+		39	c-section	+	-	+
P7	-	-	-		34	c-section	+	-	+
P8	-	-	+	Cystic hygroma at 12 weeks of gestation; vertebral block at 22 weeks of gestation; IUGR	38	Vaginal	+	-	-
P9	-	-	-	IUGR	41	Vaginal	-	-	-
P10	-	-	-	IUGR	37	Vaginal	-	-	-
P11	-	-	-		38	Vaginal	+	-	-
P12	+	-	-		37	c-section	+	-	+

IUGR: intra uterine growth restriction; NICU: neonatal intensive care unit.

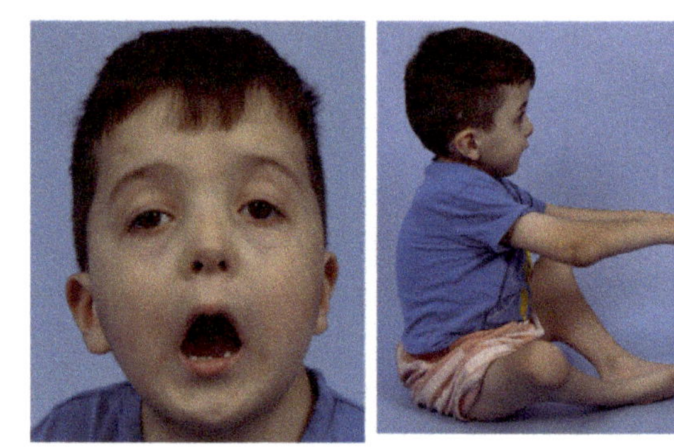
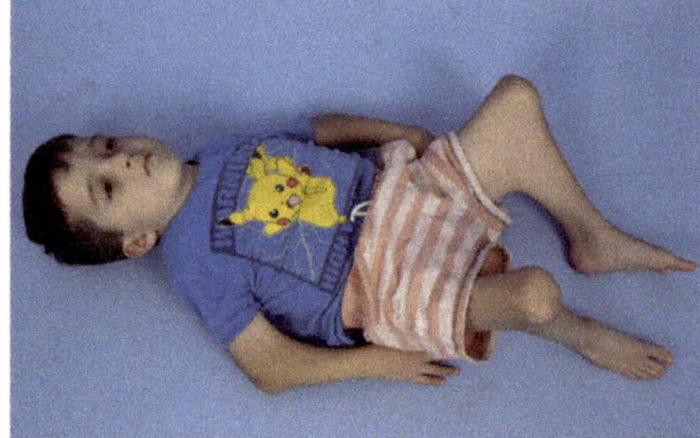

Figure 2. Patient 1 with micrognathia, downslanting palpebral fissures, lowset and posteriorly rotated ears, and pterygia apparent in the antecubital and knee popliteal regions.

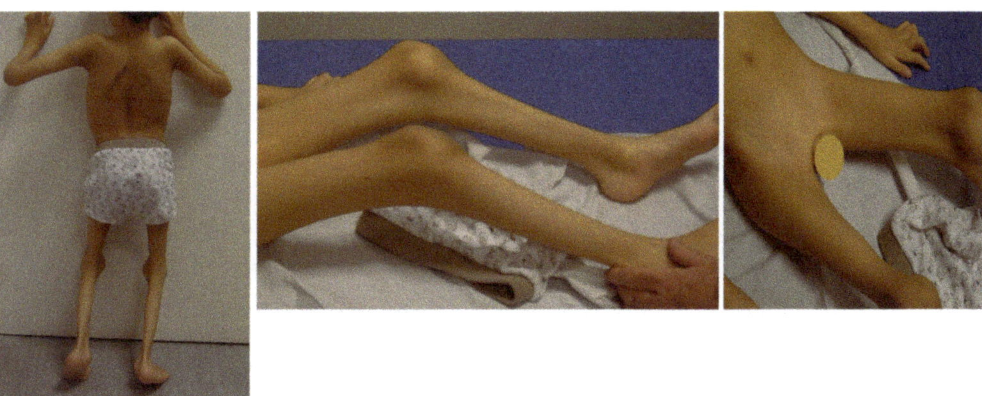

Figure 3. Patient 10 with mild to moderate axillary, popliteal, and inguinal pterygia.

Scoliosis was present in all but one patient who was only nine months of age at the last assessment. Four patients had malignant curves that had developed either during infancy or possibly prenatally (prenatal scoliosis). The level of the major curve was at the thoracolumbar spine in two of these patients, the other two patients had left low thoracic curves. These curves all included the pelvis, causing moderate to severe pelvic obliquity (18–50°). One patient had a mild but sweeping kyphosis including the pelvis, two had apical mid thoracic hyperkyphosis, whereas the last had thoracic hypokyphosis but with a thoracolumbar kyphosis. All four patients with malignant curves between 70° and 120° underwent non-fusion spinal instrumentation at 2, 4, 7 and 10 years of age in which expandable implants were placed to help control the curve during growth. Spine X-rays show a malignant curve in Figure 4.

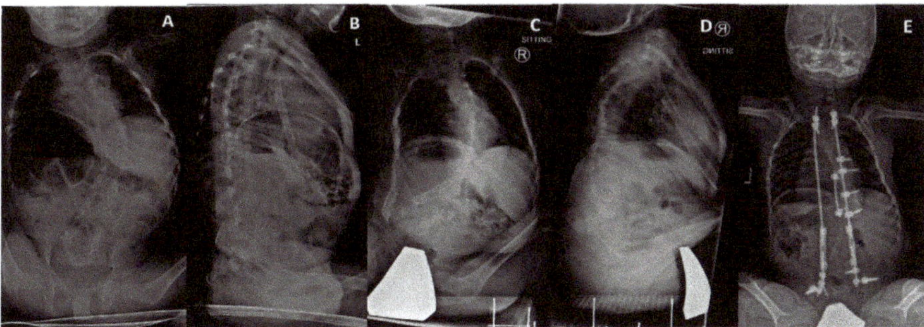

Figure 4. Patient 1 has compound heterozygous pathogenic *MYH3* variants. Curve was first noted at 13 months of age. (**A**,**B**) Posterior-anterior and lateral spine radiographs at 34 months of age showing a 68° curve with mild pelvic obliquity, and mild thoracolumbar kyphosis. (**C**) At 5 years old, the curve is 94°, with 33° pelvic obliquity and (**D**) more pronounced low thoracic kyphosis extending into the lumbar spine. (**E**) Latest follow-up at 7 years old, following halo gravity traction and a non-fusion spinal instrumentation at age 5.

Severe curves were noted in three patients, with curves progressing more slowly than the malignant curves, so that surgery could be delayed. Two patients had radiographic findings of curves at 3 and 11 months of age, whereas the third had periodic radiographs which did not demonstrate a curve until 6 years of age. All three patients had a right thoracic curve pattern, with two having thoracic hypokyphosis, which was confluent with the lumbar lordosis, the third had a thoracolumbar kyphosis. One patient was undergoing brace treatment, another had a non-fusion spinal instrumentation procedure at 11 years of age, and the last had a formal spinal fusion at 12 years of age (Figure 5).

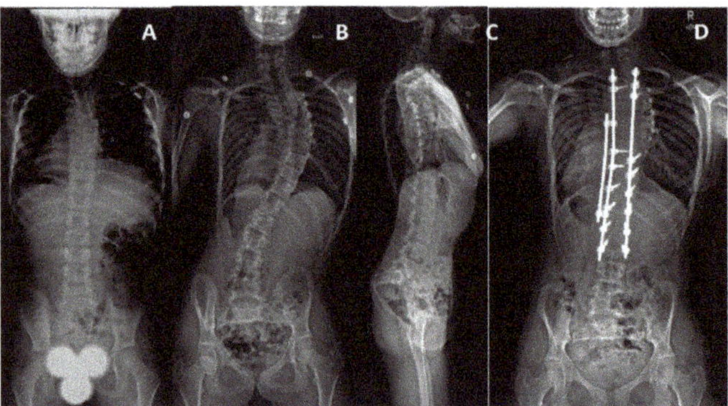

Figure 5. Patient 3 has compound heterozygous *CHRNG* variants and severe curve behavior. Curve first noted at 6 years of age. (**A**) Standing posterior-anterior spine films at 6 years old demonstrate a 19° right-sided thoracic curve. (**B,C**) A 71° curve with a lordosis spanning the thoracic and lumbar spine. (**D**) Spinal fusion at 13 years of age.

Moderate curves were found in two patients, both detected during the first year of life. Both were left sided curves, one a thoracolumbar curve with hyperkyphosis and hypolordosis, the other a lumbar curve with thoracic hypokyphosis. One patient underwent serial casting at 2 years of age, while the curve of the other did not progress past 25° until 12 years of age, and therefore was not treated. Two patients, 5 and 8 years of age at last follow up, had mild curves. They both had a forward lean on the standing sagittal radiographs with flexed hips and mild lumbar hyperlordosis. Figure 6 shows spine X-rays of a child with a moderate curve.

Thus, a total of five patients had undergone a non-fusion spinal instrumentation procedure. Interestingly, post-operative follow-up demonstrated that each of these patients had spontaneous fusions of vertebral levels, also called autofusions, in at least one uninstrumented vertebral interspace, depicted in Figure 7.

3.2. Molecular Analysis

Whole genome or whole exome sequencing was performed in 8 of the 12 patients and their parents, when available. Parents were clinically unaffected, except in family F6, as described earlier. Pathogenic recessive variants in *CHRNG* were found in five patients, all of whom had compound heterozygous variants (Table 4). Among the five different *CHRNG* variants found in these individuals, four led to premature termination codons and one represented an in-frame duplication (p.Trp98_Leu100dup) that had been described before [11]. According to family history, one other patient (P4) had compound heterozygous *CHRNG* mutations, but no detailed information was available and no DNA sample could be obtained.

Pathogenic variants in *MYH3* were found in two patients. One patient had a dominant de novo missense variant (p.Leu1204Pro), which affects the tail region of *MYH3*. This variant is not present in gnomAD and is predicted to be pathogenic by nine different prediction algorithms. Two other missense variants leading to substitutions of amino acids in the *MYH3* tail domain by proline residues have been described in the literature [12]. The other patient with pathogenic *MYH3* variants was compound heterozygous for a known recurrent splice variant (c.-9+1G > A) [12] and a novel missense variant (p.Ala183Pro). This missense is not present in gnomAD and affects the head region of MYH3, a domain where pathogenic missense variants are frequently observed [12].

A number of novel variants in genes reported to be associated with AMC were identified in patient P7, whose mother and maternal aunt have features of Escobar and scoliosis. Further validation of these variants is required given the novelty, and thus were not reported in this paper.

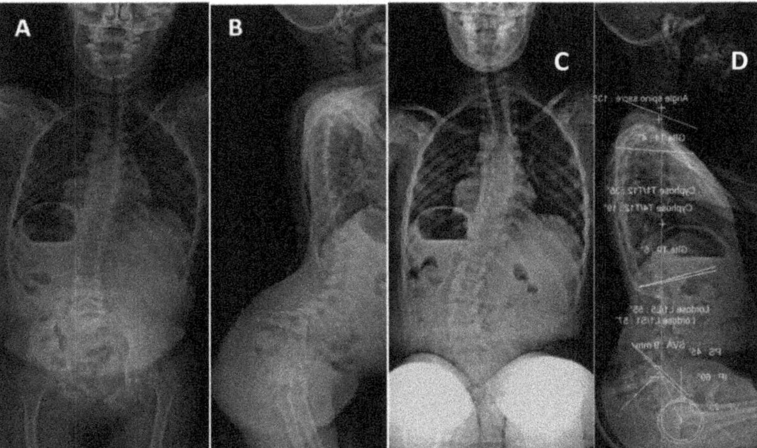

Figure 6. Patient 10 has compound heterozygous *CHRNG* variants, and a moderate scoliosis, first identified during infancy. (**A**) Posterior-anterior supine film at 10 years old, demonstrating 15° right thoracic and 18° left lumbar curves, and lateral view (**B**) Showing mild thoracic hypokyphosis and hip flexion contractures driving lumbar hyperlordosis. At this size, the curves would be considered mild. (**C,D**) Posterior-anterior and lateral sitting spine films at 12 years of age, showing that the curve had mildly progressed to 28°, reaching criteria for a moderate curve. The sagittal profile continues to show mild thoracic hypokyphosis.

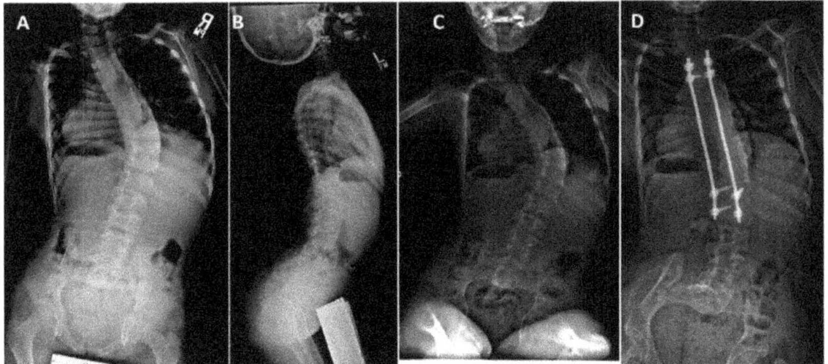

Figure 7. Patient 5 with compound heterozygous *CHRNG* variants, and severe scoliosis, first identified at 3 months of age. Spine bracing started at 24 months old. (**A**) Posterior-anterior and lateral standing films at 7 years old demonstrate a 53° right low thoracic curve, with (**B**) thoracic and lumbar lordosis on the lateral film, and flexed hips indicating hip flexion contractures. (**C**) Sitting spine films at 11 years old with curve at 66°, before non-fusion spinal instrumentation. (**D**) Autofusion along the length of the instrumented spine is seen 3 years after rod replacement at 17 years of age.

Table 4. Genotype information.

Patient	Family	Test	Gene	Variant 1	Variant 2	Inheritance
P1	F1	WES/ WGS	*MYH3*	c.-9+1G > A	c.547G > C p.Ala183Pro	AR
P2	F2	WGS	*CHRNG*	c.459dupA p.Val154SerfsTer24	c.753_754delCT p.Val253AlafsTer44	AR
P3	F2	WGS	*CHRNG*	c.459dupA p.Val154SerfsTer24	c.753_754delCT p.Val253AlafsTer44	AR
P4	F3	-	*CHRNG* *	Unavailable	Unavailable	AR
P5	F4	WGS	*CHRNG*	c.459dupA p.Val154SerfsTer24	c.401_402delCT p.Pro134ArgfsTer43	AR
P6	F5	WGS	*CHRNG*	c.459dupA p.Val154SerfsTer24	c.639_643delCAAGA p.Lys214AlafsTer82	AR
P7	F6	WGS		Negative		Dominant
P8	F7	AMC gene panel	*MYH3*	c.3611T > C p.Leu1204Pro	-	De novo
P9	F8	Single gene sequencing	*CHRNG*	c.202C > T p.Arg68Ter	c.292_300dup p.Trp98_Leu100dup	AR
P10	F9	WES		Negative		Unknown
P11	F10	WES		Pending		Unknown
P12	F11	WES		Pending		Unknown

WES: whole exome sequencing; WGS: whole genome sequencing. * According to family history, detailed information not available.

4. Discussion

Here we describe 12 individuals with variable pterygia, mild to severe flexion contractures of several joints and spine anomalies. In six of the patients, the disorder was caused by biallelic *CHRNG* mutations, one patient had an apparently dominant de novo mutation in *MYH3* and one patient was compound heterozygous for *MYH3* variants. Scoliosis was highly prevalent in our patient cohort and extremely severe in some. Intriguingly, all patients undergoing non-fusion spinal instrumentation subsequently developed fusion in at least one vertebral segment that had not been directly touched by the surgical intervention.

Biallelic loss of function variants in *CHRNG* are a well-established cause of Escobar syndrome [3,4]. Similar to previous studies, we observed that the recurrent frameshift c.459dup mutations in several unrelated patients [9]. In addition, we found that one of our patients had compound heterozygous *MYH3* variants, including the previously characterized splice variant c.-9+1G > A in the 5′ untranslated region of the gene [12,13]. Interestingly the MYH3 c.-9+1G > A variant was initially described in individuals with spondylocarpotarsal synostosis syndrome, but the reported phenotype also included webbing, contractures and scoliosis [12] and thus had considerable overlap with the clinical characteristics of our patients. Moreover, *MYH3* mutations are associated with distal arthrogryposis type 8. Based on recommendations by Biesecker et al. [14] it might be useful to use the broad term *MYH3*-related disorder to encompass these different conditions.

Multiple pterygia are seen in many types of lethal forms of AMC [15–18], but only three genes have been associated with non-lethal MPS or Escobar syndrome, namely *CHRNG*, *MYH3*, and *TPM2*. One explanation might be the exclusive or predominant expression of these genes during development and their lack or diminished expression postnatally, whereas other genes seem to be pre- and postnatally expressed. Indeed, the *CHRNG* encoded gamma-subunit of AchR, as a developmental subunit, stops being expressed at the end of the second and start of the third trimester [19], a timepoint from which the postnatal/adult CHRNE-encoded epsilon subunit is incorporated into the AchR. That most certainly also explains why individuals with *CHRNG*-related MPS do not present with clinical or electromyographic myasthenic symptoms after birth. In the same line, the *MYH3* encoded embryonic myosin heavy chain, despite its role on fiber type, fiber number, and muscle fiber differentiation [20], is much less expressed in human postnatal muscle [21]. This holds also true for *TPM2* [22].

The development of multiple pterygia probably reflects the spectrum of the most severe consequence and earliest onset of fetal akinesia. In this regard, pathogenic variants in *CHRNG*, *MYH3* and *TPM2* described to date have been associated with loss of function mutations. In *CHRNG*, nonsense, splice site and missense mutations alike have been shown to abolish AchR expression at the sarcolemma [3,4], thus probably completely abolishing neuromuscular transmission during embryonic and fetal development. Of note, the AchR is still expressed even in case of loss of function mutations in other genes involved in the maintenance and function of the neuromuscular junction, such as *RAPSN* or *MUSK* [9,23,24], leading to a severe fetal akinesia deformation sequence phenotype, but without pterygia.

Scoliosis in children with Escobar syndrome is extremely prevalent, with most curves eventually progressing to treatment. Other authors have reported prevalence rates of scoliosis in patients with multiple pterygium syndrome between 32% and 93% [25–28]. Of the 12 patients covered in this report, only three did not yet require any scoliosis treatment, one of whom is only 9 months old, the other two are under 10 years. Four of the patients had a spine deformity that behaved beyond what is typically considered severe, therefore we named curves that were greater than 50° before 4 years of age "malignant". The patients with malignant curves also had more severe limb involvement, most commonly including severe hip and knee flexion contractures with pterygium. Most malignant curves were first detected in infancy, some even in the neonatal time period, suggesting that some were actually prenatal curves, curves that reflected the patients' unchanging intrauterine position due to severe fetal akinesia. Importantly, other patients also had curves detected during their first year of life, but did not develop malignant curves, varying between mild and severe. The malignant curves all had pelvic obliquity near or greater than 20°, an indication of the uncompensated nature of these curves. Two of the patients in the severe category initially had their curves detected before their first birthday, but the third patient (Patient 3) was routinely monitored with serial radiographs, and no curve was detected prior to 6 years of age. All three of patients with severe curves underwent bracing of their spines, which allowed further growth prior to needing surgical stabilization of their curves. One of the shortcomings of the classification system we used is that a patient may progress from one class to another as they grow and their curves progress, and that the classification was based on a very small group of subjects. It is likely that some of the mild and moderate curves will progress to severe prior to the patient reaching skeletal maturity. We felt, though, that it provided some structure with which to analyze and group the subjects.

We expect that larger studies will provide insights, allowing for improvements of the scoliosis classification system used in this manuscript. A molecular diagnosis was confirmed in seven of our patients, two of which were found to have a *MYH3* gene mutation; the remaining five had a *CHRNG* mutation. The two patients with *MYH3* mutations both had malignant curves. The two other patients with malignant curves both had *CHRNG* gene mutations, of which one, Patient 2, was the older sibling of Patient 3, with the exact same mutation, but Patient 3 had a severe scoliosis. We have two other sibling pairs that we could not include in this study, since they did not have complete data available. One brother–sister pair both had malignant curves that started in infancy, whereas the brother of the other brother–sister pair had a malignant curve that started in infancy yet his sister's curve did not become severe (crossing 50°) until after 10 years of age. Curve patterns varied modestly in our cohort. Two of the malignant curves were thoracolumbar extending to the pelvis, one left-sided the other right, whereas the other two were low thoracic apex curves, also extending nearly to the pelvis, both left-sided. Three of the spines had thoracic hyperkyphosis, but the fourth had a lordotic thoracic spine with a mild thoracolumbar kyphosis. The severe curves were right sided mid- or lower thoracic-apex curves, with relatively lordotic thoracic spines, two patients had a mild thoracolumbar kyphosis. The mild and moderate curves in general were a combination of right thoracic and/or left lumbar curves, and usually a lumbar hyperlordosis. This variability complements the findings of Margalit and colleagues [29], who found three of their nine patients had right-sided thoracolumbar curves, and the rest had left-sided thoracolumbar curves. They did not describe the patterns of sagittal appearance of their patients. Coalescence of vertebral levels the spine appears to be common in patients with Escobar syndrome, particularly in the severe and malignant curves. Margalit et al. [29] noted the same on pre-operative computer tomography (CT) of their patients.

In our cohort, it is clear that spontaneous vertebral fusions occur, particularly in the severe and malignant curves, best seen in the children undergoing non-fusion spinal instrumentations. In these spines, progressive intervertebral fusions are seen both anteriorly and posteriorly in the uninstrumented section of the spine, suggesting either that lack of intervertebral motion, or the distraction of the space, leads to the fusion. We did not identify vertebral abnormalities or lack of vertebral segmentation in the films of our patients under 2 years of age, although we did not have CT scans and details could be difficult to visualize on the films. Therefore, we were unable to resolve if any of the intervertebral fusions were congenital failure of segmentation, but we suspect that most, if not all, were due to postnatal spontaneous fusions. Clearly, patients with Escobar syndrome need to be carefully monitored for the development of scoliosis, and aggressively treated to postpone the need for surgical intervention. Joo et al. [28] noted a tethered cord or a syrinx in 4 of their 16 patients. Although only one of our patients needed detethering of their spinal cord (Patient 5), treating physicians need to be vigilant for such possibilities. Both bracing and serial spine casting appear to be beneficial to some extent in controlling the curve and allowing further growth for at least a few years. Patients with spine-induced pelvic obliquity, particularly those apparent in infancy, likely have malignant curves. This seems to be particularly true for patients with *MYH3* mutations underlying their Escobar syndrome. Parents need to be informed about the challenging nature of the curve, and that non-fusion spinal instrumentation will be needed to try to modulate the curve during growth. Patients undergoing non-fusion spinal instrumentation are likely to experience spontaneous fusions of their spine, which may limit the amount of expansion possible during the child's growing years. Spine balance must be a priority at the initial implantation of the expandable device, as a formal fusion may not be necessary due to the spontaneous fusions, so long as spine balance is satisfactory. Conversely, if the spine is not well balanced, a formal fusion after a non-fusion spinal instrumentation will be very challenging due to the fusions.

5. Conclusions

In conclusion, we found a unique spine phenotype in these patients with MPS caused by *CHRNG* and *MYH3* mutations. More detailed characterization using 3D imaging may help further refine this spine phenotype in patients with MPS.

Author Contributions: Conceptualization, N.D.-O., K.D., K.S., P.F.G. and H.v.B.; methodology, N.D.-O., K.D., F.R., A.G.G., K.S., P.F.G. and H.v.B.; data acquisition, N.D.-O., K.D., F.R., G.B., T.N.B., A.G.G., R.H., X.L., K.S., P.F.G. and H.v.B.; analysis, N.D.-O., K.D., F.R., G.B., T.N.B., A.G.G., R.H., X.L., K.S., P.F.G. and H.v.B.; resources, N.D.-O., K.D., F.R., G.B., T.N.B., A.G.G., R.H., X.L., K.S., P.F.G. and H.v.B.; writing—original draft preparation, N.D.-O., K.D., T.N.B., A.G.G., X.L., K.S., P.F.G. and H.v.B.; writing—review and editing, N.D.-O., K.D., F.R., G.B., T.N.B., A.G.G., R.H., X.L., K.S., P.F.G. and H.v.B.; funding acquisition, N.D.-O., K.D., F.R., R.H., P.F.G. and H.v.B. All authors have read and agreed to the published version of the manuscript.

Funding: This research was funded by Shriners Hospitals for Children for the pediatric arthrogryposis multiplex congenita registry at the Montreal and Philadelphia hospitals (grant 79150) to N.D.-O., R.H., F.R., G.B., P.F.G. and H.v.B., and from the Direction pour la Recherche et l'Innovation (DRCI) of the Grenoble Alpes University Hospital Center (Pediatric and Adult Registry for Arthrogryposis

Multiplex Congenita—project PARART) to K.D. Funding was obtained from the Research reported in this publication was also supported in part by the Eunice Kennedy Shriver National Institute of Child Health & Human Development of the National Institutes of Health under Award Number R03HD099516 to P.F.G. N.D.-O. holds a clinical research scholar award from the Fonds de la Recherche en Santé du Québec. We gratefully acknowledge the support of the Malika Ray, Asok K. Ray, M.D., FRCS/(Edin) Initiative for Child Health Research.

Institutional Review Board Statement: The study was conducted according to the guidelines of the Declaration of Helsinki, and approved by the Institutional Review Board of McGill University's Faculty of Medicine, Canada (protocol A08-M30-19B on 4 October 2019), WCG Institutional Review Board, USA (protocol CAN1903 on 30 November 2019) and by the Institutional Review Board of Grenoble Alpes University Hospital Center, France (protocol 2205066v0 on 16 December 2019).

Informed Consent Statement: Informed consent was obtained from all subjects involved in the study. Written informed consent has been obtained from the patients to publish this paper.

Data Availability Statement: The data presented in this study are available on request from the corresponding authors. The data are not publicly available due to the ethical and privacy nature of the data.

Acknowledgments: The authors are grateful to Vasiliki Betty Darsaklis from the Shriners Hospital for Children in Montreal, Canada for her assistance with data extraction and the Molecular Biology Facility of the Grenoble University Hospital, France for technical assistance.

Conflicts of Interest: The authors declare no conflict of interest. The funders had no role in the design of the study; in the collection, analyses, or interpretation of data; in the writing of the manuscript, or in the decision to publish the results.

References

1. Dahan-Oliel, N.; Cachecho, S.; Barnes, D.; Bedard, T.; Davison, A.M.; Dieterich, K.; Donohoe, M.; Faҫara, A.; Hamdy, R.; Hjartarson, H.T.; et al. International multidisciplinary collaboration toward an annotated definition of arthrogryposis multiplex congenita. *Am. J. Med. Genet. Part C Semin. Med. Genet.* **2019**, *181*, 288–299. [CrossRef] [PubMed]
2. Chen, H. Multiple Pterygium Syndrome. In *Atlas of Genetic Diagnosis and Counseling*, 2nd ed.; Springer: New York, NY, USA; Dordrecht, The Netherland; Heidelberg, Germany; London, UK, 2012; pp. 1479–1485. [CrossRef]
3. Hoffmann, K.; Muller, J.S.; Stricker, S.; Megarbane, A.; Rajab, A.; Lindner, T.H.; Cohen, M.; Chouery, E.; Adaimy, L.; Ghanem, I.; et al. Escobar syndrome is a prenatal myasthenia caused by disruption of the acetylcholine receptor fetal γ subunit. *Am. J. Hum. Genet.* **2006**, *79*, 303–312. [CrossRef] [PubMed]
4. Morgan, N.V.; Brueton, L.A.; Cox, P.; Greally, M.T.; Tolmie, J.; Pasha, S.; Aligianis, I.A.; van Bokhoven, H.; Marton, T.; Al-Gazali, L.; et al. Mutations in the embryonal subunit of the acetylcholine receptor (CHRNG) cause lethal and Escobar variants of multiple pterygium syndrome. *Am. J. Hum. Genet.* **2006**, *79*, 390–395. [CrossRef] [PubMed]
5. Prontera, P.; Vogt, J.; McKeown, C.; Sensi, A. Familial multiple pterygium syndrome (MPS) is not associated with CHRNG gene mutation. *Am. J. Med. Genet. Part A* **2007**, *143A*, 1129. [CrossRef] [PubMed]
6. Chong, J.X.; Burrage, L.C.; Beck, A.E.; Marvin, C.T.; McMillin, M.J.; Shively, K.M.; Harrell, T.M.; Buckingham, K.J.; Bacino, C.A.; Jain, M.; et al. Autosomal-Dominant Multiple Pterygium Syndrome Is Caused by Mutations in MYH3. *Am. J. Hum. Genet.* **2015**, *96*, 841–849. [CrossRef] [PubMed]
7. Zieba, J.; Zhang, W.; Chong, J.X.; Forlenza, K.N.; Martin, J.H.; Heard, K.; Grange, D.K.; Butler, M.G.; Kleefstra, T.; Lachman, R.S.; et al. A postnatal role for embryonic myosin revealed by MYH3 mutations that alter TGFβ signaling and cause autosomal dominant spondylocarpotarsal synostosis. *Sci. Rep.* **2017**, *7*, 41803. [CrossRef]
8. Komolkin, I.; Ulrich, E.V.; Agranovich, O.E.; van Bosse, H. Treatment of Scoliosis Associated with Arthrogryposis Multiplex Congenita. *J. Pediatric Orthop.* **2017**, *37* (Suppl. 1), S24–S26. [CrossRef] [PubMed]
9. Vogt, J.; Harrison, B.J.; Spearman, H.; Cossins, J.; Vermeer, S.; ten Cate, L.N.; Morgan, N.V.; Beeson, D.; Maher, E.R. Mutation analysis of CHRNA1, CHRNB1, CHRND, and RAPSN genes in multiple pterygium syndrome/fetal akinesia patients. *Am. J. Hum. Genet.* **2008**, *82*, 222–227. [CrossRef]
10. Ravenscroft, G.; Clayton, J.S.; Faiz, F.; Sivadorai, P.; Milnes, D.; Cincotta, R.; Moon, P.; Kamien, B.; Edwards, M.; Delatycki, M.; et al. Neurogenetic fetal akinesia and arthrogryposis: Genetics, expanding genotype-phenotypes and functional genomics. *J. Med. Genet.* **2020**. [CrossRef]
11. Carrera-García, L.; Natera-de Benito, D.; Dieterich, K.; de la Banda, M.; Felter, A.; Inarejos, E.; Codina, A.; Jou, C.; Roldan, M.; Palau, F.; et al. CHRNG-related nonlethal multiple pterygium syndrome: Muscle imaging pattern and clinical, histopathological, and molecular genetic findings. *Am. J. Med. Genet. Part A* **2019**, *179*, 915–926. [CrossRef]

12. Cameron-Christie, S.R.; Wells, C.F.; Simon, M.; Wessels, M.; Tang, C.; Wei, W.; Takei, R.; Aarts-Tesselaar, C.; Sandaradura, S.; Sillence, D.O.; et al. Recessive Spondylocarpotarsal Synostosis Syndrome Due to Compound Heterozygosity for Variants in MYH3. *Am. J. Hum. Genet.* **2018**, *102*, 1115–1125. [CrossRef]
13. Hakonen, A.H.; Lehtonen, J.; Kivirikko, S.; Keski-Filppula, R.; Moilanen, J.; Kivisaari, R.; Almusa, H.; Jakkula, E.; Saarela, J.; Avela, K.; et al. Recessive MYH3 variants cause "Contractures, pterygia, and variable skeletal fusions syndrome 1B" mimicking Escobar variant multiple pterygium syndrome. *Am. J. Med. Genet. Part A* **2020**, *182*, 2605–2610. [CrossRef] [PubMed]
14. Biesecker, L.G.; Adam, M.P.; Alkuraya, F.S.; Amemiya, A.R.; Bamshad, M.J.; Beck, A.E.; Bennett, J.T.; Bird, L.M.; Carey, J.C.; Chung, B.; et al. A dyadic approach to the delineation of diagnostic entities in clinical genomics. *Am. J. Hum. Genet.* **2021**, *108*, 8–15. [CrossRef] [PubMed]
15. Laquérriere, A.; Maluenda, J.; Camus, A.; Fontenas, L.; Dieterich, K.; Nolent, F.; Zhou, J.; Monnier, N.; Latour, P.; Gentil, D.; et al. Mutations in CNTNAP1 and ADCY6 are responsible for severe arthrogryposis multiplex congenita with axoglial defects. *Hum. Mol. Genet.* **2014**, *23*, 2279–2289. [CrossRef]
16. Ravenscroft, G.; Thompson, E.M.; Todd, E.J.; Yau, K.S.; Kresoje, N.; Sivadorai, P.; Friend, K.; Riley, K.; Manton, N.D.; Blumbergs, P.; et al. Whole exome sequencing in foetal akinesia expands the genotype-phenotype spectrum of GBE1 glycogen storage disease mutations. *Neuromuscul. Disord.* **2013**, *23*, 165–169. [CrossRef]
17. McKie, A.B.; Alsaedi, A.; Vogt, J.; Stuurman, K.E.; Weiss, M.M.; Shakeel, H.; Tee, L.; Morgan, N.V.; Nikkels, P.G.; van Haaften, G.; et al. Germline mutations in RYR1 are associated with foetal akinesia deformation sequence/lethal multiple pterygium syndrome. *Acta Neuropathol. Commun.* **2014**, *2*, 148. [CrossRef]
18. Zhou, X.; Zhou, J.; Wei, X.; Yao, R.; Yang, Y.; Deng, L.; Zou, G.; Wang, X.; Yang, Y.; Duan, T.; et al. Value of Exome Sequencing in Diagnosis and Management of Recurrent Non-immune Hydrops Fetalis: A Retrospective Analysis. *Front. Genet.* **2021**, *12*, 616392. [CrossRef] [PubMed]
19. Hesselmans, L.F.; Jennekens, F.G.; Van den Oord, C.J.; Veldman, H.; Vincent, A. Development of innervation of skeletal muscle fibers in man: Relation to acetylcholine receptors. *Anat. Rec.* **1993**, *236*, 553–562. [CrossRef] [PubMed]
20. Agarwal, M.; Sharma, A.; Kumar, P.; Kumar, A.; Bharadwaj, A.; Saini, M.; Kardon, G.; Mathew, S.J. Myosin heavy chain-embryonic regulates skeletal muscle differentiation during mammalian development. *Development* **2020**, *147*, dev184507. [CrossRef] [PubMed]
21. Racca, A.W.; Beck, A.E.; McMillin, M.J.; Korte, F.S.; Bamshad, M.J.; Regnier, M. The embryonic myosin R672C mutation that underlies Freeman-Sheldon syndrome impairs cross-bridge detachment and cycling in adult skeletal muscle. *Hum. Mol. Genet.* **2015**, *24*, 3348–3358. [CrossRef] [PubMed]
22. Laurent-Winter, C.; Soussi-Yanicostas, N.; Butler-Browne, G.S. Biphasic expression of slow myosin light chains and slow tropomyosin isoforms during the development of the human quadriceps muscle. *FEBS Lett.* **1991**, *280*, 292–296. [CrossRef]
23. Tan-Sindhunata, M.B.; Mathijssen, I.B.; Smit, M.; Baas, F.; de Vries, J.I.; van der Voorn, J.P.; Kluijt, I.; Hagen, M.A.; Blom, E.W.; Sistermans, E.; et al. Identification of a Dutch founder mutation in MUSK causing fetal akinesia deformation sequence. *Eur. J. Hum. Genet.* **2015**, *23*, 1151–1157. [CrossRef] [PubMed]
24. Wilbe, M.; Ekvall, S.; Eurenius, K.; Ericson, K.; Casar-Borota, O.; Klar, J.; Dahl, N.; Ameur, A.; Annerén, G.; Bondeson, M.L. MuSK: A new target for lethal fetal akinesia deformation sequence (FADS). *J. Med. Genet.* **2015**, *52*, 195–202. [CrossRef] [PubMed]
25. Escobar, V.; Bixler, D.; Gleiser, S.; Weaver, D.D.; Gibbs, T. Multiple pterygium syndrome. *Am. J. Dis. Child.* **1978**, *132*, 609–611. [CrossRef] [PubMed]
26. Hall, J.G.; Reed, S.; Rosenbaum, K.N.; Gershanik, J.; Chen, H.; Wilson, K.M. Limb pterygium syndromes: A review and report of eleven patients. *Am. J. Med. Genet.* **1982**, *12*, 377–409. [CrossRef]
27. Angsanuntsukh, C.; Oto, M.; Holmes, L.; Rogers, K.J.; King, M.M.; Donohoe, M.; Kumar, S.J. Congenital vertical talus in multiple pterygium syndrome. *J. Pediatr. Orthop.* **2011**, *31*, 564–569. [CrossRef] [PubMed]
28. Joo, S.; Rogers, K.J.; Donohoe, M.; King, M.M.; Kumar, S.J. Prevalence and patterns of scoliosis in children with multiple pterygium syndrome. *J. Pediatr. Orthop.* **2012**, *32*, 190–195. [CrossRef]
29. Margalit, A.; Sponseller, P.D.; McCarthy, R.E.; Pawelek, J.B.; McCullough, L.; Karlin, L.I.; Shirley, E.D.; Schwend, R.M.; Samdani, A.F.; Akbarnia, B.A.; et al. Growth-Friendly Spine Surgery in Escobar Syndrome. *J. Pediatric Orthop.* **2019**, *39*, e506–e513. [CrossRef] [PubMed]

genes

Article

Novel *FGFR1* Variants Are Associated with Congenital Scoliosis

Shengru Wang [1,2,†], Xiran Chai [1,3,†], Zihui Yan [1,2,3,4,†], Sen Zhao [1,2,3,4], Yang Yang [1,2], Xiaoxin Li [2,4,5], Yuchen Niu [2,4,5], Guanfeng Lin [1,2], Zhe Su [1,2], Zhihong Wu [1,2,4,5], Terry Jianguo Zhang [1,2,4,*,‡] and Nan Wu [1,2,4]

[1]   Department of Orthopedic Surgery, State Key Laboratory of Complex Severe and Rare Diseases, Peking Union Medical College Hospital, Peking Union Medical College and Chinese Academy of Medical Sciences, Beijing 100730, China; wangshengru@foxmail.com (S.W.); chaixiran1124@163.com (X.C.); yanzihui1127@126.com (Z.Y.); zhaosen830@163.com (S.Z.); kaido137@hotmail.com (Y.Y.); richard_lingf@foxmail.com (G.L.); suesu0092@126.com (Z.S.); orthoscience@126.com (Z.W.); dr.wunan@pumch.cn (N.W.)
[2]   Key Laboratory of Big Data for Spinal Deformities, Chinese Academy of Medical Sciences, Beijing 100730, China; ldx217@yeah.net (X.L.); nhtniuyuchen@126.com (Y.N.)
[3]   Graduate School of Peking Union Medical College, Beijing 100005, China
[4]   Beijing Key Laboratory for Genetic Research of Skeletal Deformity, Beijing 100730, China
[5]   Medical Research Center, State Key Laboratory of Complex Severe and Rare Diseases, Peking Union Medical College Hospital, Peking Union Medical College and Chinese Academy of Medical Sciences, Beijing 100000, China
*   Correspondence: jgzhang_pumch@yahoo.com
†   These authors contributed equally to this work.
‡   Current address: The Department of Orthopedic Surgery, Beijing Key Laboratory for Genetic Research of Skeletal Deformity, State Key Laboratory of Complex Severe and Rare Diseases, All at Peking Union Medical College Hospital, Peking Union Medical College and Chinese Academy of Medical Sciences, No. 1 Shuaifuyuan, Beijing 100730, China.

Citation: Wang, S.; Chai, X.; Yan, Z.; Zhao, S.; Yang, Y.; Li, X.; Niu, Y.; Lin, G.; Su, Z.; Wu, Z.; et al. Novel *FGFR1* Variants Are Associated with Congenital Scoliosis. *Genes* **2021**, *12*, 1126. https://doi.org/10.3390/genes12081126

Academic Editors: Roel Ophoff, Nancy Hadley-Miller, Philip F. Giampietro and Cathy L. Raggio

Received: 1 April 2021
Accepted: 14 July 2021
Published: 24 July 2021

Publisher's Note: MDPI stays neutral with regard to jurisdictional claims in published maps and institutional affiliations.

Abstract: *FGFR1* encodes a transmembrane cytokine receptor, which is involved in the early development of the human embryo and plays an important role in gastrulation, organ specification and patterning of various tissues. Pathogenic *FGFR1* variants have been associated with Kallmann syndrome and hypogonadotropic hypogonadism. In our congenital scoliosis (CS) patient series of 424 sporadic CS patients under the framework of the Deciphering disorders Involving Scoliosis and COmorbidities (DISCO) study, we identified four unrelated patients harboring *FGFR1* variants, including one frameshift and three missense variants. These variants were predicted to be deleterious by in silico prediction and conservation analysis. Signaling activities and expression levels of the mutated protein were evaluated in vitro and compared to that of the wild type (WT) *FGFR1*. As a result, the overall protein expressions of c.2334dupC, c.2339T>C and c.1261A>G were reduced to 43.9%, 63.4% and 77.4%, respectively. By the reporter gene assay, we observed significantly reduced activity for c.2334dupC, c.2339T>C and c.1261A>G, indicating the diminished FGFR1 signaling pathway. In conclusion, *FGFR1* variants identified in our patients led to only mild disruption to protein function, caused milder skeletal and cardiac phenotypes than those reported previously.

Keywords: *FGFR1* (Fibroblast growth factor receptor 1); genetic variations; congenital scoliosis

1. Introduction

The Fibroblast growth factor receptor 1 (*FGFR1*) gene encodes a transmembrane cytokine receptor, which comprises an extracellular region of three immunoglobulin-like domains (D1, D2 and D3), a transmembrane helix and a cytoplasmic tyrosine kinase domain [1]. Although different isoforms have different tissue expression and varied affinity to FGFs, *FGFR1-IIIc*, spliced through the use of exon 8B, is the predominant isoform that carries out most of the functions of the *FGFR1* gene [2].

The downstream signaling of *FGFR1* is activated by the dimerization and activation of the receptor and autophosphorylation of the tyrosine kinase domains. These downstream signaling pathways include the mitogen activated protein kinases (MAPK), the phosphatidylinositide 3 kinase/AKT (PI3K/AKT) and the phospholipase C γ (PLC) [1,3]. *FGFR1*-related signaling pathways are involved in the early development of the human embryo, and thus play an important role in gastrulation, organ specification and patterning of many tissues [4].

Many *FGFR1* mutations have been identified in both Kallmann syndrome and isolated hypogonadotropic hypogonadism (IHH) [5–9]. *FGFR1* loss-of-function mutations were also reported to be found in Kallmann syndrome patients with skeletal phenotypes, including oligodactyly, hemivertebrae and butterfly vertebrae [10] and *FGFR1* signaling was reported to be important for different stages of osteoblast maturation [11]. Mice models with *FGFR1* variants presented various skeletal phenotypes, especially vertebral malformation from cervical vertebrae to lumbar vertebrae, making *FGFR1* a candidate gene for congenital scoliosis [12]. However, whether *FGFR1* is associated with vertebral malformations in human remains unknown.

In this study, we analyzed variants of *FGFR1* identified in a cohort of congenital scoliosis (CS) and performed in vitro experiments to determine the effects of these variants on the protein function.

2. Materials and Methods

2.1. Human Subjects

A total of 424 sporadic Han Chinese probands who received a diagnosis of congenital scoliosis (CS) were consecutively collected into the cohort between 2009 and 2018 at Peking Union Medical College Hospital (PUMCH) under the the framework of the Deciphering disorders Involving Scoliosis and COmorbidities (DISCO, http://discostudy.org/, accessed on 15 March 2021) study. Demographic information, physical examination results, clinical symptoms on presentation, and a detailed medical history were obtained from each proband. Clinical diagnoses were confirmed by radiology imaging. Blood was obtained from all the probands and whole exome sequencing (WES) was performed.

A total of 942 Han Chinese individuals without evidence of congenital scoliosis or other congenital malformations from the DISCO project served as in-house controls. All in-house controls provided their blood for DNA analysis and signed written informed consent.

2.2. Bioinformatic Analysis and Mutation Interpretation

WES data processing was performed using the Peking Union Medical college hospital Pipeline (PUMP) [13,14] developed in-house. Computational prediction tools (Genomic Evolutionary Rate Profiling [GERP] [15], Combined Annotation Dependent Depletion [CADD PHRED-score, GRCh37-v1.6] [16], Sorting Intolerant Form Tolerant [SIFT] [17], Polyphen-2 [18], and MutationTaster [19]) were used to predict the conservation and pathogenicity of candidate variants. All variants were compared against population genomic databases such as the 1000 Genomes Project (http://www.internationalgenome.org/, accessed on 15 March 2021), the NHLBI GO Exome Sequencing Project (ESP) Exome Variant Server (http://evs.gs.washington.edu/EVS/, accessed on 15 March 2021) and the genome Aggregation Database (gnomAD, http://gnomad.broadinstitute.org/, accessed on 15 March 2021).

Candidate variants in *FGFR1* were extracted and filtered using the following criteria:

(1) Truncating (nonsense, frameshift, splice acceptor/donor) variants or missense variants/inframe indels with a CADD score ≥ 20;
(2) Absent from population genomic databases listed above.

2.3. Site-Directed Mutagenesis

Plasmids of pcDNA3.1+ with N-terminal myc-tagged WT and mutant *FGFR1c* cDNA (NM_023110.2) were acquired from Beijing Hitrobio Biotechnology. The mutant constructs were sequenced on both strands to verify nucleotide changes.

2.4. Receptor Expression and Maturation Studies

2.4.1. Endoglycosidase Digestion

Endoglycosidase assays were performed as previously published [8]. In brief, COS-7 cells (Cell Resource Center, Peking Union Medical College, Beijing, China) with 60–70% confluence were transiently transfected with 300 ng of plasmid containing myc-tagged WT or mutated *FGFR1* cDNA in 6-well plates using Lipofectamine 3000 reagent (Thermo Fisher Scientific, Waltham, MA, USA). Forty-eight hours post transfection, cells were washed with phosphate-buffered saline (PBS), and then, lysed with 100 μL of RIPA buffer (Thermo Fisher Scientific, Waltham, MA, USA) containing 1× protease inhibitor (Solarbio, Beijing, China). For deglycosylation analysis, all protein lysates were diluted to 10 μg/μL, and 9 μL of diluted lysate (90 μg of total protein) was subjected to PNGasef and EndoH digestion according to the manufacturer's recommendations (New England Biolab, Ipswich, MA, USA).

2.4.2. Western Analysis

Untreated or endoglycosidase-treated samples were resolved on gels under reducing conditions and then, subjected to Western analysis using an anti-myc primary antibody (clone 4A6, 1:1000, Upstate Biotechnology, Inc., Lake Placid, NY, USA) and a goat anti-mouse horseradish peroxidase-conjugated secondary antibody (1:5000, Bioss, Edinburgh, UK). Immunoreactivity was visualized using Western Lighting chemiluminescence reagent (Beyotime, Wuhan, China). To control for equal loading, blots were stripped using Restore Western Blot Stripping Buffer (Applygen Technologies, Beijing, China) and reprobed with horseradish peroxidase-conjugated anti-β-actin antibody (1:5000, Proteintech, Rosemont, IL, USA). FGFR1 and β-actin immunoreactivity were quantified by densitometry using an automatic chemiluminescence imaging system (Tanon, Shanghai, China). Overall expression levels of WT and mutant receptors were determined from the PNGase-treated samples and were normalized to their respective β-actin levels. The ratio between mutant and WT was reported. For receptor maturation studies, the upper (mature) and lower (immature) band densities were determined individually from the EndoH-treated samples, and the percent of mature fraction (maturation level) was calculated as overall protein divided by matured protein. The maturation levels of four variants were compared with the WT group, i.e., maturation ratio. Endoglycosidase and Western experiments were repeated three times.

2.4.3. FGF Reporter Gene Assay

The activation of downstream signaling pathways by wild type and mutated *FGFR1* constructs was interrogated using the luciferase-based reporter assay; the osteocalcin FGF response element (OCFRE) reporter reports the activity of the MAPK pathway downstream of FRS2α signaling [9]. In detail, L6 myoblasts (Cell Resource Center, Peking Union Medical College, Beijing, China), which are largely devoid of endogenous FGFRs and FGFs, were maintained in DMEM-H containing penicillin (100 U/L), streptomycin (100 ug/L), and 10% fetal calf serum. Transient transfections were performed at 60–70% cell confluency in 24-well plates with 300 ng of plasmid containing WT or mutant *FGFR1* cDNA, 400 ng of osteocalcin FGF response element-pGL3 plasmid, and 10 ng of pRL plasmid using Lipofectamine 3000 reagent (Thermo Fisher Scientific, Waltham, MA, USA). After 24 h of serum starvation, cells were treated for 16 h with FGF18 (10-8 M) in DMEM-H containing 0.1% BSA. The cells were lysed with passive lysis buffer (Promega, Madison, WI, USA), and assayed for luciferase activity using a Promega luciferase assay system. Experiments were performed in triplicate and repeated at least three times. Results of each experiment were normalized to the WT and the mean values of three experiments were calculated.

2.4.4. Statistical Analyses

The frequency of candidate variants of *FGFR1* was compared between the control group and the CS group using the Fisher Exact Test. Luciferase activities and overall expression levels were normalized to WT (set as 100%) and mean values of mutant versus WT from all three experiments were compared using one-way ANOVA and Dunnett's multiple comparisons test. All charts were drawn and analyzed using GraphPad Prism 7 and $p < 0.05$ was considered significant for all analyses.

3. Results

3.1. Mutation and Phenotype Analyses

In the 424 sporadic CS patients, 79 patients (18.6%) were found to have a molecular diagnosis by pathogenic genetic variants, as previously reported [13]. From the probands who remained undiagnosed, four likely deleterious heterozygous variants of *FGFR1*, including one frameshift variant and three rare missense variants (c.2334dupC; c.2339T>C; c.1107G>A; c.1261A>G), were identified (Table 1), presenting a significant mutational burden as compared with the in-house controls (one candidate variant in 942 control individuals, $p = 0.035$, Fisher Exact Test). The authenticity of all variants was validated by manual review of BAM files using the Integrative Genomics Viewer (http://igv.org, accessed on 15 March 2021).

Table 1. Demographic, phenotypic and variant information of four patients in our series. All variants' nomenclatures were based on the *FGFR1* transcript NM_023110.2. All positions were aligned to GRCh37/hg19.

	Patient #1	Patient #2	Patient #3	Patient #4
Sex	Female	Female	Male	Male
Age of onset	11	4	0	1
CS type	Failure of segmentation	Mixed defects	Failure of formation	Failure of formation
Vertebral malformation	T6-T10 Spine fusion	T9 Hemivertebrae, T8 Butterfly vertebrae	T10 Hemivertebrae	T10 Hemivertebrae
Associated anomalies	Mitral valve prolapse; Fusion of 9th and 10th ribs	9th, 10th and 12th ribs absent	None	None
Variant nomenclature	c.2334dupC(p.Ser779GlnfsTer21)	c.2339T>C(p.Phe780Ser)	c.1107G>A(p.Met369Ile)	c.1261A>G(p.Ile421Val)
Mutation type	Frameshift	Missense	Missense	Missense
Position	Chr8_38271280	Chr8_38271276	Chr8_38277228	Chr8_38277074
1000G_ASN_AF	0	0	0	0
gnomAD_EAS_AF	0	0	0	0
ESP6500_AF	0	0	0	0
MutationTaster	NA	1	0.999	1
SIFT	NA	0.53	0.25	0.02
Polyphen2	NA	0.948	0.174	0.481
LRT	NA	0	0	0
CADD PHRED-score	32	24.2	22.8	22.0

AF, allele frequency; pLI, probability of loss-of function intolerance; Ref, reference.

Patient #1 is a 13-year-old female with T6-10 segmentation defect (Figure 1a,b), fused left 9-10 ribs and mitral valve prolapse. She has a heterozygous duplication of nucleotide 2334 (c.2334dupC; p.Ser779GlnfsTer21). The variant was mapped in the intracellular region and post-translational phosphorylation site of the FGFR1 protein (Figure 2a) and was predicted by the NMD (nonsense-mediated decay) Prediction Tool (https://nmdpredictions.shinyapps.io/, accessed on 15 March 2021) to be located in the NMD-incompetent region (Figure 2b), suggesting that the variant is unlikely to cause nonsense-mediated decay. The variant was not found in population genomic databases, such as 1000G, ESP6500 and gnomAD. The CADD PHRED score of this variant is 32, indicating the deleteriousness of this variant.

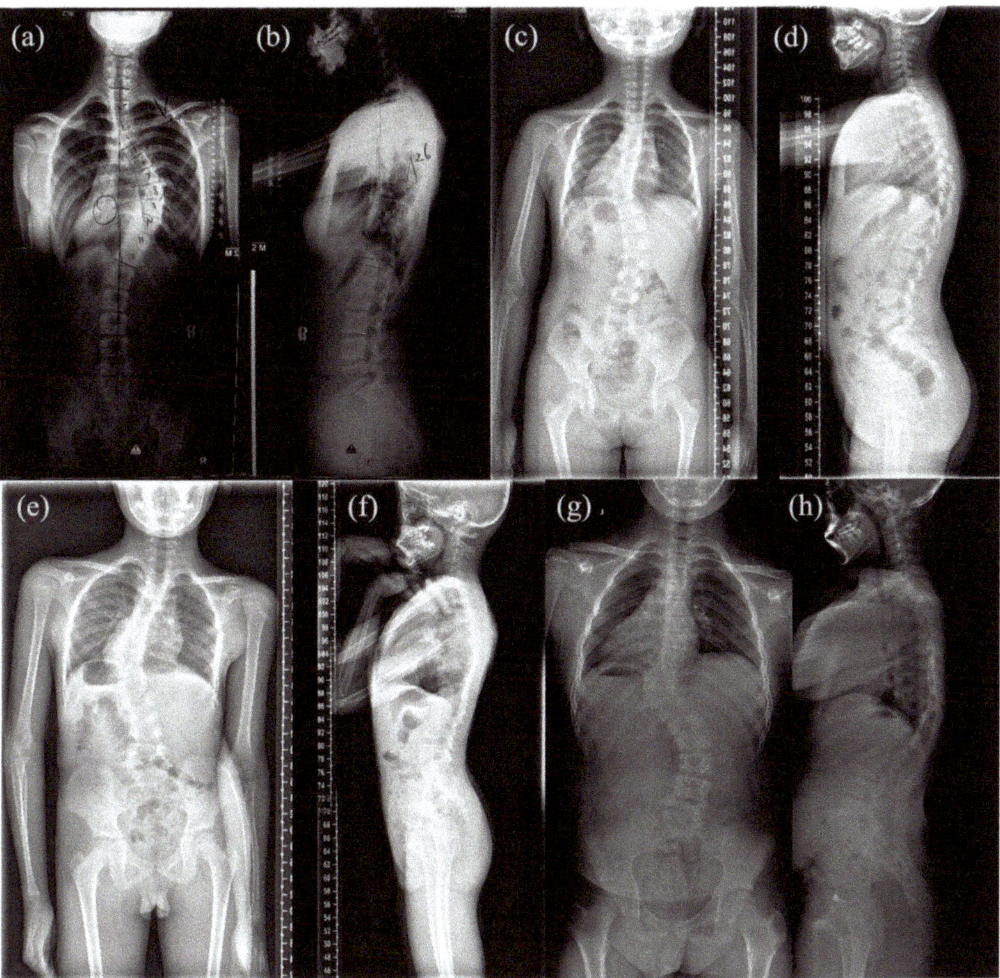

Figure 1. Anteroposterior and lateral spinal X-ray of four patients: (**a,b**) Anteroposterior and lateral spinal X-ray of patient #1; (**c,d**) Anteroposterior and lateral spinal X-ray of patient #2; (**e,f**) Anteroposterior and lateral spinal X-ray of patient #3; (**g,h**) Anteroposterior and lateral spinal X-ray of patient #4.

Patient #2, a 4-year-old female, has T9 hemivertebrae, T8 butterfly vertebrae, and three ribs absent (Figure 1c,d). She has an *FGFR1* missense variant c.2339T>C (p.Phe780Ser). This variant was also mapped in the intracellular region and post-translational phosphorylation sites of the FGFR1 protein (Figure 2a). It was not found in most population databases, such as 1000 G, gnomAD and ESP6500. The variant was highly conservative across different vertebral species (Figure 2c). In silico prediction had contradictory results (tolerant or benign for SIFT, pathogenic for MutationTaster, Polyphen2, LRT and CADD PHRED-score).

Patient #3 is a male newborn affected with a T10 hemivertebrae (Figure 1e,f) with a missense variant c.1107G>A (p.Met369Ile). It is a novel mutation according to all population databases. In silico predictions were tolerant or benign for SIFT and Polyphen2, but deleterious for MutationTaster, LRT and CADD PHRED score.

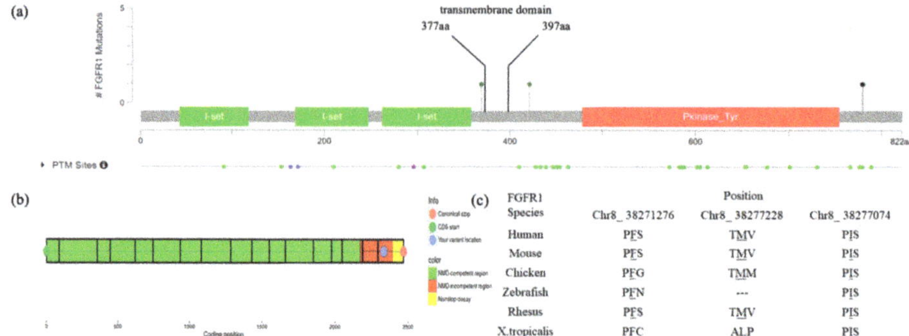

FGFR1 Species	Chr8_38271276	Position Chr8_38277228	Chr8_38277074
Human	PFS	TMV	PIS
Mouse	PFS	TMV	PIS
Chicken	PFG	TMM	PIS
Zebrafish	PFN	---	PIS
Rhesus	PFS	TMV	PIS
X.tropicalis	PFC	ALP	PIS

Figure 2. Mapping and conservation analysis of four variants: (**a**) Mapping of four *FGFR1* variants revealed that c.2334dupC and c.2339T>C are located in the intracellular region and post-translational phosphorylation sites of the FGFR1 protein, whereas c.1261A>G is located in the transmembrane region and close to the post-translational phosphorylation site; (**b**) The result of NMD prediction of c.2334dupC showed that it is located in the NMD-incompetent region; (**c**) Mutation loci of the three missense variants (c.2339T>C, c.1261A>G and c.1107G>A) are highly conservative across different species.

Patient #4 is a male newborn who presents T10 hemivertebrae (Figure 1g,h). This patient has a novel missense variant (c.1261A>G; p.Ile421Val), which was mapped in the transmembrane region and close to the post-translational phosphorylation site of FGFR1 protein (Figure 2a). This variant is highly conservative among different vertebral species (Figure 2c). It was predicted to be deleterious by SIFT, MutationTaster, LRT and CADD PHRED score.

3.2. Functional Characterization of FGFR1 Variants

3.2.1. Western Analysis

To identify the influences of these four variants on the function of the FGFR1 protein, we evaluated overall protein expression and maturation of the different *FGFR1* variants compared to WT. Endoglycosidase digestion and Western blotting analysis showed two immunoreactive-specific bands for WT FGFR1 at 140 and 120 kDa, corresponding to a differently N-glycosylated receptor. These two bands were reduced to a single lower molecular weight band following peptide N-glycosidase (PNGase) digestion to remove all types of N-linked carbohydrate chains. Treatment with endoglycosidase H (EndoH), which only removes high mannose N-linked sugars, merely affects the immature form (120 kDa), leaving the fully glycosylated mature form (140 kDa) intact. Thus, maturation rate can be calculated by dividing the band of 140 kDa from EndoH-treated samples into the band of 100 kDa. Overall expression level was quantified by measuring bands from PNGase-treated samples and normalized to the WT group (set as 100%). The overall expression of the frameshift variant was decreased to 43.9% compared with that of WT ($p = 0.06$), and those of three missense variants were reduced to 63.4% ($p < 0.01$), 82.8% ($p = 0.887$), and 77.4% ($p = 0.743$), respectively (Figure 3). As for maturation analysis, densitometric analysis revealed that 29.1% of the WT FGFR1 protein was expressed as a mature form (Figure 3). Consistent with our mapping analysis predicting that all four variants are not localized in the FGFR1 functional ectodomain, these mutant receptors showed no difference in the level of protein maturation, compared to WT (Figure 3).

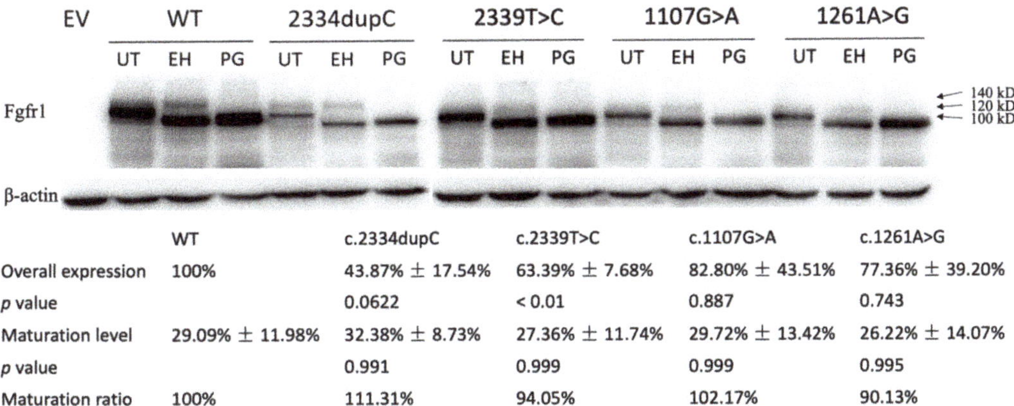

	WT	c.2334dupC	c.2339T>C	c.1107G>A	c.1261A>G
Overall expression	100%	43.87% ± 17.54%	63.39% ± 7.68%	82.80% ± 43.51%	77.36% ± 39.20%
p value		0.0622	< 0.01	0.887	0.743
Maturation level	29.09% ± 11.98%	32.38% ± 8.73%	27.36% ± 11.74%	29.72% ± 13.42%	26.22% ± 14.07%
p value		0.991	0.999	0.999	0.995
Maturation ratio	100%	111.31%	94.05%	102.17%	90.13%

Figure 3. Western blot analysis of COS-7 cells transiently transfected with WT or mutant *FGFR1* constructs reveal diminished protein expression levels of c.2334dupC, c.2339T>C, c.1107G>A and c.1261A>G. Overall expression was significantly decreased in all four variants, especially in c.2334dupC. No difference in protein maturation process was detected using a receptor deglycosylation. EV = empty vector, WT = wild type, UT = untreated, EH = EndoH-treated, PG = PNGase-f-treated.

3.2.2. FGF Reporter Gene Assay

To assess the influence of the four *FGFR1* variants on the receptor functionality, we first used the FGF-responsive reporter osteocalcin FGF response element-luciferase in L6 myoblasts, which acts downstream of the MAPK pathway (Figure 4). FGF18 is included in the FGF8 subfamily, which is expressed during somitogenesis and is essential for the morphogenesis of many tissues. In the *FGF18* knockout mice model, skeletal phenotypes have been detected [20], indicating an important role of FGF18 signaling in skeletal development. Previously, FGF18 was found to be expressed in and required for osteogenesis and chondrogenesis [21–25]. Compared to WT FGFR1, the receptor signaling capacity of the truncating variant (c.2334dupC) was reduced by 20.7% ($p < 0.05$, Figure 4). The responses of missense variants (c.2339T>C, c.1261A>G) were also significantly reduced by 26.6% and 28.8%, respectively ($p < 0.01$, Figure 4). These results indicated the diminished signaling pathway of FGFR1 activated by FGF18.

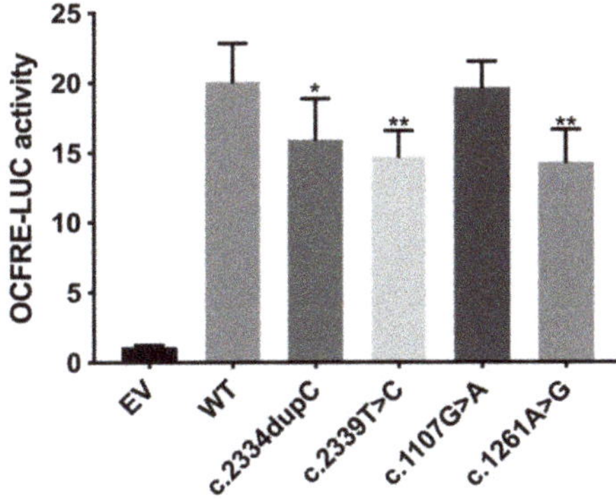

Figure 4. FGF reporter gene assay showing reduced signaling capacity of c.2334dupC, c.2339T>C and c.1261A>G. L6 myoblasts were transiently transfected with OCFRE-luciferase reporter together with wild type (WT) or mutant constructs and then treated with 10-8M FGF18. The average receptor signaling capacities of c.2334dupC, c.2339T>C and c.1261A>G were reduced by 20.7%, 26.6% and 28.8%, respectively. EV = empty vector. * *p* value < 0.05, ** *p* value < 0.01.

4. Discussion

In this study, we identified four pathogenic variants, namely one frameshift and three missense variations in patients with congenital scoliosis. The frameshift variant, c.2334dupC, found in a patient with vertebral segmentation defects and mitral valve prolapse, was the first *FGFR1* variant to be associated with spinal malformations and heart defects. Previous mouse models with *FGFR1* mutations were found to have malformations in both vertebrae and the heart [12], suggesting that *FGFR1* variants were associated with skeletal and cardiac abnormalities. Functional studies of this frameshift variant showed that this variant decreases overall protein expression compared with that of WT with a trend to significance but left protein maturation intact (Figure 3). The decreased overall protein expression of this variant might contribute to the diminished luciferase activity, suggesting a diminished signaling function induced by this variant (Figure 4). As the frameshift variant was located in the NMD-incompetent region, we proposed that this truncating variant did not lead to nonsense-mediated mRNA decay but only mildly affected the protein expression, and thus, merely resulted in mild skeletal and cardiac phenotypes.

As for the three missense variants, all of them were predicted to be deleterious by MutationTaster, LRT and CADD, but had different predictions by SIFT and Polyphen2. Two missense variants (c.2339T>C and c.1261A>G) were highly conservative across a wide range of vertebral species, suggesting them to be deleterious variants. Functional studies revealed that all missense variants had reduced overall protein expression, but only the decrease in c.2339T>C was statistically significant (Figure 3). Further luciferase assay indicated significantly reduced luciferase reporter activities (c.2339T>C and c.1261A>G), and thus, had diminished signaling functions (Figure 4). As the OCFRE reporter used in luciferase assays reports the activity of the MAPK pathway downstream of FRS2α signaling, we can conclude that c.2334dupC, c.2339T>C and c.1261A>G diminish the MAPK pathway.

Western blotting and maturation assay of the missense variant (c.1107G>A) showed a slightly decreased overall protein expression and normal maturation level. However, luciferase assay indicated that this variant has similar luciferase activity compared to WT, suggesting a normal effect on downstream signaling of this variant. As most proteins are redundant regarding their expression level, a minor decrease in expression level might not impact normal function. The missense variant c.1107G>A has an 82.8% expression level and a normal maturation ratio and thus, the matured protein of c.1107G>A is decreased to 84.6% compared to WT (82.8% times 102.17%), while matured protein of the other three variants is decreased to 48.8% for c.2334dupC, 59.6% for c.2339T>A and 69.7% for c.1261A>G compared to WT. Therefore, we proposed that matured protein with less than 70%~80% of WT could not be compensated by the redundant expression and might lead to the diminished signaling function indicated by the luciferase assay.

Furthermore, these three hypomorphic variants, including c.2334dupC, c.2339T>C and c.1261A>G, were mapped around post-translational modification sites and may affect protein phosphorylation, which plays an important role in normal protein function. Ying et al. [26] reported a patient with cryptorchidism, micropenis, strabismus, and hypopsia, who was diagnosed with nIHH. The patient had a de novo mutation in *FGFR1* (c.2008G>A), which induced a post-translational modification defect, including defective glycosylation and impaired trans-autophosphorylation. This study revealed the significance of post-translational modification of FGFR1. Based on in silico analysis and functional study results, we believe these three hypomorphic variants (c.2334dupC, c.2339T>C and c.1261A>G) of *FGFR1* may be associated with spinal defects in our patients. As for patient #3 with c.1107G>A, we propose that his skeletal defects are caused by other unknown genetic or environmental factors.

Pathogenic loss-of-function variants of the *FGFR1* gene were reported to be involved in patients with Kallmann Syndrome, including hypogonadotropic hypogonadism and anosmia [5,10,27], and isolated HH [6–9]. Patients with *FGFR1* mutations also presented with skeletal phenotypes [7,9,10], including oligodactyly on both feet, fusion of metacarpal bones, hemivertebrae, butterfly vertebrae and split hand/foot malformation.

In our cohort, a broad range of skeletal phenotypes were observed, as one patient had failure of segmentation, one patient had mixed defects and two patients had failure of formation. This is consistent with previous studies of *FGFR1* pathogenic variants, in which patients with HH can present a varied spectrum of reproductive phenotypes and non-reproductive phenotypes [8,10]. Furthermore, different patients carrying identical *FGFR1* mutations were observed to exhibit largely variable expressivity of reproductive phenotypes [8]. *FGFR1* signaling is involved in the determination of mesodermal cell fates and regional patterning of the mesoderm during gastrulation [28], and thus, affects organ specification. For the *FGFR1* signaling pathway, different organ systems respond to ligand binding with discrepant patterns [29], and several distinct downstream pathways, such as Erk1/2, Frs2, Crk proteins and Plcγ, are involved [30]. Given the broad function of *FGFR1* in embryo development, wide crosslink with other signaling pathways, tissue-specific response patterns and different downstream pathways, it is reasonable that patients with *FGFR1* mutations can present distinct phenotypes affecting different organ systems. However, the detailed mechanisms through which *FGFR1* mutations lead to different diseases need to be further studied and clarified.

In previous studies, patients with different FGFR1 domains affected have been revealed to present different phenotype spectra. The variants found in these patients all impair the functional domain of FGFR1 protein, including exon 1U, which is located around multiple transcription factor-binding sites, the FRS2α-binding domain and the tyrosine kinase domain [7,9]. Compared to these studies, patients in our cohort only presented mild spine and heart defects. As none of our variants were located in the functional region of the FGFR1 protein or led to severe damage to protein structure, we hypothesized that mild variants in our patients can only result in mild phenotypes compared with those in previous studies.

5. Conclusions

In conclusion, we found four *FGFR1* variants in our CS cohorts—one frameshift variant (c.2334dupC) and three missense variants (c.2339T>C; c.1107G>A; c.1261A>G). Functional studies revealed diminished signaling function and reduced protein expression in three of them (c.2334dupC; c.2339T>C; c.1261A>G). These variants in our patients only caused mild damage to the protein expression, and thus, resulted in mild skeletal and cardiac phenotypes, compared to those in previous studies.

Author Contributions: Conceptualization, S.W., X.C., Z.Y., N.W. and T.J.Z.; methodology, X.C., S.Z. and X.L.; software, S.Z. and X.L.; validation, S.W., S.Z., Y.N. and G.L.; formal analysis, X.C., Z.Y. and Z.S.; investigation, X.C.; resources, S.W. and Y.N.; data curation, X.C., Z.Y. and S.Z.; writing—original draft preparation, X.C., S.W., Z.Y; writing—review and editing, Z.Y., S.Z., G.L., Y.Y. and N.W.; visualization, X.C.; supervision, Y.Y., N.W., Z.W. and T.J.Z.; project administration, N.W., Z.W. and T.J.Z.; funding acquisition, N.W., Z.W. and T.J.Z. All authors have read and agreed to the published version of the manuscript.

Funding: This research was funded in part by the National Natural Science Foundation of China (81902178 to S.W., 81972037 to J.Z., 81822030 and 82072391 to N.W., 81930068 and 81772299 to Z.W.), Beijing Natural Science Foundation (No. L192015 to J.Z., JQ20032 to N.W., 7191007 to Z.W.), Capital's Funds for Health Improvement and Research (2020-4-40114 to N.W.), Tsinghua University-Peking Union Medical College Hospital Initiative Scientific Research Program, Non-profit Central Research Institute Fund of Chinese Academy of Medical Sciences (No. 2019PT320025), and the PUMC Youth Fund & the Fundamental Research Funds for the Central Universities (No. 3332019021 to Shengru W.).

Institutional Review Board Statement: Approval for the study was obtained from the ethics committee at Peking Union Medical College Hospital (JS-098).

Informed Consent Statement: Written informed consent was provided by each participant.

Data Availability Statement: Data are available upon reasonable request. The datasets analyzed during the current study are available from the corresponding author on reasonable request.

Acknowledgments: We thank the patients and their families who participated in this research. We also thank the Nanjing Geneseq Technology Inc. for technical help in sequencing, and Ekitech Technology Inc. for technical support in database and data management. We also thank the Deciphering disorders Involving Scoliosis and Comorbidities study.

Conflicts of Interest: The authors declare no conflict of interest.

References

1. Mohammadi, M.; Olsen, S.K.; Ibrahimi, O.A. Structural basis for fibroblast growth factor receptor activation. *Cytokine Growth Factor Rev.* **2005**, *16*, 107–137. [CrossRef]
2. Partanen, J.; Schwartz, L.; Rossant, J. Opposite phenotypes of hypomorphic and Y766 phosphorylation site mutations reveal a function for Fgfr1 in anteroposterior patterning of mouse embryos. *Genes Dev.* **1998**, *12*, 2332–2344. [CrossRef]
3. Miraoui, H.; Marie, P.J. Fibroblast growth factor receptor signaling crosstalk in skeletogenesis. *Sci. Signal.* **2010**, *3*, re9. [CrossRef]
4. Itoh, N.; Ornitz, D.M. Fibroblast growth factors: From molecular evolution to roles in development, metabolism and disease. *J. Biochem.* **2011**, *149*, 121–130. [CrossRef]
5. Goncalves, C.; Bastos, M.; Pignatelli, D.; Borges, T.; Aragues, J.M.; Fonseca, F.; Pereira, B.D.; Socorro, S.; Lemos, M.C. Novel FGFR1 mutations in Kallmann syndrome and normosmic idiopathic hypogonadotropic hypogonadism: Evidence for the involvement of an alternatively spliced isoform. *Fertil. Steril.* **2015**, *104*, 1261–1267 e1261. [CrossRef] [PubMed]
6. Koika, V.; Varnavas, P.; Valavani, H.; Sidis, Y.; Plummer, L.; Dwyer, A.; Quinton, R.; Kanaka-Gantenbein, C.; Pitteloud, N.; Sertedaki, A.; et al. Comparative functional analysis of two fibroblast growth factor receptor 1 (FGFR1) mutations affecting the same residue (R254W and R254Q) in isolated hypogonadotropic hypogonadism (IHH). *Gene* **2013**, *516*, 146–151. [CrossRef]
7. Ohtaka, K.; Fujisawa, Y.; Takada, F.; Hasegawa, Y.; Miyoshi, T.; Hasegawa, T.; Miyoshi, H.; Kameda, H.; Kurokawa-Seo, M.; Fukami, M.; et al. FGFR1 Analyses in Four Patients with Hypogonadotropic Hypogonadism with Split-Hand/Foot Malformation: Implications for the Promoter Region. *Hum. Mutat.* **2017**, *38*, 503–506. [CrossRef]
8. Raivio, T.; Sidis, Y.; Plummer, L.; Chen, H.; Ma, J.; Mukherjee, A.; Jacobson-Dickman, E.; Quinton, R.; Van Vliet, G.; Lavoie, H.; et al. Impaired fibroblast growth factor receptor 1 signaling as a cause of normosmic idiopathic hypogonadotropic hypogonadism. *J. Clin. Endocrinol. Metab.* **2009**, *94*, 4380–4390. [CrossRef] [PubMed]
9. Villanueva, C.; Jacobson-Dickman, E.; Xu, C.; Manouvrier, S.; Dwyer, A.A.; Sykiotis, G.P.; Beenken, A.; Liu, Y.; Tommiska, J.; Hu, Y.; et al. Congenital hypogonadotropic hypogonadism with split hand/foot malformation: A clinical entity with a high frequency of FGFR1 mutations. *Genet. Med.* **2015**, *17*, 651–659. [CrossRef]
10. Jarzabek, K.; Wolczynski, S.; Lesniewicz, R.; Plessis, G.; Kottler, M.L. Evidence that FGFR1 loss-of-function mutations may cause variable skeletal malformations in patients with Kallmann syndrome. *Adv. Med. Sci.* **2012**, *57*, 314–321. [CrossRef] [PubMed]
11. Jacob, A.L.; Smith, C.; Partanen, J.; Ornitz, D.M. Fibroblast growth factor receptor 1 signaling in the osteo-chondrogenic cell lineage regulates sequential steps of osteoblast maturation. *Dev. Biol.* **2006**, *296*, 315–328. [CrossRef] [PubMed]
12. Bult, C.J.; Blake, J.A.; Smith, C.L.; Kadin, J.A.; Richardson, J.E.; The Mouse Genome Database Group. Mouse Genome Database (MGD) 2019. *Nucleic Acids Res.* **2019**, *47*, D801–D806. [CrossRef] [PubMed]
13. Zhao, S.; Zhang, Y.; Chen, W.; Li, W.; Wang, S.; Wang, L.; Zhao, Y.; Lin, M.; Ye, Y.; Lin, J.; et al. Diagnostic yield and clinical impact of exome sequencing in early-onset scoliosis (EOS). *J. Med. Genet.* **2021**, *58*, 41–47. [CrossRef]
14. Chen, N.; Zhao, S.; Jolly, A.; Wang, L.; Pan, H.; Yuan, J.; Chen, S.; Koch, A.; Ma, C.; Tian, W.; et al. Perturbations of genes essential for Mullerian duct and Wolffian duct development in Mayer-Rokitansky-Kuster-Hauser syndrome. *Am. J. Hum. Genet.* **2021**, *108*, 337–345. [CrossRef]
15. Davydov, E.V.; Goode, D.L.; Sirota, M.; Cooper, G.M.; Sidow, A.; Batzoglou, S. Identifying a high fraction of the human genome to be under selective constraint using GERP++. *PLoS Comput. Biol.* **2010**, *6*, e1001025. [CrossRef] [PubMed]
16. Kircher, M.; Witten, D.M.; Jain, P.; O'Roak, B.J.; Cooper, G.M.; Shendure, J. A general framework for estimating the relative pathogenicity of human genetic variants. *Nat. Genet.* **2014**, *46*, 310–315. [CrossRef] [PubMed]
17. Vaser, R.; Adusumalli, S.; Leng, S.N.; Sikic, M.; Ng, P.C. SIFT missense predictions for genomes. *Nat. Protoc.* **2016**, *11*, 1–9. [CrossRef]
18. Adzhubei, I.A.; Schmidt, S.; Peshkin, L.; Ramensky, V.E.; Gerasimova, A.; Bork, P.; Kondrashov, A.S.; Sunyaev, S.R. A method and server for predicting damaging missense mutations. *Nat. Methods* **2010**, *7*, 248–249. [CrossRef]
19. Schwarz, J.M.; Cooper, D.N.; Schuelke, M.; Seelow, D. MutationTaster2: Mutation prediction for the deep-sequencing age. *Nat. Methods* **2014**, *11*, 361–362. [CrossRef]
20. Horton, W.A.; Degnin, C.R. FGFs in endochondral skeletal development. *Trends Endocrinol. Metab. TEM* **2009**, *20*, 341–348. [CrossRef]
21. Hung, I.H.; Schoenwolf, G.C.; Lewandoski, M.; Ornitz, D.M. A combined series of Fgf9 and Fgf18 mutant alleles identifies unique and redundant roles in skeletal development. *Dev. Biol.* **2016**, *411*, 72–84. [CrossRef]
22. Lewandoski, M.; Sun, X.; Martin, G.R. Fgf8 signalling from the AER is essential for normal limb development. *Nat. Genet.* **2000**, *26*, 460–463. [CrossRef] [PubMed]
23. Liu, Z.; Lavine, K.J.; Hung, I.H.; Ornitz, D.M. FGF18 is required for early chondrocyte proliferation, hypertrophy and vascular invasion of the growth plate. *Dev. Biol.* **2007**, *302*, 80–91. [CrossRef]

24. Liu, Z.; Xu, J.; Colvin, J.S.; Ornitz, D.M. Coordination of chondrogenesis and osteogenesis by fibroblast growth factor 18. *Genes Dev.* **2002**, *16*, 859–869. [CrossRef]
25. Ohbayashi, N.; Shibayama, M.; Kurotaki, Y.; Imanishi, M.; Fujimori, T.; Itoh, N.; Takada, S. FGF18 is required for normal cell proliferation and differentiation during osteogenesis and chondrogenesis. *Genes Dev.* **2002**, *16*, 870–879. [CrossRef] [PubMed]
26. Ying, H.; Sun, Y.; Wu, H.; Jia, W.; Guan, Q.; He, Z.; Gao, L.; Zhao, J.; Ji, Y.; Li, G.; et al. Posttranslational Modification Defects in Fibroblast Growth Factor Receptor 1 as a Reason for Normosmic Isolated Hypogonadotropic Hypogonadism. *Oxid. Med. Cell Longev.* **2020**, *2020*, 2358719. [CrossRef] [PubMed]
27. Dode, C.; Levilliers, J.; Dupont, J.M.; De Paepe, A.; Le Du, N.; Soussi-Yanicostas, N.; Coimbra, R.S.; Delmaghani, S.; Compain-Nouaille, S.; Baverel, F.; et al. Loss-of-function mutations in FGFR1 cause autosomal dominant Kallmann syndrome. *Nat. Genet.* **2003**, *33*, 463–465. [CrossRef]
28. Yamaguchi, T.P.; Harpal, K.; Henkemeyer, M.; Rossant, J. fgfr-1 is required for embryonic growth and mesodermal patterning during mouse gastrulation. *Genes Dev.* **1994**, *8*, 3032–3044. [CrossRef]
29. Hajihosseini, M.K.; Lalioti, M.D.; Arthaud, S.; Burgar, H.R.; Brown, J.M.; Twigg, S.R.; Wilkie, A.O.; Heath, J.K. Skeletal development is regulated by fibroblast growth factor receptor 1 signalling dynamics. *Development* **2004**, *131*, 325–335. [CrossRef]
30. Brewer, J.R.; Molotkov, A.; Mazot, P.; Hoch, R.V.; Soriano, P. Fgfr1 regulates development through the combinatorial use of signaling proteins. *Genes Dev.* **2015**, *29*, 1863–1874. [CrossRef]

 *genes*

MDPI

Article

Whole Exome Sequencing of 23 Multigeneration Idiopathic Scoliosis Families Reveals Enrichments in Cytoskeletal Variants, Suggests Highly Polygenic Disease

Elizabeth A. Terhune [1], Cambria I. Wethey [1], Melissa T. Cuevas [1], Anna M. Monley [1,2], Erin E. Baschal [1], Morgan R. Bland [1], Robin Baschal [1,2], G. Devon Trahan [3], Matthew R. G. Taylor [4], Kenneth L. Jones [3,5] and Nancy Hadley Miller [1,2,*]

[1] Department of Orthopedics, University of Colorado Anschutz Medical Campus, Aurora, CO 80045, USA; elizabeth.a.terhune@cuanschutz.edu (E.A.T.); cambria.wethey@cuanschutz.edu (C.I.W.); Melissa.cuevas@cuanschutz.edu (M.T.C.); anna.monley@cuanschutz.edu (A.M.M.); erin.baschal@cuanschutz.edu (E.E.B.); morganrbland@gmail.com (M.R.B.); robin.baschal@cuanschutz.edu (R.B.)
[2] Musculoskeletal Research Center, Children's Hospital Colorado, Aurora, CO 80045, USA
[3] Department of Pediatrics, University of Colorado Anschutz Medical Campus, Aurora, CO 80045, USA; devon.trahan@cuanschutz.edu (G.D.T.); ken.jones@ouhsc.edu (K.L.J.)
[4] Department of Medicine, Adult Medical Genetics Program, University of Colorado Anschutz Medical Campus, Aurora, CO 80045, USA; matthew.taylor@cuanschutz.edu
[5] Department of Cell Biology, University of Oklahoma Health Science Center, Oklahoma City, OK 73104, USA
* Correspondence: Nancy.hadley-miller@cuanschutz.edu

Citation: Terhune, E.A.; Wethey, C.I.; Cuevas, M.T.; Monley, A.M.; Baschal, E.E.; Bland, M.R.; Baschal, R.; Trahan, G.D.; Taylor, M.R.G.; Jones, K.L.; et al. Whole Exome Sequencing of 23 Multigeneration Idiopathic Scoliosis Families Reveals Enrichments in Cytoskeletal Variants, Suggests Highly Polygenic Disease. *Genes* **2021**, *12*, 922. https://doi.org/10.3390/genes12060922

Academic Editor: Cathy L. Raggio

Received: 18 May 2021
Accepted: 10 June 2021
Published: 16 June 2021

Publisher's Note: MDPI stays neutral with regard to jurisdictional claims in published maps and institutional affiliations.

Abstract: Adolescent idiopathic scoliosis (AIS) is a lateral spinal curvature >10° with rotation that affects 2–3% of healthy children across populations. AIS is known to have a significant genetic component, and despite a handful of risk loci identified in unrelated individuals by GWAS and next-generation sequencing methods, the underlying etiology of the condition remains largely unknown. In this study, we performed exome sequencing of affected individuals within 23 multigenerational families, with the hypothesis that the occurrence of rare, low frequency, disease-causing variants will co-occur in distantly related, affected individuals. Bioinformatic filtering of uncommon, potentially damaging variants shared by all sequenced family members revealed 1448 variants in 1160 genes across the 23 families, with 132 genes shared by two or more families. Ten genes were shared by >4 families, and no genes were shared by all. Gene enrichment analysis showed an enrichment of variants in cytoskeletal and extracellular matrix related processes. These data support a model that AIS is a highly polygenic disease, with few variant-containing genes shared between affected individuals across different family lineages. This work presents a novel resource for further exploration in familial AIS genetic research.

Keywords: idiopathic scoliosis; exome sequencing; spine; polygenic; variants; musculoskeletal disease; cytoskeleton; extracellular matrix

1. Introduction

Adolescent idiopathic scoliosis (AIS) is a structural lateral spinal curvature $\geq 10°$ that affects 2–3% of healthy children [1], with females at the greatest risk for severe progression [2–4]. In individuals with severe progressive curvatures, life-long problems of cosmetic deformity, respiratory compromise, back pain, and degenerative disease often arise and in many cases require surgical intervention, placing a significant economic burden upon our healthcare system [5]. AIS is known to be highly heritable, however, our knowledge of its etiology is severely limited, both in terms of the individuals at risk for curve initiation and those likely to experience curve progression. Understanding key risk variants or genetic pathways leading to AIS holds the potential to improve patient care by way of targeted clinical treatments or prognostics for detecting AIS risk or risk of curvature progression.

Multiple studies support the genetic foundation of AIS, with sibling recurrence risks reported to be near 18%, and heritability estimates of approximately 87.5% [6–8]. To understand the genetics of AIS, traditional approaches including genome wide association studies (GWAS), exome sequencing, and familial linkage studies have been applied to multiple populations. GWAS of unrelated individuals with AIS have identified potential common risk alleles, notably those in or near *LBX1* [9–23] and *GPR126/ADGRG6* [23–27], which are the most well-replicated genetic findings to date across populations. Next-generation sequencing studies (i.e., whole exome sequencing) both within families and unrelated individuals with AIS have identified *rare* variants in extracellular matrix genes that may contribute to the AIS phenotype [28–30]. Recent genetic and functional studies have led to varied hypotheses of AIS etiology, including dysfunction within neuroinflammatory pathways [31,32], the cartilage matrisome [33], cilia, the cytoskeleton [34–38], or the vestibular system [39–43]. However, the inability thus far to relate specific genetic variants to the biology of AIS and the wide variation in which AIS presents are indicative of the complex heterogeneity of this disorder.

Disease susceptibility variants for complex diseases may collectively be common in the general population, but specific variants may be rare within families or affected individuals. Family studies present unique advantages over case-control studies, as they may reveal rare disease-associated variants enriched within the family that are amenable to targeted sequencing. Sequencing studies of multiple large families can reveal multiple variants within a gene or molecular pathway that contribute to the disease phenotype [44,45].

Within our laboratory, a pilot exome sequencing study of five AIS families identified an enrichment of damaging uncommon variants in cilia, extracellular matrix, and cytoskeletal genes, although no specific variants or genes were found to be present across all families [46].

This led us to propose the hypothesis that the development of AIS may be due to damaging variants within a specific set of pathways or molecular classes, rather than being driven by just a few select 'AIS genes'. In this study, we expand upon our previous work and present exome sequencing of affected individuals from 23 AIS families, interpreted using gene enrichment analyses to identify overrepresented functional categories. We then investigate specific genes containing variants in multiple families via genotyping confirmation in additional affected and unaffected members of the family to assess how closely these variants track with the AIS phenotype.

2. Materials and Methods

An overview of the methodology for this study—including subject enrollment, sample collection, sequencing, and bioinformatic filtering strategies—are provided in Figure 1.

2.1. Subjects

Study subjects were enrolled as previously described [29,38,46]. Inclusion in the sequencing pool required a standing anteroposterior spinal radiograph showing $\geq 10°$ curvature by the Cobb method with pedicle rotation, and no evidence of congenital deformity or other co-existing genetic disorders [47–49].

Blood samples were collected from all study subjects as described previously [38]. DNA was then extracted from whole blood using standard phenol chloroform protocols or the QIA-GEN Gentra PureGene Blood Kit. DNA quality was verified by Qubit (Invitrogen, Waltham, MA, USA), agarose gel electrophoresis and Nanodrop (Thermo Scientific, Waltham, MA, USA).

2.2. Family Selection for Exome Sequencing

Twenty-three large families were selected through a tiered process wherein each pedigree was reviewed and evaluated by five project experts. Selection was based on the number of affected individuals in the family, the severity of their scoliosis curvatures, and the estimated genetic relationship between enrolled, affected individuals. Families with other musculoskeletal conditions or AIS from multiple sides of the family were excluded.

Three to five affected individuals were selected for sequencing per family based on the degree of hypothesized genetic distance between them, availability of high-quality DNA, and severity of spinal curvature.

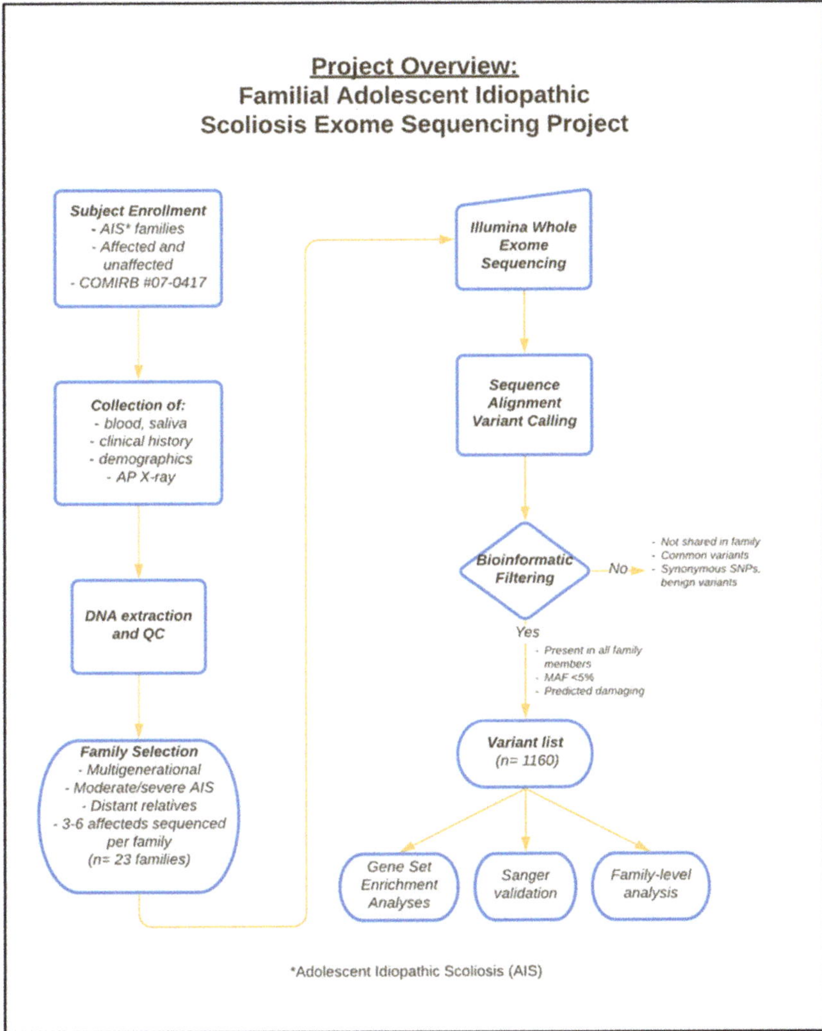

Figure 1. Project overview of subject enrollment, sample collection and extraction, exome sequencing, and bioinformatic filtering strategy.

Pedigrees for these families and a summary of clinical information for all sequenced individuals are provided in Supplementary Files S1 and S2. On the pedigrees, degree of spinal curvature as measured by Cobb angle is indicated in numbers below affected individuals (i.e., 20D). Individuals with known double curves have both listed (i.e., 26/30D), as are any triple curvatures. All individuals with listed curvatures were examined radiographically, and any found to be unaffected on physical exam are labeled as "confirmed negative". Proband for each family is indicated with an arrow.

2.3. Whole Exome Sequencing

Exome capture was completed using 1 µg of genomic DNA from 86 individuals across 23 families using the Agilent SureSelect Human V5 (51 Mb) exon capture kit. Samples were sequenced with a 2 × 100 bp run on an Illumina HiSeq 2500 at the Otogenetics Corporation facility in Atlanta, GA, with a minimum average coverage of 50X guaranteed per sample.

2.4. Bioinformatic Filtering

Whole exome reads were aligned to GRCh38 and variants were identified with Free-Bayes as previously described [38,46]. Candidate variants were filtered by SnpEff (version 4.1g) [50] along with custom scripts to retain only non-synonymous SNPs, coding indels, and variants affecting splice sites. Known artifacts and variants whose frequency was greater than 0.05 in the ExAC database (r0.3) [51] were also stripped. If the variant was annotated in the dbNSFP database (version 3.0) [52,53], it was retained only if at least one of the included variant prediction algorithms (SIFT, Polyphen2, LRT, MutationTaster) scored it as "damaging", signifying that the resulting change to the encoded protein had a predicted functional consequence. Variants that were not shared by all sequenced members of the family were not retained. Variants with a Minor Allele Frequency (MAF) < 0.05 that remained after the above filters were applied were retained for further analysis in separate gene sets. This MAF threshold was intentionally set higher than typical rare variant thresholds to account for the prevalence of the disease (2–3% of the general population), working under the hypothesis that low frequency variants may contribute to the high prevalence of this disease.

2.5. Genotyping

Variants appearing in multiple families that were present in the GO functional categories for "cytoskeleton" or "extracellular matrix" (or related terms) were prioritized for genotyping. Additional enrolled affected and unaffected family members were sequenced at the variant site by the Sanger method to establish whether the variant segregated with the AIS phenotype.

PCR was conducted in 20 µL reactions containing 10 µL Premix D (Epicentre Biotechnologies, Madison, WI, USA), 0.2 µL Taq Polymerase (Sigma, St. Louis, MO, USA), 60 ng genomic DNA, and 10 µM Forward and Reverse Primers. PCR reactions were run on a SimpliAmp Thermocycler (Fisher Scientific, Waltham, MA, USA) with a touchdown PCR protocol [38]. Primer sequences were obtained from Integrated DNA Technologies and are provided in Supplementary Files S2. Sanger sequencing was performed by Quintara Biosciences and chromatograms were analyzed using the CodonCode Aligner v9.0 (Codon-Code Corporation, Centerville, MA, USA, https://www.codoncode.com/index.htm (accessed on 16 June 2021)).

Pedigrees for each sequenced family are provided in Supplementary Files S1 and were created using PedigreeXP software (PC Pal, https://www.pedigreexp.com (accessed on 16 June 2021)).

2.6. Gene Set Overrepresentation Analyses

Gene set overrepresentation analyses (GOA) were performed on MAF < 0.05 gene lists with duplicate genes removed (*n* = 1160) from the combined list from each family. Both DAVID and EnrichR websites were used, as described below.

DAVID: DAVID (Database for Annotation, Visualization and Integrated Discovery) v6.8 was used to identify the significant GO terms and clusters in each gene list (https://david.ncifcrf.gov, accessed on 20 April 2021) [54,55]. The input gene list of 1160 genes resulted in 1146 DAVID IDs. Functional annotation clustering was used on our dataset with default settings and the GOTERM_All and GOTERM_CC_Direct annotation categories.

Enrichr: The same input gene lists as used for DAVID were used in Enrichr, a gene enrichment software developed by the Ma'ayan laboratory [56,57] (https://maayanlab.cloud/Enrichr/, accessed on 27 April 2021). Additionally, family-specific gene lists were separately inputted into EnrichR. We report results from the 2018 GO term Cellular Component and KEGG pathways. Volcano plots and charts were generated with Appyter (https://appyters.maayanlab.cloud/#/, accessed on 27 April 2021).

3. Results

To identify rare and low frequency variants associated with familial AIS, we performed whole exome sequencing on 23 multigenerational IS families (3–5 individuals per family, 86 individuals in total). Pedigrees for all families are provided in Supplementary Files S1, and clinical information for sequenced individuals is provided in Supplementary Files S2. Whole exome sequencing was performed using DNA extracted from whole blood, as described in the Methods. Illumina HiSeq reads were mapped to the human reference genome (hg39), and a minimum of 50X average coverage was obtained for each sample.

We then filtered the list of variants, requiring that each retained variant was present in all sequenced members of the family, was predicted to be damaging, and had an ExAC minor allele frequency (MAF) of <0.05. This MAF filter thus included both low-frequency (MAF 1–5%) and rare variants (MAF < 1%). These filters resulted in a total of 1448 variants in 1160 genes across the 23 families, with 11 to 128 variants identified in each family (median = 51 variants). Figure 2 provides a summary of variant information identified across all families. Nonsynonymous single nucleotide polymorphisms (SNPs) constituted the majority of filtered variants (88%, $n = 1281$), followed by non-frameshift deletions (4%, $n = 52$), non-frameshift insertions (2%, $n = 34$), frameshift deletions (2%, $n = 25$), stop gains (2%, $n = 22$), frameshift insertions (2%, $n = 22$), and splice sites (1%, $n = 12$). About half (48%) of variants had a minor allele frequency (MAF) of 0.01–0.05, 22% had an MAF 0.001–0.01, 18% had an MAF < 0.001, and 12% were predicted to be novel. Chromosomes with the largest number of variants were chromosome 1 ($n = 150$) and chromosome 19 ($n = 119$). Full details of each variant, sorted by family, are provided in Supplementary Files S3.

We next searched for genes containing variants in multiple families, with the hypothesis that shared genes would be more likely to contribute to the IS phenotype. Most variant-containing genes ($n = 1028$ of 1160 total, 88.6%) were specific to only one family. 132 genes were shared by at least 2 families, 38 were shared between 3+ families, and 10 were shared by 4+ families. No genes were shared among all families. Table 1 provides a list of genes found within multiple families. Of this list, four genes had a previously observed association with either idiopathic, degenerative, or infantile scoliosis (*DLL3* [58,59], *AHNAK* [60], *TTN* [61], *ANKRD11* [62]).

Although few genes were shared across families, we hypothesized that AIS families may share an enrichment of variants in specific functional categories. To identify categories of genes which were overrepresented with damaging variants within the AIS families, we performed gene set overrepresentation analysis (GOA) of our resulting filtered gene lists. We entered the MAF < 0.05 gene lists from individual families and combined data from all families into DAVID to identify overrepresented gene ontologies (GO Terms) (Table 2). The most overrepresented GO Cell Component categories across all families using DAVID were "microtubule" ($p = 1.62 \times 10^4$, 1.98 fold enrichment), "slit diaphragm" ($p = 8.88 \times 10^4$, 10.66 fold enrichment), and "intermediate filament" ($p = 8.88 \times 10^4$, 2.57 fold enrichment). Categories within extracellular matrix (ECM) GO terms, including "proteinaceous extracellular matrix" and "collagen trimer" also showed mild enrichments (Table 2) and the top enriched KEGG term was "ECM receptor activation" (Figure 2E). A full list of all GO functional categories is provided in Supplementary Files S2. We also used GOA to analyze an overrepresentation of cytogenic bands, as specific chromosomal regions have been linked with IS development by linkage analyses [63–71]. The most overrepresented band was 16p13.3 ($p < 0.001$), as described in Supplementary Files S2. This region was previously identified as significant in familial linkage analyses of a large AIS familial cohort. A neighboring region, distal chromosome 16p11.2 duplication, has been recently identified as a significant risk factor for severe AIS.

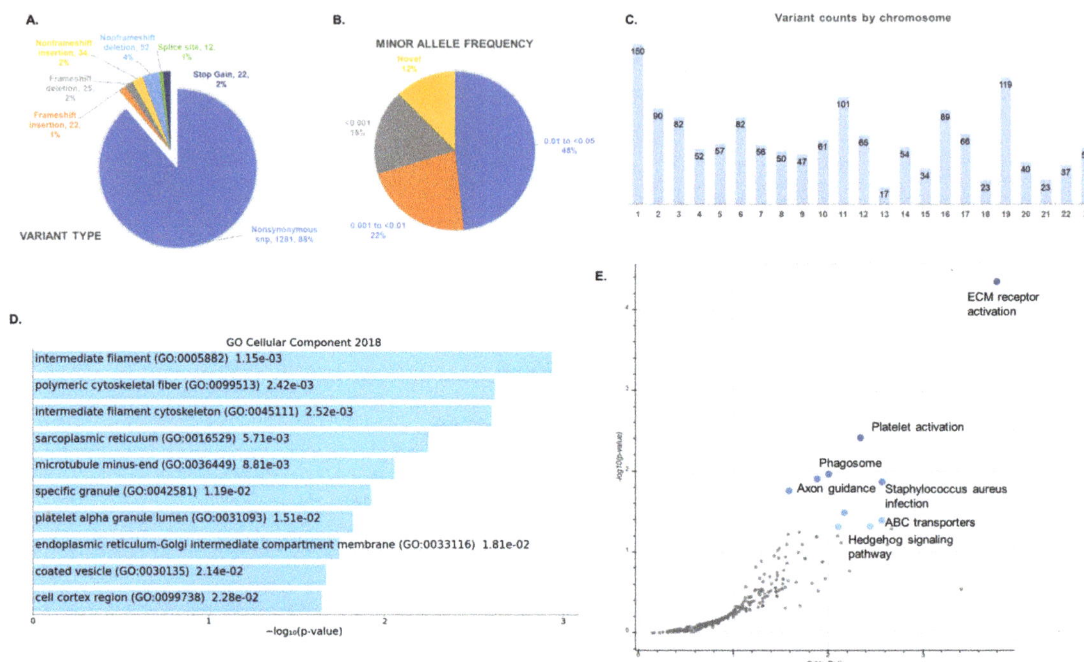

Figure 2. Summary information for total gene list (*n* = 1160) of familial AIS-associated variants across families with a minor allele frequency (MAF) < 0.05 passing bioinformatic filtering, as described in the Methods. (**A**) Variant type with *n* variants and % of total provided. (**B**) Minor allele frequency of all variants using ExAC. (**C**) Variant counts by chromosome. (**D**) Top GO Cellular Component, 2018 terms using EnrichR. See Table 2 for top GO terms using DAVID. (**E**) Volcano plot of top enriched KEGG terms over expected.

Table 1. Genes containing variants in multiple idiopathic scoliosis (IS) families. Genes with the same variant identified across families are indicated with an *. Citations are given for genes previously associated with scoliosis (including idiopathic, degenerative, or infantile). Full variant results by family are provided in Supplementary Files S3.

N Families Containing Variant	Gene Names
2 families	*TMEM52, GCFC2, RTP5, SLC10A6, SAMD9L, PTPRD, ASCL1, GALC, CD34, ALPPL2, FAM189B, NTRK1, CHIT1, DCBLD2, COL6A5, ATG9B, NUTM2F, NME3, GFAP, KRTAP10-5, ZNF644, FLG, OR6P1, CSMD1, UQCRB, LMNTD2, THAP11, MLLT1, OR14A2, VWA3B, CFC1B, MSH3, AKAP3, GNB1L, RAB36, FGD1, THOC3, TNXB, AHNAK2, XYLT1, APOL3, CACFD1, RPL3L, PIGT, CTNNA3, LRIT1, TYMP, ASXL2, GPAT2, AARD, MGA, LGALS9C, C18orf65, AHCY, CSMD2, VPS41, BHLHE22, CCDC68, SYNJ1, HMGCS2, SYN2, DNASE1L3, CASP12, FSIP2, TET2, FAM153B, FAM153A, PFAS, PDZD2, KLHL32, ANKRD18A, MCM8, NPHP4, USP32, COL21A1, SSTR5, CEACAM21, FAM26F, OR4A5, KRT81, BDNF, SLC1A7, SPTA1, TNK2, PTPRE, CYB5R2, MTUS2, FANCA, KATNAL2, OTC, GPIHBP1, CLEC18B, SLC22A31, SETDB1*
3 families	*GPRIN1, SKIV2L, PRR23D1, TBC1D26, KIR3DL1, TTN* [61], *KRTAP4-3, DLL3* [58,59], *OBSCN, KCNQ5, CCDC168, PRR25, ANKRD11* [62], *HRC, GPR179, USP26, FCGBP, KCNN3, CACNA1H, ANKRD30B, KIAA1875, MRC1, ACOT4, KIAA0556, ERCC6L, DYSF, CEP170, AHNAK* [60]
4 families	*POM121, PKD1L2, TMPRSS13, KRT2, ZNF717, PDE4DIP*
≥5 families	*AL589743.1, TPRX1, FAM47A, WIPF3*

Table 2. Overrepresented Gene Ontology, Cellular Component terms using the combined Minor Allele Frequency (MAF) < 0.05 gene list from all families (input n = 1160). The genes under the terms "cytoplasm" and "plasma membrane" are limited to 50 for brevity. The full list of enriched GO Terms (all GO annotation sets) is provided in Supplementary Files S2.

Term	Count	%	p Value	Genes	Fold Enrichment
microtubule	36	3.14	1.62×10^4	*INVS, DNAH1, DNAH7, TUBAL3, DCTN1, DNAH6, IQGAP1, CAMSAP2, GOLGA2, TEKT1, DVL1, KIF13B, TEKT2, CEP170, KIF21B, FSD1, KIF1A, PCNT, DYNC2H1, DNAH14, MTUS2, KCNAB2, SHROOM1, HAUS5, EML2, DLG1, SYNJ1, KATNAL2, INCENP, KIF26A, KIFC1, FEZ1, EHHADH, TTLL11, EIF3A, GAS8*	1.98
slit diaphragm	5	0.44	6.761×10^4	*TRPC6, KIRREL2, NPHS1, MAGI2, IQGAP1*	10.66
intermediate filament	17	1.48	8.88×10^4	*FLG, DSP, KRTAP13-4, KRT2, KRTAP26-1, KRTAP27-1, KRT79, KRT10, GFAP, KRT28, KRT37, PKP2, KRTAP6-1, SYNC, NES, KRT6A, PRPH*	2.57
Z disc	17	1.48	1.42×10^3	*PPP1R12A, SYNPO2, AHNAK2, ATP2B4, SLC4A1, NEB, ANK3, ADRA1A, IGFN1, RYR3, MYPN, TTN, OBSCN, HRC, PDE4B, SYNC, CRYAB*	2.46
spindle	17	1.48	1.85×10^3	*INVS, DIDO1, SPAG8, NUMA1, DCTN1, DCTN3, HEPACAM2, HAUS5, EML2, INCENP, KIFC1, NUP85, CLTCL1, ANKRD53, CEP170, E4F1, KBTBD8*	2.40
cytoplasm	347	30.28	3.04×10^3	**RPL5, ANKLE1, MTRR, HDAC10, RGSL1, ABCA12, BACH2, WDR87, ENDOV, TBK1, CYP2D7, MPRIP, PPP4R2, C7ORF31, FAM65C, PLCE1, TRIM26, RTTN, KIF21B, ADGB, SDS, BSX, KRT2, MTUS2, AFAP1, SCRIB, LCE1E, MAPK8IP2, EML2, PPP1R3G, INPP4B, KATNAL2, TEP1, FEZ1, ZSCAN26, WDPCP, TTLL11, FLYWCH1, ALPK2, PFN1, RIN1, TRIB2, KPRP, SPTBN5, PARPBP, DHX8, PCDH15, C1ORF198, AGAP1, XPC*	1.13
lateral plasma membrane	10	0.87	3.37×10^3	*CEACAM1, DLG1, KCNB1, MTCL1, ABCC6, PTPRO, DVL1, ANK3, IQGAP1, NKD2*	3.22
cytoskeleton	36	3.14	3.73×10^3	*TENM1, DRC7, MTCL1, EVPL, PPL, GPHN, FGD1, SYNE1, KIAA0556, CNN2, AKAP12, EPB41L5, SGCD, EPB41L2, TNKS1BP1, PLEK2, NPHP4, CTNNA3, TRIM67, DSP, VASP, FARP2, AFAP1, FRMD4A, KCNAB2, PTPN13, ARHGAP24, APBB1IP, UBXN11, FRMD7, TRIP10, FILIP1, NF2, RIN1, PFN1, TRIB2*	1.66
neuron projection	25	2.18	6.37×10^3	*TENM1, GPI, TENM2, TENM3, AHCY, TENM4, PTPRO, IQGAP1, STON2, PARK2, DVL1, CADM1, DTNBP1, ATP2B4, ANK3, PTPN13, SSTR5, SYNJ1, FRMD7, RAPGEF2, FAS, ATP13A2, NF2, PFN1, CPEB2*	1.80
centrosome	38	3.32	1.08×10^2	*ERCC6L, NUMA1, DCTN1, VPS4B, DCTN3, HEPACAM2, CDC14A, RPGR, TTC28, CAMSAP2, PPP4R2, TMEM67, C7ORF31, CHEK1, PDE4B, CEP170, NPHP4, RTTN, NEK3, CCDC141, PCNT, ZNF322, DCAF13, LRRCC1, PPP1R12A, CEP135, DYX1C1, CCDC116, IFT140, CEP131, KLHL21, MTUS2, PDE4DIP, HAUS5, ALS2, CROCC, FEZ1, ALMS1*	1.52

Table 2. *Cont.*

Term	Count	%	*p* Value	Genes	Fold Enrichment
axon	23	2.01	1.11×10^2	*NTRK1, EPHA5, CNTNAP2, TENM3, KCNB1, PTPRO, DAB2IP, DTNBP1, IQGAP1, IGSF9, ROBO1, SPTA1, ALCAM, ALS2, FEZ1, DVL1, KIF13B, CHRNA10, KIF21B, NEK3, LDLRAP1, CRYAB, VPS16*	1.77
spectrin	4	0.35	1.29×10^2	*SPTA1, SPTBN5, EPB41L2, SPTBN2*	7.58
sarcoplasmic reticulum	7	0.61	1.30×10^2	*ATP2A3, ITPR1, CLEC18B, MRVI1, ANK3, XDH, RYR3*	3.51
plasma membrane	272	23.73	1.48×10^2	**RGSL1, SLC4A1, ABCA12, ABRA, SLC4A5, PIEZO1, PLCE1, GPR179, DYNC2H1, EPHA5, IL15RA, UNC5A, OR1J4, KCNK13, CACNA2D2, SCRIB, ANK3, SYTL5, PHKA2, HSPG2, SYTL3, KEL, FEZ1, OR6C6, KCNQ5, WDPCP, RIN1, EPHA1, OR5AK2, CFB, TAS2R42, SLC22A1, PCDH15, MGST2, C2CD4A, OR10T2, XPC, PCDH12, IQGAP1, MST1R, PPL, IGSF9, OR52L1, ART1, EPB41L5, NCSTN, EPB41L2, PLCG2, KCNN3, APOB*	1.13
proteinaceous extracellular matrix	25	2.18	2.55×10^2	*FBN2, TNXB, LAMC3, ADAMTS12, ADAMTS10, ADAMTS14, GPC1, SLIT2, MUC4, IMPG1, TECTB, AMBN, FN1, MMP10, MMP16, CILP, COL4A3, COL4A6, COL8A2, COL21A1, COL9A3, COL6A5, COL9A2, MATN3, MATN2*	1.59
adherens junction	8	0.70	2.56×10^2	*EPHA5, EPB41L5, CEACAM1, TNK2, PKP2, FMN1, CTNNA3, NF2*	2.73
sarcolemma	11	0.96	2.65×10^2	*DLG1, SGCD, AHNAK, KCNB1, AHNAK2, DYSF, DTNBP1, ANK3, SYNC, SLC8B1, RYR3*	2.21
presynapse	9	0.79	3.25×10^2	*SYT3, SYNJ1, DVL1, SYT15, SCRIB, STON2, SYTL5, SYTL3, PARK2*	2.40
apical plasma membrane	26	2.27	3.54×10^2	*MTCL1, PTPRO, SIPA1L3, SLC4A5, PARD6B, DSTYK, SLCO2B1, CD34, DUOX2, GPIHBP1, SPTBN2, SLC22A11, MUC17, SLC10A2, ABCC6, AKR1A1, FN1, NRG1, ABCA7, CEACAM1, CDHR2, RAPGEF2, CHRFAM7A, WDPCP, SLC26A4, KCNK1*	1.52
collagen trimer	11	0.96	4.25×10^2	*MSR1, COL27A1, COL7A1, COL4A3, COL8A2, SAAL1, COL4A6, COL21A1, COL9A2, COL6A5, OTOL1*	2.04

Next, we performed GOA on each family individually, to determine whether the cytoskeletal functional categories seen in the combined gene list were driven by a single family or many families. Overall, a cytoskeletal GO term was detected among the significantly enriched terms in every family except Family H, J, and O (Supplementary Files S2). Every family possessed at least one variant in a cytoskeletal gene.

We next selected variants of interest for genotyping to determine whether the variant segregated with the AIS phenotype within each relevant family. Specific variants were genotyped if they were present in functional categories of interest (cytoskeletal or extracellular matrix GO terms), and if additional affected and unaffected members of the family were enrolled in our study. Ten genes (*ANKRD11, COL21A1, COL6A5, FGD1, NPHP4, OBSCN, TNXB, CTNNA3, NTRK1*, and *PDE4DIP*) containing multiple variants across families were genotyped. Of these, variants within *TNXB, CTNNA3, NTRK1*, and *PDE4DIP* showed segregation of the variant with the IS phenotype in additional affected and unaffected family members (Supplementary Files S2). *TNXB* (tenascin XB) localizes to the major histocompatibility region on chromosome 6 and encodes an ECM glycoprotein and has been associated to classic Ehlers-Danlos syndrome (EDS), a connective tissue disorder with scoliosis as one characteristic of the phenotype [72]. The *TNXB* variants within our dataset (Family G: TNXB:NM_019105:exon24:c.C8192G:p.P2731R; Family S: TNXB:NM_019105:exon12:c.G4444A: p.V1482M) do not appear in the EDS variant database to date [73–75]. *CTNNA3* (Catenin α 3) encodes a protein within the vinculin family of cell-cell adhesion proteins, and has been linked with certain types of cardiomyopathy [76]. *NTRK1* (Neurotrophic Receptor Tyrosine Kinase 1) encodes a protein that binds neurotrophin peptide and participates within neuronal cell differentiation and specification of neuronal cell subtypes [77]. *PDE4DIP* encodes myomegalin, which anchors phosphodiesterase 4D to the Golgi/centrosome regions. An isoform of myomegalin was recently shown to form a complex with AKAP9 and CDK5RAP to link the pericentrosomal complex to the microtubule-nucleating complex [78]. The remaining variants showed no segregation or only partial segregation of the variant with the AIS phenotype.

4. Discussion

In this study, we report an enrichment of predicted damaging variants in cytoskeletal and extracellular matrix (ECM) functional categories within adolescent idiopathic scoliosis (AIS) families. Additionally, this cohort showed minimal overlap in specific genes across families. Our results support and add to the growing evidence that AIS is a highly polygenic disorder [7,8,28,79,80] in which multiple variants of variable effect size, potentially in combination with epigenetic and environmental phenomenon [81], contribute to the disease phenotype.

The ECM is a dynamic molecular network of proteoglycans, glycoproteins, minerals, and related proteins that plays a critical role within musculoskeletal tissues [82]. The ECM provides critical structural networks and scaffolding to tissues, and contributes to cellular signaling, growth, and repair [83]. Studies of genes encoding the "matrisome", the proteins composing and relating to the ECM, have identified dozens of causal ECM mutations for connective tissue and musculoskeletal disorders [83]. A polygenic burden of rare variants in musculoskeletal collagen genes was linked to AIS risk in a large case-control cohort [28], and our group previously showed mild variant enrichments in our pilot exome sequencing study of five AIS families [46]. Candidate ECM genes including *HSPG2* [29] and *FBN2* [30] have also been linked to AIS in specific cohorts. Several collagen genes (*COL8A2, COL4A3, COL6A5, COL27A1, COL7A1, COL21A1, COL9A2, COL9A3, COL4A6*) and perlecan (*HSPG2*) appeared in our families (Supplementary Files S3). Candidate studies of ECM genes in AIS [63,84–86] were launched, in part, because scoliosis is a common phenotype of monogenic connective tissue disorders including Marfan and Ehlers-Danlos syndromes (EDS) [72,74]. *TNXB* (tenascin XB) variants, which have been associated with classic EDS, were found in two families in our current cohort as well as one family in our previous study [46]. The *TNXB* variant within Family G (TNXB:NM_019105:exon24:c.C8192G:p.P2731R) appeared in the Leiden Open Variation Database for EDS but was predicted as "benign" (https://databases.lovd.nl/shared/variants/0000313458#00021614, accessed on 4 May 2021). Functional data will be required to determine the pathogenicity of these variants. In a recent review, Wise et al. proposed that genetic variants affecting the cartilage matrisome, specifically of the intravertebral disc, may be involved in a subset of AIS cases [33]. Our results provide support to polygenic variants in ECM genes as a contributing factor to AIS in some family lineages.

The cytoskeleton comprises actin, microtubule, and intermediate filament networks and is responsible for a multitude of functions including molecular transport, cellular stability, molecular signaling, cell migration, and cell division [87]. The cytoskeleton and ECM are intricately connected within musculoskeletal tissues and play important roles in mechanotransduction, tissue stability and response to biomechanical loading [88,89]. Every family within our dataset had at least one cytoskeletal genetic variant, and 20/23 families showed an enriched cytoskeletal GO term (Supplementary Files S2). Variants in the centriolar protein *POC5* have been associated with scoliosis through human and animal studies [90–92]. The cytoskeletal kinesin *kif6* was shown to be necessary for proper spine development in zebrafish [34], and mutations in *kif6* also appeared in a zebrafish ENU genetic screen for scoliosis [93].

Defects in cilia, microtubule-based projections critical for cell signaling and fluid flow, have been linked to scoliosis in animal models [31,34,35,37,94,95]. A variant in the ciliary kinesin *KIF7* was found within one AIS family and specific *KIF7* mutations produced scoliosis in zebrafish [38]. Additionally, an overrepresentation of cilia variants was observed in our pilot exome sequencing cohort [46]. However, functional roles of the cytoskeleton and ECM in the development of IS have not yet been demonstrated in humans.

This study contains several limitations. Our familial cohort is relatively small and our ability to understand the impact of rare or low frequency variants in relation to the expression of a complex genetic disease, such as AIS, is limited [96]. Low frequency variants frequently lie outside the scope of large statistical association studies, and therefore may contribute to the "missing heritability" that accompanies many complex traits [97]. The majority of our discovered variants are heterozygous, and we do not know without functional testing whether these variants have a true dominant-negative effect on the resulting protein. To confirm our observed functional pathway enrichments, validate the presence of specific genetic variants, and obtain adequate statistical power, a larger cohort of AIS families must be sequenced. Additional functional studies of specific variants, with particular focus on those most likely to be damaging (i.e., stopgain, frameshift mutations) will be required to demonstrate causality [93].

Collectively, our results provide support to the hypothesis of ECM and cytoskeletal involvement in AIS etiology through what is, to our knowledge, the largest sequencing study of AIS families to date. These results suggest that there are many specific genes that can collectively increase disease risk, although there may be affected pathways that are shared across families. We hypothesize that (1) individual AIS families harbor low frequency mutations in different functional categories, resulting in different subsets of AIS, and/or (2) individuals with AIS require mutations in multiple gene categories (i.e., the cytoskeleton and ECM), resulting in mild dysfunction across several molecular pathways, to cause disease. Specifically, these results suggest that mild mutations in cytoskeletal or ECM genes may play a role in AIS etiology. Ultimately, this work will assist in the ability to predict the onset of adolescent idiopathic scoliosis and the risk of progressive disease and, thus, lead to the development of more personalized treatments for individuals with AIS.

5. Conclusions

This work presents a novel set of candidate variants found in affected individuals from 23 AIS families. In agreement with our previous WES study of five AIS families, we observed an overrepresentation of variants in cytoskeletal and ECM functional categories, with few specific genes shared across families. Overall, this work paints a picture of AIS genetic etiology as highly polygenic and specific to individual family lineages. An analytical approach that integrates data from family-based sequencing with genetic association studies, an understanding of study population variation, population stratification and genetic heterogeneity, and advances in clinical phenotyping will enhance our ability to define the genetic complexity of this disorder.

Supplementary Materials: The following are available online at https://www.mdpi.com/article/10.3390/genes12060922/s1: File S1: Family Pedigrees. Degree of spinal curvature is indicated with a D (degrees). Individuals marked "confirmed positive" were confirmed as affected with AIS but were unable to provide us with an X-ray for measurement. Exome sequenced individuals are indicated. File S2: Table S1: Clinical information from all exome sequenced individuals; Table S2: Enriched cytobands from MAF < 0.05 gene list; Table S3: Genotyping of genes in multiple families and in functional categories of interest; Table S4: Full Gene Ontology (GO) Term enrichment results

($p < 0.01$) for all functional categories using DAVID; File S3: Detail of all filtered variants, sorted by family. All filtered variants had a minor allele frequency (MAF) < 0.05, were present in all sequenced members of the family, and were predicted to be damaging to the resulting protein by at least one algorithm. Details of the exome sequencing strategy and bioinformatic filtering process are provided in the Methods.

Author Contributions: Conceptualization: N.H.M., K.L.J., M.R.G.T., E.E.B. and E.A.T.; Methodology: C.I.W., M.T.C., A.M.M., G.D.T., M.R.B., E.E.B. and R.B.; Software: G.D.T. and K.L.J.; Validation: C.I.W.; Formal analysis: K.L.J., G.D.T. and E.A.T.; Investigation: N.H.M.; Resources: N.H.M.; Data curation: N.H.M. and E.A.T.; Writing—original draft preparation: E.A.T.; Writing—review and editing: N.H.M. and A.M.M.; Visualization: E.A.T.; Supervision: N.H.M., E.E.B. and E.A.T.; Project administration: N.H.M.; Funding acquisition: N.H.M. and E.E.B. All authors have read and agreed to the published version of the manuscript.

Funding: Funding for NHM's laboratory and this study was provided by NIH/NIAMS R01AR068292. Additional funds were provided by Children's Hospital Colorado.

Institutional Review Board Statement: Written informed consent was obtained from study subjects who were enrolled in accordance with protocols approved by the Johns Hopkins School of Medicine Institutional Review Board and the University of Colorado Anschutz Medical Campus Institutional Review Board (Colorado Multiple Institutional Review Board, Studies #06-1161 and #07-0417). All procedures involving human participants were performed in accordance with the ethical standards of these institutional review boards, the 1964 Declaration of Helsinki and its later amendments, or comparable ethical standards.

Informed Consent Statement: Informed consent was obtained from all subjects involved in the study.

Data Availability Statement: The authors affirm that all data necessary for confirming the conclusions of this article are represented fully within the article and its supplementary files, including the complete lists of filtered variants for each family.

Acknowledgments: Patient data was collected and managed using REDCap (Research Electronic Data Capture) tools hosted at the University of Colorado Anschutz Medical Campus. REDCap is a secure, web-based application designed to support data capture for research studies, providing: (1) an intuitive interface for validated data entry; (2) audit trails for tracking data manipulation and export procedures; (3) automated export procedures for seamless data downloads to common statistical packages; and (4) procedures for importing data from external sources. We thank Lee Niswander and Bruce Appel for their helpful discussions regarding the overall design of this study, as well as assistance selecting families for sequencing. Above all, we thank the study subjects and their families for contributing their time, samples, and information to this research effort.

Conflicts of Interest: The authors declare no conflict of interest.

References

1. Asher, M.A.; Burton, D.C. Adolescent idiopathic scoliosis: Natural history and long term treatment effects. *Scoliosis* **2006**, *1*, 2. [CrossRef] [PubMed]
2. Kane, W.J.; Moe, J.H. A scoliosis-prevalence survey in Minnesota. *Clin. Orthop. Relat. Res.* **1970**, *69*, 216–218. [CrossRef]
3. Weinstein, S.L. Adolescent idiopathic scoliosis: Prevalence and natural history. In *The Pediatric Spine: Principles and Practice*; Weinstein, S.L., Ed.; Raven Press: New York, NY, USA, 1994; pp. 463–478.
4. Weinstein, S.L.; Zavala, D.C.; Ponseti, I.V. Idiopathic scoliosis: Long-term follow-up and prognosis in untreated patients. *J. Bone Jt. Surg. Am. Vol.* **1981**, *63*, 702–712. [CrossRef]
5. Bozzio, A.E.; Hu, X.; Lieberman, I.H. Cost and Clinical Outcome of Adolescent Idiopathic Scoliosis Surgeries-Experience From a Nonprofit Community Hospital. *Int. J. Spine Surg.* **2019**, *13*, 474–478. [CrossRef]
6. Tang, N.L.; Yeung, H.Y.; Lee, K.M.; Hung, V.W.; Cheung, C.S.; Ng, B.K.; Kwok, R.; Guo, X.; Qin, L.; Cheng, J.C. A relook into the association of the estrogen receptor [alpha] gene (PvuII, XbaI) and adolescent idiopathic scoliosis: A study of 540 Chinese cases. *Spine* **2006**, *31*, 2463–2468. [CrossRef] [PubMed]
7. Khanshour, A.M.; Wise, C.A. The Genetic Architecture of Adolescent Idiopathic Scoliosis. In *Pathogenesis of Idiopathic Scoliosis*; Machida, M., Weinstein, S.L., Dubousset, J., Eds.; Springer: Tokyo, Japan, 2018; pp. 51–74. [CrossRef]
8. Ward, K.; Ogilvie, J.; Argyle, V.; Nelson, L.; Meade, M.; Braun, J.; Chettier, R. Polygenic inheritance of adolescent idiopathic scoliosis: A study of extended families in Utah. *Am. J. Med. Genet A* **2010**, *152A*, 1178–1188. [CrossRef] [PubMed]

9. Takahashi, Y.; Kou, I.; Takahashi, A.; Johnson, T.A.; Kono, K.; Kawakami, N.; Uno, K.; Ito, M.; Minami, S.; Yanagida, H.; et al. A genome-wide association study identifies common variants near LBX1 associated with adolescent idiopathic scoliosis. *Nat. Genet.* **2011**, *43*, 1237–1240. [CrossRef] [PubMed]

10. Fan, Y.H.; Song, Y.Q.; Chan, D.; Takahashi, Y.; Ikegawa, S.; Matsumoto, M.; Kou, I.; Cheah, K.S.; Sham, P.; Cheung, K.M.; et al. SNP rs11190870 near LBX1 is associated with adolescent idiopathic scoliosis in southern Chinese. *J. Hum. Genet* **2012**, *57*, 244–246. [CrossRef] [PubMed]

11. Jiang, H.; Qiu, X.; Dai, J.; Yan, H.; Zhu, Z.; Qian, B.; Qiu, Y. Association of rs11190870 near LBX1 with adolescent idiopathic scoliosis susceptibility in a Han Chinese population. *Eur. Spine J. Off. Publ. Eur. Spine Soc. Eur. Spinal Deform. Soc. Eur. Sect. Cerv. Spine Res. Soc.* **2013**, *22*, 282–286. [CrossRef]

12. Gao, W.; Peng, Y.; Liang, G.; Liang, A.; Ye, W.; Zhang, L.; Sharma, S.; Su, P.; Huang, D. Association between common variants near LBX1 and adolescent idiopathic scoliosis replicated in the Chinese Han population. *PLoS ONE* **2013**, *8*, e53234. [CrossRef]

13. Londono, D.; Kou, I.; Johnson, T.A.; Sharma, S.; Ogura, Y.; Tsunoda, T.; Takahashi, A.; Matsumoto, M.; Herring, J.A.; Lam, T.P.; et al. A meta-analysis identifies adolescent idiopathic scoliosis association with LBX1 locus in multiple ethnic groups. *J. Med. Genet* **2014**, *51*, 401–406. [CrossRef]

14. Chen, S.; Zhao, L.; Roffey, D.M.; Phan, P.; Wai, E.K. Association of rs11190870 near LBX1 with adolescent idiopathic scoliosis in East Asians: A systematic review and meta-analysis. *Spine J. Off. J. N. Am. Spine Soc.* **2014**, *14*, 2968–2975. [CrossRef]

15. Liang, J.; Xing, D.; Li, Z.; Chua, S.; Li, S. Association Between rs11190870 Polymorphism Near LBX1 and Susceptibility to Adolescent Idiopathic Scoliosis in East Asian Population: A Genetic Meta-Analysis. *Spine* **2014**, *39*, 862–869. [CrossRef]

16. Zhu, Z.; Tang, N.L.; Xu, L.; Qin, X.; Mao, S.; Song, Y.; Liu, L.; Li, F.; Liu, P.; Yi, L.; et al. Genome-wide association study identifies new susceptibility loci for adolescent idiopathic scoliosis in Chinese girls. *Nat. Commun.* **2015**, *6*, 8355. [CrossRef]

17. Chettier, R.; Nelson, L.; Ogilvie, J.W.; Albertsen, H.M.; Ward, K. Haplotypes at LBX1 have distinct inheritance patterns with opposite effects in adolescent idiopathic scoliosis. *PLoS ONE* **2015**, *10*, e0117708. [CrossRef]

18. Grauers, A.; Wang, J.; Einarsdottir, E.; Simony, A.; Danielsson, A.; Akesson, K.; Ohlin, A.; Halldin, K.; Grabowski, P.; Tenne, M.; et al. Candidate gene analysis and exome sequencing confirm LBX1 as a susceptibility gene for idiopathic scoliosis. *Spine J. Off. J. N. Am. Spine Soc.* **2015**, *15*, 2239–2246. [CrossRef]

19. Liu, S.; Wu, N.; Zuo, Y.; Zhou, Y.; Liu, J.; Liu, Z.; Chen, W.; Liu, G.; Chen, Y.; Chen, J.; et al. Genetic Polymorphism of LBX1 Is Associated With Adolescent Idiopathic Scoliosis in Northern Chinese Han Population. *Spine* **2017**, *42*, 1125–1129. [CrossRef] [PubMed]

20. Nada, D.; Julien, C.; Samuels, M.E.; Moreau, A. A Replication Study for Association of LBX1 Locus With Adolescent Idiopathic Scoliosis in French-Canadian Population. *Spine* **2018**, *43*, 172–178. [CrossRef]

21. Li, Y.L.; Gao, S.J.; Xu, H.; Liu, Y.; Li, H.L.; Chen, X.Y.; Ning, G.Z.; Feng, S.Q. The association of rs11190870 near LBX1 with the susceptibility and severity of AIS, a meta-analysis. *Int. J. Surg.* **2018**, *54*, 193–200. [CrossRef] [PubMed]

22. Man, G.C.; Tang, N.L.; Chan, T.F.; Lam, T.P.; Li, J.W.; Ng, B.K.; Zhu, Z.; Qiu, Y.; Cheng, J.C. Replication Study for the Association of GWAS-associated Loci with Adolescent Idiopathic Scoliosis Susceptibility and Curve Progression in a Chinese Population. *Spine* **2018**. [CrossRef]

23. Kou, I.; Takahashi, Y.; Johnson, T.A.; Takahashi, A.; Guo, L.; Dai, J.; Qiu, X.; Sharma, S.; Takimoto, A.; Ogura, Y.; et al. Genetic variants in GPR126 are associated with adolescent idiopathic scoliosis. *Nat. Genet.* **2013**, *45*, 676–679. [CrossRef] [PubMed]

24. Xu, J.F.; Yang, G.H.; Pan, X.H.; Zhang, S.J.; Zhao, C.; Qiu, B.S.; Gu, H.F.; Hong, J.F.; Cao, L.; Chen, Y.; et al. Association of GPR126 gene polymorphism with adolescent idiopathic scoliosis in Chinese populations. *Genomics* **2015**, *105*, 101–107. [CrossRef] [PubMed]

25. Qin, X.; Xu, L.; Xia, C.; Zhu, W.; Sun, W.; Liu, Z.; Qiu, Y.; Zhu, Z. Genetic Variant of GPR126 Gene is Functionally Associated With Adolescent Idiopathic Scoliosis in Chinese Population. *Spine* **2017**, *42*, E1098–E1103. [CrossRef] [PubMed]

26. Kou, I.; Watanabe, K.; Takahashi, Y.; Momozawa, Y.; Khanshour, A.; Grauers, A.; Zhou, H.; Liu, G.; Fan, Y.H.; Takeda, K.; et al. A multi-ethnic meta-analysis confirms the association of rs6570507 with adolescent idiopathic scoliosis. *Sci. Rep.* **2018**, *8*, 11575. [CrossRef]

27. Haller, G.; Alvarado, D.; McCall, K.; Yang, P.; Cruchaga, C.; Harms, M.; Goate, A.; Willing, M.; Morcuende, J.A.; Baschal, E.; et al. A polygenic burden of rare variants across extracellular matrix genes among individuals with adolescent idiopathic scoliosis. *Hum. Mol. Genet.* **2016**, *25*, 202–209. [CrossRef]

28. Baschal, E.E.; Wethey, C.I.; Swindle, K.; Baschal, R.M.; Gowan, K.; Tang, N.L.; Alvarado, D.M.; Haller, G.E.; Dobbs, M.B.; Taylor, M.R.; et al. Exome sequencing identifies a rare HSPG2 variant associated with familial idiopathic scoliosis. *G3* **2014**, *5*, 167–174. [CrossRef]

29. Buchan, J.G.; Alvarado, D.M.; Haller, G.E.; Cruchaga, C.; Harms, M.B.; Zhang, T.; Willing, M.C.; Grange, D.K.; Braverman, A.C.; Miller, N.H.; et al. Rare variants in FBN1 and FBN2 are associated with severe adolescent idiopathic scoliosis. *Hum. Mol. Genet.* **2014**, *23*, 5271–5282. [CrossRef]

30. Rose, C.D.; Pompili, D.; Henke, K.; Van Gennip, J.L.M.; Meyer-Miner, A.; Rana, R.; Gobron, S.; Harris, M.P.; Nitz, M.; Ciruna, B. SCO-Spondin Defects and Neuroinflammation Are Conserved Mechanisms Driving Spinal Deformity across Genetic Models of Idiopathic Scoliosis. *Curr. Biol.* **2020**. [CrossRef]

31. Van Gennip, J.L.M.; Boswell, C.W.; Ciruna, B. Neuroinflammatory signals drive spinal curve formation in zebrafish models of idiopathic scoliosis. *Sci. Adv.* **2018**, *4*, eaav1781. [CrossRef]

32. Wise, C.A.; Sepich, D.; Ushiki, A.; Khanshour, A.M.; Kidane, Y.H.; Makki, N.; Gurnett, C.A.; Gray, R.S.; Rios, J.J.; Ahituv, N.; et al. The cartilage matrisome in adolescent idiopathic scoliosis. *Bone Res.* **2020**, *8*, 13. [CrossRef]

33. Buchan, J.G.; Gray, R.S.; Gansner, J.M.; Alvarado, D.M.; Burgert, L.; Gitlin, J.D.; Gurnett, C.A.; Goldsmith, M.I. Kinesin family member 6 (kif6) is necessary for spine development in zebrafish. *Dev. Dyn. Off. Publ. Am. Assoc. Anat.* **2014**, *243*, 1646–1657. [CrossRef] [PubMed]

34. Grimes, D.T.; Boswell, C.W.; Morante, N.F.; Henkelman, R.M.; Burdine, R.D.; Ciruna, B. Zebrafish models of idiopathic scoliosis link cerebrospinal fluid flow defects to spine curvature. *Science* **2016**, *352*, 1341–1344. [CrossRef] [PubMed]

35. Hayes, M.; Gao, X.; Yu, L.X.; Paria, N.; Henkelman, R.M.; Wise, C.A.; Ciruna, B. ptk7 mutant zebrafish models of congenital and idiopathic scoliosis implicate dysregulated Wnt signalling in disease. *Nat. Commun.* **2014**, *5*, 4777. [CrossRef] [PubMed]

36. Konjikusic, M.J.; Yeetong, P.; Boswell, C.W.; Lee, C.; Roberson, E.C.; Ittiwut, R.; Suphapeetiporn, K.; Ciruna, B.; Gurnett, C.A.; Wallingford, J.B.; et al. Mutations in Kinesin family member 6 reveal specific role in ependymal cell ciliogenesis and human neurological development. *PLoS Genet.* **2018**, *14*, e1007817. [CrossRef] [PubMed]

37. Terhune, E.A.; Cuevas, M.T.; Monley, A.M.; Wethey, C.I.; Chen, X.; Cattel, M.V.; Bayrak, M.N.; Bland, M.R.; Sutphin, B.; Devon Trahan, G.; et al. Mutations in KIF7 Implicated in Idiopathic Scoliosis in Humans and Axial Curvatures in Zebrafish. *Hum. Mutat* **2020**. [CrossRef]

38. Carry, P.M.; Duke, V.R.; Brazell, C.J.; Stence, N.; Scholes, M.; Rousie, D.L.; Hadley Miller, N. Lateral semi-circular canal asymmetry in females with idiopathic scoliosis. *PLoS ONE* **2020**, *15*, e0232417. [CrossRef]

39. Hawasli, A.H.; Hullar, T.E.; Dorward, I.G. Idiopathic scoliosis and the vestibular system. *Eur. Spine J. Off. Publ. Eur. Spine Soc. Eur. Spinal Deform. Soc. Eur. Sect. Cerv. Spine Res. Soc.* **2015**, *24*, 227–233. [CrossRef]

40. Pialasse, J.P.; Descarreaux, M.; Mercier, P.; Blouin, J.; Simoneau, M. The Vestibular-Evoked Postural Response of Adolescents with Idiopathic Scoliosis Is Altered. *PLoS ONE* **2015**, *10*, e0143124. [CrossRef]

41. Haumont, T.; Gauchard, G.C.; Lascombes, P.; Perrin, P.P. Postural instability in early-stage idiopathic scoliosis in adolescent girls. *Spine* **2011**, *36*, E847–E854. [CrossRef]

42. Simoneau, M.; Lamothe, V.; Hutin, E.; Mercier, P.; Teasdale, N.; Blouin, J. Evidence for cognitive vestibular integration impairment in idiopathic scoliosis patients. *BMC Neurosci.* **2009**, *10*, 102. [CrossRef]

43. Mathieson, I.; McVean, G. Differential confounding of rare and common variants in spatially structured populations. *Nat. Genet.* **2012**, *44*, 243–246. [CrossRef]

44. McClellan, J.; King, M.C. Genetic heterogeneity in human disease. *Cell* **2010**, *141*, 210–217. [CrossRef]

45. Baschal, E.E.; Terhune, E.A.; Wethey, C.I.; Baschal, R.M.; Robinson, K.D.; Cuevas, M.T.; Pradhan, S.; Sutphin, B.S.; Taylor, M.R.G.; Gowan, K.; et al. Idiopathic Scoliosis Families Highlight Actin-Based and Microtubule-Based Cellular Projections and Extracellular Matrix in Disease Etiology. *G3* **2018**, *8*, 2663–2672. [CrossRef]

46. Shands, A.R.; Eisberg, H.B. The incidence of scoliosis in the state of Delaware; a study of 50,000 minifilms of the chest made during a survey for tuberculosis. *J. Bone Jt. Surg. Am. Vol.* **1955**, *37*, 1243–1249. [CrossRef]

47. Kane, W.J. Scoliosis prevalence: A call for a statement of terms. *Clin. Orthop. Relat. Res.* **1977**, *126*, 43–46. [CrossRef]

48. Armstrong, G.W.; Livermore, N.B., 3rd; Suzuki, N.; Armstrong, J.G. Nonstandard vertebral rotation in scoliosis screening patients. Its prevalence and relation to the clinical deformity. *Spine (Phila Pa 1976)* **1982**, *7*, 50–54. [CrossRef]

49. Cingolani, P.; Platts, A.; Wang le, L.; Coon, M.; Nguyen, T.; Wang, L.; Land, S.J.; Lu, X.; Ruden, D.M. A program for annotating and predicting the effects of single nucleotide polymorphisms, SnpEff: SNPs in the genome of Drosophila melanogaster strain w1118; iso-2; iso-3. *Fly* **2012**, *6*, 80–92. [CrossRef] [PubMed]

50. Lek, M.; Karczewski, K.J.; Minikel, E.V.; Samocha, K.E.; Banks, E.; Fennell, T.; O'Donnell-Luria, A.H.; Ware, J.S.; Hill, A.J.; Cummings, B.B.; et al. Analysis of protein-coding genetic variation in 60,706 humans. *Nature* **2016**, *536*, 285–291. [CrossRef] [PubMed]

51. Liu, X.; Jian, X.; Boerwinkle, E. dbNSFP: A lightweight database of human nonsynonymous SNPs and their functional predictions. *Hum. Mutat* **2011**, *32*, 894–899. [CrossRef]

52. Liu, X.; Wu, C.; Li, C.; Boerwinkle, E. dbNSFP v3.0: A One-Stop Database of Functional Predictions and Annotations for Human Nonsynonymous and Splice-Site SNVs. *Hum. Mutat* **2016**, *37*, 235–241. [CrossRef]

53. Huang da, W.; Sherman, B.T.; Lempicki, R.A. Systematic and integrative analysis of large gene lists using DAVID bioinformatics resources. *Nat. Protoc* **2009**, *4*, 44–57. [CrossRef] [PubMed]

54. Huang da, W.; Sherman, B.T.; Lempicki, R.A. Bioinformatics enrichment tools: Paths toward the comprehensive functional analysis of large gene lists. *Nucleic Acids Res.* **2009**, *37*, 1–13. [CrossRef] [PubMed]

55. Chen, E.Y.; Tan, C.M.; Kou, Y.; Duan, Q.; Wang, Z.; Meirelles, G.V.; Clark, N.R.; Ma'ayan, A. Enrichr: Interactive and collaborative HTML5 gene list enrichment analysis tool. *BMC Bioinform.* **2013**, *14*, 128. [CrossRef]

56. Kuleshov, M.V.; Jones, M.R.; Rouillard, A.D.; Fernandez, N.F.; Duan, Q.; Wang, Z.; Koplev, S.; Jenkins, S.L.; Jagodnik, K.M.; Lachmann, A.; et al. Enrichr: A comprehensive gene set enrichment analysis web server 2016 update. *Nucleic Acids Res.* **2016**, *44*, W90–W97. [CrossRef] [PubMed]

57. Giampietro, P.F.; Raggio, C.L.; Reynolds, C.; Ghebranious, N.; Burmester, J.K.; Glurich, I.; Rasmussen, K.; McPherson, E.; Pauli, R.M.; Shukla, S.K.; et al. DLL3 as a candidate gene for vertebral malformations. *Am. J. Med. Genet. A* **2006**, *140*, 2447–2453. [CrossRef]

58. Loomes, K.M.; Stevens, S.A.; O'Brien, M.L.; Gonzalez, D.M.; Ryan, M.J.; Segalov, M.; Dormans, N.J.; Mimoto, M.S.; Gibson, J.D.; Sewell, W.; et al. Dll3 and Notch1 genetic interactions model axial segmental and craniofacial malformations of human birth defects. *Dev. Dyn. Off. Publ. Am. Assoc. Anat.* **2007**, *236*, 2943–2951. [CrossRef]

59. Shen, N.; Chen, N.; Zhou, X.; Zhao, B.; Huang, R.; Liang, J.; Yang, X.; Chen, M.; Song, Y.; Du, Q. Alterations of the gut microbiome and plasma proteome in Chinese patients with adolescent idiopathic scoliosis. *Bone* **2019**, *120*, 364–370. [CrossRef] [PubMed]
60. Liu, X.Y.; Wang, L.; Yu, B.; Zhuang, Q.Y.; Wang, Y.P. Expression Signatures of Long Noncoding RNAs in Adolescent Idiopathic Scoliosis. *Biomed. Res. Int.* **2015**, *2015*, 276049. [CrossRef]
61. Shin, J.H.; Ha, K.Y.; Jung, S.H.; Chung, Y.J. Genetic predisposition in degenerative lumbar scoliosis due to the copy number variation. *Spine* **2011**, *36*, 1782–1793. [CrossRef]
62. Miller, N.H.; Mims, B.; Child, A.; Milewicz, D.M.; Sponseller, P.; Blanton, S.H. Genetic analysis of structural elastic fiber and collagen genes in familial adolescent idiopathic scoliosis. *J. Orthop. Res.* **1996**, *14*, 994–999. [CrossRef]
63. Wise, C.A.; Barnes, R.; Gillum, J.; Herring, J.A.; Bowcock, A.M.; Lovett, M. Localization of susceptibility to familial idiopathic scoliosis. *Spine* **2000**, *25*, 2372–2380. [CrossRef] [PubMed]
64. Justice, C.M.; Miller, N.H.; Marosy, B.; Zhang, J.; Wilson, A.F. Familial idiopathic scoliosis: Evidence of an X-linked susceptibility locus. *Spine* **2003**, *28*, 589–594. [CrossRef]
65. Miller, N.H.; Justice, C.M.; Marosy, B.; Doheny, K.F.; Pugh, E.; Zhang, J.; Dietz, H.C., 3rd; Wilson, A.F. Identification of candidate regions for familial idiopathic scoliosis. *Spine* **2005**, *30*, 1181–1187. [CrossRef]
66. Miller, N.H.; Marosy, B.; Justice, C.M.; Novak, S.M.; Tang, E.Y.; Boyce, P.; Pettengil, J.; Doheny, K.F.; Pugh, E.W.; Wilson, A.F. Linkage analysis of genetic loci for kyphoscoliosis on chromosomes 5p13, 13q13.3, and 13q32. *Am. J. Med. Genet. A* **2006**, *140*, 1059–1068. [CrossRef] [PubMed]
67. Ocaka, L.; Zhao, C.; Reed, J.A.; Ebenezer, N.D.; Brice, G.; Morley, T.; Mehta, M.; O'Dowd, J.; Weber, J.L.; Hardcastle, A.J.; et al. Assignment of two loci for autosomal dominant adolescent idiopathic scoliosis to chromosomes 9q31.2-q34.2 and 17q25.3-qtel. *J. Med. Genet.* **2008**, *45*, 87–92. [CrossRef]
68. Esposito, T.; Uccello, R.; Caliendo, R.; Di Martino, G.F.; Gironi Carnevale, U.A.; Cuomo, S.; Ronca, D.; Varriale, B. Estrogen receptor polymorphism, estrogen content and idiopathic scoliosis in human: A possible genetic linkage. *J. Steroid Biochem. Mol. Biol* **2009**, *116*, 56–60. [CrossRef]
69. Gurnett, C.A.; Alaee, F.; Bowcock, A.; Kruse, L.; Lenke, L.G.; Bridwell, K.H.; Kuklo, T.; Luhmann, S.J.; Dobbs, M.B. Genetic linkage localizes an adolescent idiopathic scoliosis and pectus excavatum gene to chromosome 18 q. *Spine* **2009**, *34*, E94–E100. [CrossRef] [PubMed]
70. Marosy, B.; Justice, C.M.; Vu, C.; Zorn, A.; Nzegwu, N.; Wilson, A.F.; Miller, N.H. Identification of susceptibility loci for scoliosis in FIS families with triple curves. *Am. J. Med. Genet. A* **2010**, *152A*, 846–855. [CrossRef]
71. Meester, J.A.N.; Verstraeten, A.; Schepers, D.; Alaerts, M.; Van Laer, L.; Loeys, B.L. Differences in manifestations of Marfan syndrome, Ehlers-Danlos syndrome, and Loeys-Dietz syndrome. *Ann. Cardiothorac Surg.* **2017**, *6*, 582–594. [CrossRef]
72. Lao, Q.; Mallappa, A.; Rueda Faucz, F.; Joyal, E.; Veeraraghavan, P.; Chen, W.; Merke, D.P. A TNXB splice donor site variant as a cause of hypermobility type Ehlers-Danlos syndrome in patients with congenital adrenal hyperplasia. *Mol. Genet. Genom. Med.* **2021**, *9*, e1556. [CrossRef]
73. Rymen, D.; Ritelli, M.; Zoppi, N.; Cinquina, V.; Giunta, C.; Rohrbach, M.; Colombi, M. Clinical and Molecular Characterization of Classical-Like Ehlers-Danlos Syndrome Due to a Novel TNXB Variant. *Genes* **2019**, *10*, 843. [CrossRef]
74. Miller, W.L.; Merke, D.P. Tenascin-X, Congenital Adrenal Hyperplasia, and the CAH-X Syndrome. *Horm Res. Paediatr* **2018**, *89*, 352–361. [CrossRef]
75. McNally, E.; MacLeod, H.; Dellefave-Castillo, L. *Arrhythmogenic Right Ventricular Cardiomyopathy*; Adam, M.P., Ardinger, H.H., Pagon, R.A., Wallace, S.E., Bean, L.J.H., Mirzaa, G., Amemiya, A., Eds.; GeneReviews®: Seattle, WA, USA, 1993.
76. Cocco, E.; Scaltriti, M.; Drilon, A. NTRK fusion-positive cancers and TRK inhibitor therapy. *Nat. Rev. Clin. Oncol.* **2018**, *15*, 731–747. [CrossRef]
77. Bouguenina, H.; Salaun, D.; Mangon, A.; Muller, L.; Baudelet, E.; Camoin, L.; Tachibana, T.; Cianferani, S.; Audebert, S.; Verdier-Pinard, P.; et al. EB1-binding-myomegalin protein complex promotes centrosomal microtubules functions. *Proc. Natl. Acad. Sci. USA* **2017**, *114*, E10687–E10696. [CrossRef]
78. Kruse, L.M.; Buchan, J.G.; Gurnett, C.A.; Dobbs, M.B. Polygenic threshold model with sex dimorphism in adolescent idiopathic scoliosis: The Carter effect. *J. Bone Jt. Surg. Am. Vol.* **2012**, *94*, 1485–1491. [CrossRef] [PubMed]
79. Grauers, A.; Einarsdottir, E.; Gerdhem, P. Genetics and pathogenesis of idiopathic scoliosis. *Scoliosis Spinal Disord.* **2016**, *11*, 45. [CrossRef] [PubMed]
80. Perez-Machado, G.; Berenguer-Pascual, E.; Bovea-Marco, M.; Rubio-Belmar, P.A.; Garcia-Lopez, E.; Garzon, M.J.; Mena-Molla, S.; Pallardo, F.V.; Bas, T.; Vina, J.R.; et al. From genetics to epigenetics to unravel the etiology of adolescent idiopathic scoliosis. *Bone* **2020**, *140*, 115563. [CrossRef]
81. Theocharis, A.D.; Skandalis, S.S.; Gialeli, C.; Karamanos, N.K. Extracellular matrix structure. *Adv. Drug Deliv Rev.* **2016**, *97*, 4–27. [CrossRef]
82. Lamande, S.R.; Bateman, J.F. Genetic Disorders of the Extracellular Matrix. *Anat. Rec.* **2020**, *303*, 1527–1542. [CrossRef] [PubMed]
83. Wang, S.; Qiu, Y.; Ma, Z.; Xia, C.; Zhu, F.; Zhu, Z. Expression of Runx2 and type X collagen in vertebral growth plate of patients with adolescent idiopathic scoliosis. *Connect. Tissue Res.* **2010**, *51*, 188–196. [CrossRef]
84. Carr, A.J.; Ogilvie, D.J.; Wordsworth, B.P.; Priestly, L.M.; Smith, R.; Sykes, B. Segregation of structural collagen genes in adolescent idiopathic scoliosis. *Clin. Orthop. Relat. Res.* **1992**, *274*, 305–310.
85. Montanaro, L.; Parisini, P.; Greggi, T.; Di Silvestre, M.; Campoccia, D.; Rizzi, S.; Arciola, C.R. Evidence of a linkage between matrilin-1 gene (MATN1) and idiopathic scoliosis. *Scoliosis* **2006**, *1*, 21. [CrossRef]

86. Hohmann, T.; Dehghani, F. The Cytoskeleton-A Complex Interacting Meshwork. *Cells* **2019**, *8*, 362. [CrossRef]
87. Humphrey, J.D.; Dufresne, E.R.; Schwartz, M.A. Mechanotransduction and extracellular matrix homeostasis. *Nat. Rev. Mol. Cell Biol.* **2014**, *15*, 802–812. [CrossRef] [PubMed]
88. Klein-Nulend, J.; Bacabac, R.G.; Bakker, A.D. Mechanical loading and how it affects bone cells: The role of the osteocyte cytoskeleton in maintaining our skeleton. *Eur. Cell Mater.* **2012**, *24*, 278–291. [CrossRef]
89. Patten, S.A.; Margaritte-Jeannin, P.; Bernard, J.C.; Alix, E.; Labalme, A.; Besson, A.; Girard, S.L.; Fendri, K.; Fraisse, N.; Biot, B.; et al. Functional variants of POC5 identified in patients with idiopathic scoliosis. *J. Clin. Investig.* **2015**, *125*, 1124–1128. [CrossRef] [PubMed]
90. Xu, L.; Sheng, F.; Xia, C.; Li, Y.; Feng, Z.; Qiu, Y.; Zhu, Z. Common Variant of POC5 Is Associated With the Susceptibility of Adolescent Idiopathic Scoliosis. *Spine* **2018**, *43*, E683–E688. [CrossRef] [PubMed]
91. Hassan, A.; Parent, S.; Mathieu, H.; Zaouter, C.; Molidperee, S.; Bagu, E.T.; Barchi, S.; Villemure, I.; Patten, S.A.; Moldovan, F. Adolescent idiopathic scoliosis associated POC5 mutation impairs cell cycle, cilia length and centrosome protein interactions. *PLoS ONE* **2019**, *14*, e0213269. [CrossRef] [PubMed]
92. Gray, R.S.; Gonzalez, R.; Ackerman, S.D.; Minowa, R.; Griest, J.F.; Bayrak, M.N.; Troutwine, B.; Canter, S.; Monk, K.R.; Sepich, D.S.; et al. Postembryonic screen for mutations affecting spine development in zebrafish. *Dev. Biol.* **2021**, *471*, 18–33. [CrossRef] [PubMed]
93. Sternberg, J.R.; Prendergast, A.E.; Brosse, L.; Cantaut-Belarif, Y.; Thouvenin, O.; Orts-Del'Immagine, A.; Castillo, L.; Djenoune, L.; Kurisu, S.; McDearmid, J.R.; et al. Pkd2l1 is required for mechanoception in cerebrospinal fluid-contacting neurons and maintenance of spine curvature. *Nat. Commun.* **2018**, *9*, 3804. [CrossRef] [PubMed]
94. Zhang, X.; Jia, S.; Chen, Z.; Chong, Y.L.; Xie, H.; Feng, D.; Wu, X.; Song, D.Z.; Roy, S.; Zhao, C. Cilia-driven cerebrospinal fluid flow directs expression of urotensin neuropeptides to straighten the vertebrate body axis. *Nat. Genet.* **2018**, *50*, 1666–1673. [CrossRef] [PubMed]
95. Bomba, L.; Walter, K.; Soranzo, N. The impact of rare and low-frequency genetic variants in common disease. *Genome Biol.* **2017**, *18*, 77. [CrossRef]
96. Manolio, T.A.; Collins, F.S.; Cox, N.J.; Goldstein, D.B.; Hindorff, L.A.; Hunter, D.J.; McCarthy, M.I.; Ramos, E.M.; Cardon, L.R.; Chakravarti, A.; et al. Finding the missing heritability of complex diseases. *Nature* **2009**, *461*, 747–753. [CrossRef] [PubMed]
97. Harris, P.A.; Taylor, R.; Thielke, R.; Payne, J.; Gonzalez, N.; Conde, J.G. Research electronic data capture (REDCap)–a metadata-driven methodology and workflow process for providing translational research informatics support. *J. Biomed. Inf.* **2009**, *42*, 377–381. [CrossRef] [PubMed]

genes MDPI

Article

A Genomic Approach to Delineating the Occurrence of Scoliosis in Arthrogryposis Multiplex Congenita

Xenia Latypova [1], Stefan Giovanni Creadore [2], Noémi Dahan-Oliel [3,4], Anxhela Gjyshi Gustafson [2], Steven Wei-Hung Hwang [5], Tanya Bedard [6], Kamran Shazand [2], Harold J. P. van Bosse [5], Philip F. Giampietro [7,*] and Klaus Dieterich [8,*]

[1] Grenoble Institut Neurosciences, Université Grenoble Alpes, Inserm, U1216, CHU Grenoble Alpes, 38000 Grenoble, France; XMartin@chu-grenoble.fr
[2] Shriners Hospitals for Children Headquarters, Tampa, FL 33607, USA; SCreadore@shrinenet.org (S.G.C.); agustafson@shrinenet.org (A.G.G.); kshazand@shrinenet.org (K.S.)
[3] Shriners Hospitals for Children, Montreal, QC H4A 0A9, Canada; ndahan@shrinenet.org
[4] School of Physical & Occupational Therapy, Faculty of Medicine and Health Sciences, McGill University, Montreal, QC H3G 2M1, Canada
[5] Shriners Hospitals for Children, Philadelphia, PA 19140, USA; sthwang@shrinenet.org (S.W.-H.H.); hvanbosse@shrinenet.org (H.J.P.v.B.)
[6] Alberta Congenital Anomalies Surveillance System, Alberta Health Services, Edmonton, AB T5J 3E4, Canada; tanya.bedard@albertahealthservices.ca
[7] Department of Pediatrics, University of Illinois-Chicago, Chicago, IL 60607, USA
[8] Institut of Advanced Biosciences, Université Grenoble Alpes, Inserm, U1209, CHU Grenoble Alpes, 38000 Grenoble, France
*   Correspondence: philipg@uic.edu (P.F.G.); KDieterich@chu-grenoble.fr (K.D.)

Citation: Latypova, X.; Creadore, S.G.; Dahan-Oliel, N.; Gustafson, A.G.; Wei-Hung Hwang, S.; Bedard, T.; Shazand, K.; van Bosse, H.J.P.; Giampietro, P.F.; Dieterich, K. A Genomic Approach to Delineating the Occurrence of Scoliosis in Arthrogryposis Multiplex Congenita. *Genes* 2021, 12, 1052. https://doi.org/10.3390/genes12071052

Academic Editor: Stephen Robertson

Received: 18 May 2021
Accepted: 29 June 2021
Published: 8 July 2021

Publisher's Note: MDPI stays neutral with regard to jurisdictional claims in published maps and institutional affiliations.

Abstract: Arthrogryposis multiplex congenita (AMC) describes a group of conditions characterized by the presence of non-progressive congenital contractures in multiple body areas. Scoliosis, defined as a coronal plane spine curvature of $\geq$10 degrees as measured radiographically, has been reported to occur in approximately 20% of children with AMC. To identify genes that are associated with both scoliosis as a clinical outcome and AMC, we first queried the DECIPHER database for copy number variations (CNVs). Upon query, we identified only two patients with both AMC and scoliosis (AMC-SC). The first patient contained CNVs in three genes (*FBN2*, *MGF10*, and *PITX1*), while the second case had a CNV in *ZC4H2*. Looking into small variants, using a combination of Human Phenotype Ontogeny and literature searching, 908 genes linked with scoliosis and 444 genes linked with AMC were identified. From these lists, 227 genes were associated with AMC-SC. Ingenuity Pathway Analysis (IPA) was performed on the final gene list to gain insight into the functional interactions of genes and various categories. To summarize, this group of genes encompasses a diverse group of cellular functions including transcription regulation, transmembrane receptor, growth factor, and ion channels. These results provide a focal point for further research using genomics and animal models to facilitate the identification of prognostic factors and therapeutic targets for AMC.

Keywords: Amyoplasia; scoliosis; DECIPHER (DatabasE of genomiC variation and Phenotype in Humans using Ensemble Resources); CNV (copy number variant); DA (distal arthrogryposis); IPA (ingenuity pathway analysis); HPO (human phenotype ontology); akinesia; MYOD; IGF2

1. Introduction

Arthrogryposis multiplex congenita (AMC or interchangeably arthrogryposis) describes a group of conditions characterized by the presence of non-progressive congenital contractures in multiple body areas [1]. The congenital contractures are a result of decreased fetal movement (fetal akinesia), leading to joint fibrosis and dysplasia/lack of elasticity of the soft tissues surrounding the joint. The longer the duration and the earlier the onset of fetal akinesia, the more severe the contractures. The direct causes of fetal akinesia are

through a literature review for association with AMC, only 12 of which were found to be associated with AMC. These 12 genes were: *C12orf65, DSE, NEK9, PHGDH, PPP3CA, PSAT1, TBCD, VAMP1, CACN1E, CEP55, RFT1*and *SHPK*.

2.2. Identification of Genes Associated with Scoliosis

The same method used for AMC was applied to identify the genes associated with scoliosis using the Human Phenotype Ontology (HPO) project (identifier HP:0002650; accessed on Tuesday April 13th 2021, version hpo-web@1.7.9-hpo-obo@2021-02-08). A total of 895 genes associated with scoliosis extracted using HPO were reviewed. An additional 16 genes reported in the literature based on Perez-Machado and colleagues' 2020 paper since then were also reviewed, of which three were duplicates among the 895 genes already identified, resulting in a total of 908 genes (see Figure 1) [12].

2.3. Identification of Genes Associated with Both AMC and Scoliosis and Gene Interaction Pathway Analysis

The list of genes identified for AMC and for scoliosis were compared to identify the genes that are associated with both. Ingenuity Pathway Analysis (IPA), which represents a functional analysis of a set of identified genes, was then conducted using the IPA Ingenuity Systems QIAGEN, content version 60467501 software. A core analysis type and subsequent variant effect analysis were used to generate the outputs in each case.

2.4. Identification of Copy Number Variants (CNV) Associated with AMC and Scoliosis

In order to identify CNVs associated with both AMC and scoliosis, we queried the DECIPHER (DatabasE of genomiC variation and Phenotype in Humans using Ensemble Resources) (https://decipher.sanger.ac.uk/) database to identify reported cases with scoliosis or vertebral malformation(s) with AMC (accessed on 9 February 2021). To do so, the 444 genes associated with AMC were queried through the implementation of an in-house Selenium-based automation software package written in the Python 3.8 programming language. The data points extracted into DECIPHER included gene name, number of associated genes, DECIPHER patient number, phenotype(s)/conditions, chromosome location, start position, end position, mode of inheritance, and genotype. The resulting DECIPHER patient IDs with their associated data were then sub-sampled into identifiable cohort groups representing the phenotype(s) of interest including Arthrogryposis-like hand anomaly, Arthrogryposis Multiplex Congenita, Distal Arthrogryposis, and Scoliosis. We then filtered for DECIPHER Patient IDs containing both arthrogryposis and scoliosis. The arthrogryposis and scoliosis DECIPHER Patient IDs containing single nucleotide variants (SNVs) were removed by subtracting the start position from the end position to identify the allelic depth and kept only copy number variation (CNV). Next, we removed any duplicates within our dataset resulting in an accurate representation of the copy number variant genes associated with arthrogryposis and scoliosis for the DECIPHER Patient IDs extracted.

3. Results

Combining the initial literature search results with the HPO identified genes for AMC-SC independently yielded a total of 444 genes associated with AMC and 908 genes associated with scoliosis. When comparing these two sets of genes, 227 genes were found in common (Figure 2).

Scoliosis associated genes

Scoliosis associated genes: 681

A2ML1, ABCB7, ABCC6, ABL1, ACADS, ACP5, ACTA2, ACTL6B, ACTN2, ACVR1, ADAMTS2, ADAR, AEBP1, AGA, AGRN, AHDC1, AHI1, AHSG, AIFM1, AIMP2, AJAP1, ALDH18A1, ALDH3A2, ALG11, ALG14, ALMS1, ALS2, ALX3, AMER1, ANKRD11, ANO5, ANTXR1, APC, APC2, APTX, ARF1, ARFGEF2, ARID1A, ARID1B, ARID2, ARL13B, ARL3, ARMC9, AS1, ASH1L, ASXL2, ATAD1, ATL1, ATP6AP2, ATP6V0A2, ATP6V1E1, B3GALT6, B3GLCT, B4GALNT1, B4GALT7, B9D1, BANF1, BCL2, BCN2, BCOR, BCORL1, BMP1, BMPR1B, BPTF, BRAF, BRCA1, BRCA2, BRF1, BRIP1, C1R, C1S, CACNA1G, CACNG2, CAMK2A, CAMK2B, CAPN1, CBS, CC2D2A, CCBE1, CCDC47, CCDC8, CCM2, CCN6, CDC42, CDC45, CDH11, CDH13, CDH15, CDK19, CDKL5, CENPE, CENPJ, CENPT, CEP104, CEP120, CEP152, CEP290, CEP41, CHAMP1, CHD3, CHD7, CHRM3, CHST11, CIC, CLCF1, CLCN4, CLIC2, CLP1, CLTC, CNOT2, CNTNAP2, COG1, COG8, COL27A1, COL5A1, COL5A2, COLEC10, COLQ, COMP, COMT, COQ4, COX8A, CPLANE1, CPLX1, CREBBP, CRPPA, CSGALNACT1, CSNK2A1, CSNK2B, CSPP1, CTBP1, CTC1, CTDP1, CTSK, CUL4B, CUL7, CUX1, CYP7B1, DACT1, DCC, DDHD1, DDR2, DDX3X, DDX59, DDX6, DEAF1, DEGS1, DHX37, DKC1, DLG4, DLL1, DLL3, DLX5, DMD, DNA2, DNAJC6, DNM1L, DNMT3A, DOCK3, DOCK8, DPF2, DPM2, DPP6, DSTYK, DVL1, DVL3, DYNC2H1, DYRK1A, EED, EEF1A2, EFNB1, ELN, ELP1, EMC1, ENPP1, EP300, EPB41L1, EPHA4, ERCC4, ERCC8, ERLIN1, ERMARD, EXOC6B, EXT1, EXT2, FAM111B, FAM126A, FANCA, FANCB, FANCC, FANCD2, FANCE, FANCF, FANCG, FANCI, FANCL, FANCM, FARS2, FARSA, FARSB, FAT4, FBLN5, FBXW11, FGD4, FGF8, FGFRL1, FKBP14, FLAD1, FLCN, FLI1, FLVCR1, FMR1, FN1, FOXA2, FOXE3, FOXG1, FOXRED1, FUS, FUT8, FUZ, FXN, FZD2, GABBR2, GABRD, GALNS, GAN, GARS1, GBA2, GCH1, GDF3, GDF6, GJB1, GJC2, GLB1, GLI2, GLIS3, GMPPB, GNAQ, GNAS, GNE, GNPAT, GNPTAB, GNPTG, GORAB, GOSR2, GP1BB, GPC4, GPR126, GRIA3, GRIN2B, GTPBP2, H1-4, H19, HACE1, HERC1, HERC2, HES7, HGSNAT, HINT1, HIVEP2, HK1, HMGA2, HMGB3, HNRNPH2, HNRNPK, HNRNPU, HPDL, HPGD, HSPB8, HSPD1, HTT, HUWE1, HYLS1, IARS2, IDH1, IDH2, IDUA, IGBP1, IHH, IKBKG, IL6ST, INPP5E, INPPL1, INTS1, IPW, IQSEC2, ITCH, ITGA7, ITGB6, JPH1, KANSL1, KATNIP, KCNJ2, KCNJ6, KCNQ1OT1, KCNQ5, KDM5B, KDM6A, KIF22, KIF5A, KLC2, KLLN, KMT2C, KMT2D, KMT2E, KRAS, KRIT1, KY, LAGE3, LAMB2, LARP7, LBX1, LEMD3, LETM1, LFNG, LHX3, LIG4, LMNB2, LONP1, LOX, LRP4, LRP5, LSS, LTBP4, LZTR1, MAD2L2, MADD, MAF, MAGI1, MAP1B, MAP2K1, MAP2K2, MAP3K20, MAPK8IP3, MAPT, MARS2, MBD5, MBTPS2, MCM3AP, MECP2, MED12L, MED13, MED25, MEGF8, MEIS1, MEIS2, MEOX1, MESD, MESP2, MFAP5, MGAT2, MKRN3, MKRN3-AS1, MKS1, MLXIPL, MMP13, MOGS, MORC2, MPL, MPZ, MRAS, MRPS34, MSTO1, MTMR2, MTRR, MTTP, MVK, MYF5, MYH11, MYH7, MYL2, MYLK, MYT1L, NACC1, NCAPG2, NDN, NDP, NDUFA12, NDUFAF1, NDUFAF6, NDUFS3, NEDD4L, NEFL, NEPRO, NFIX, NGLY1, NHP2, NIN, NKAP, NKX6-2, NODAL, NONO, NOP10, NOTCH2, NOTCH2NLC, NOTCH3, NPAP1, NPHP1, NPM1, NPR2, NRAS, NSD2, NSDHL, NSUN2, NUP107, NUS1, NXN, OBSL1, OCA2, OTUD6B, P4HB, P4HTM, PACS1, PALB2, PAPSS2, PARN, PAX1, PAX6, PAX7, PCGF2, PCNT, PCYT1A, PDCD10, PDE4D, PDGFRB, PET100, PGAP2, PGAP3, PGM3, PHACTR1, PHF6, PHF8, PIBF1, PIEZO1, PIGL, PIGO, PIGQ, PIGU, PIGV, PIGW, PIGY, PIK3C2A, PIK3CA, PIK3R2, PISD, PLK4, PNKP, POLA1, POLD1, POLE, POMK, POP1, PORCN, PPP1R12A, PPP1R15B, PPP2R1A, PPP2R3C, PPP2R5D, PRDM16, PRDM5, PRKD1, PRKG1, PRPS1, PRUNE1, PSMD12, PTCH1, PTCH2, PTEN, PTPN11, PUF60, PUS1, PUS7, PWAR1, PWRN1, PYCR1, PYCR2, PYROXD1, RAB23, RAC1, RAC3, RAD51, RAD51C, RAF1, RAI1, RBBP8, RBCK1, RBM28, RBM8A, RECQL4, RERE, RFWD3, RIN2, RIPPLY2, RIT1, RLIM, RNASEH1, RNF125, RNF13, ROBO3, RPGRIP1L, RPL10, RPL11, RPL13, RPS6KA3, RRAS2, RRM2B, RSPRY1, RTEL1, RUNX2, RUSC2, SACS, SALL4, SAMD9, SBF1, SBF2, SDHB, SDHC, SDHD, SEC23A, SEC23B, SEC24D, SEMA3E, SERPINH1, SET, SETD2, SETD5, SF3B4, SFRP4, SGCA, SGMS2, SH3PXD2B, SH3TC2, SHH, SHROOM4, SIGMAR1, SIL1, SIN3A, SIX3, SLC10A7, SLC16A2, SLC1A2, SLC25A1, SLC25A21, SLC25A24, SLC25A46, SLC35A2, SLC39A14, SLC52A2, SLC52A3, SLC6A1, SLCO2A1, SLX4, SMAD3, SMAD4, SMARCA2, SMARCA4, SMARCB1, SMARCC2, SMARCD1, SMARCE1, SMC1A, SMS, SNORD115-1, SNORD116-1, SNRPB, SNRPN, SNX14, SON, SORD, SOS1, SOS2, SOX11, SOX4, SOX5, SP7, SPARC, SPATA5, SPG11, SPG7, SPRTN, SRY, SSR4, STAG1, STAG2, STAT3, STXBP1, SUFU, SURF1, SUZ12, SVBP, SYNE2, SYNGAP1, SYNJ1, SYT1, TAF1, TANC2, TBC1D20, TBC1D24, TBCE, TBCK, TBL1XR1, TBX1, TBX2, TBX4, TBX6, TCF20, TCF4, TCTN1, TCTN2, TCTN3, TELO2, TERC, TERT, TGDS, TGFB1, TGFB2, TGIF1, TINF2, TK2, TLK2, TMCO1, TMEM138, TMEM165, TMEM216, TMEM231, TMEM237, TMEM38B, TMEM43, TMEM67, TMEM94, TMTC3, TNIK, TONSL, TRAIP, TRAPPC11, TRAPPC12, TRAPPC2, TRAPPC4, TRIO, TRIP11, TRMT1, TRMT10A, TRPS1, TRRAP, TTC21B, TTI2, TTPA, TUBB3, TUBB4A, TUBGCP4, TUBGCP6, TWIST1, UBE2T, UFSP2, UNC80, UPB1, UPF3B, USB1, USP7, USP9X, VANGL1, VDR, VPS13B, WDFY3, WDR45B, WDR81, WHCR, WNT1, WRAP53, WT1, XRCC2, XYLT1, YARS2, YWHAG, ZBTB20, ZDHHC9, ZIC1, ZMIZ1, ZMYND11, ZNF423, ZNF469

AMC and scoliosis associated genes: 227

ACTA1, ACTB, ACTG1, ADAMTS10, ADGRG6, AIMP1, AKT1, ALG2, AP1S2, ARX, ASAH1, ASXL3, ATAD3A, ATN1, ATP7A, ATR, ATRX, AUTS2, B3GAT3, BAG3, BICD2, BIN1, C12orf65, CANT1, CASK, CAVIN1, CBL, CCDC22, CD96, CDON, CFL2, CHAT, CHRNA1, CHRNB1, CHRND, CHRNE, CHRNG, CHST14, CHST3, COL12A1, COL13A1, COL1A1, COL1A2, COL2A1, COL3A1, COL6A1, COL6A2, COL6A3, COLEC11, CRLF1, CRTAP, DCHS1, DCX, DES, DHCR7, DOK7, DPAGT1, DSE, DYM, DYNC1H1, EBP, ECEL1, EGR2, EMD, ERCC1, ERCC2, ERCC6, ERLIN2, EXTL3, EZH2, FBN1, FBN2, FBXL4, FGD1, FGFR1, FGFR2, FGFR3, FHL1, FKBP10, FKRP, FKTN, FLNA, FLNB, FUCA1, GAD1, GBA, GDAP1, GDF5, GFM2, GFPT1, GLE1, GPC3, GRIN1, GUSB, GZF1, HRAS, HSD17B4, HSPG2, IFIH1, IGF2, IGHMBP2, INPP5K, IRF6, KBTBD13, KCNA1, KCNH1, KIAA0586, KIF1A, KLHL41, LAMA2, LARGE1, LIFR, LMNA, LMOD3, LMX1B, MAGEL2, MAP3K7, MASP1, MED12, MEGF10, MFN2, MMP2, MTM1, MUSK, MYBPC1, MYH2, MYH3, MYMK, MYO18B, MYO9A, MYOD1, MYPN, NAA10, NALCN, NEB, NEK9, NEU1, NF1, NRXN1, NSD1, NUP88, OCRL, P3H1, PAFAH1B1, PAX3, PEX5, PEX7, PHGDH, PIEZO2, PIGS, PIGT, PLEKHG5, PLOD1, PLOD2, PLOD3, PMM2, PMP22, POLR3A, POMT1, POMT2, POR, PPIB, PPP3CA, PQBP1, PRKAR1A, PRX, PSAT1, PTDSS1, RAB18, RAB3GAP1, RAB3GAP2, RAPSN, RBM10, RET, RMRP, RNASEH2A, RNASEH2B, RNASEH2C, ROR2, RYR1, SAMHD1, SCN4A, SELENON, SETBP1, SETX, SGCG, SHOX, SKI, SLC12A6, SLC18A3, SLC26A2, SLC2A10, SLC35A3, SLC39A13, SLC5A7, SNAP25, SOX9, SPART, SPTBN4, STAC3, SYNE1, SYT2, TBCD, TBX5, TGFB3, TGFBR1, TGFBR2, TNNI2, TNNT3, TOR1A, TPM2, TPM3, TREX1, TRIP4, TRPV4, TTN, UBA1, UBE3A, VAMP1, VMA21, VPS53, WASHC5, WNT5A, ZC4H2, ZEB2, ZIC2, ZMPSTE24

AMC associated genes: 217

ABCA7, ABCC8, ACTG1P10, ADAMTSL2, ADCY6, ADSL, ADSSL1, AK9, ALG3, ANTXR2, APLNR, ASCC1, ASCC2, ASCC3, ASNS, ASPM, ASXL1, ATM, ATP1A2, ATP2B3, ATXN2, ATXN3, CACNA1, CACNA1E, CAPN3, CD24, CD6, CDK5, CEP55, CHMP1A, CHUK, CLIC3, CNTN1, CNTNAP1, COASY, COG6, COG7, COL11A2, COL7A1, CPT2, CTSA, CTSL, DHCR24, DMPK, DNM2, DOCK7, DOLK, DPM1, DRG2, DST, DYSF, EARS2, EGFLAM, EIF2AK4, EIF2S3, EMG1, ERBB3, ERCC5, ERGIC1, ESCO2, EXOSC3, EXOSC8, FAM20C, FBLN1, FBN3, FGF9, FLVCR2, FOXP3, GAA, GBE1, GCK, GJA1, GLDN, GLI1, GLI3, GLRA1, GLRB, GLUL, GRHL3, GRN, HEXA, HEXB, HLA-DRB1, HOGA1, HOXA13, HOXD13, HWE1, IBA57, IDS, IGF1, IMPAD1, INSR, ISLR2, ISPD, ITGA6, ITGB4, KAT6B, KCNJ11, KCNK9, KHL7, KIAA1109, KIF14, KIF5C, KIF7, KLHL3, KLHL40, KLKB1, L1CAM, LGI4, LMBR1, LTBP2, MECOM, MGP, MHS3, MNX1, MYBPC2, MYH14, MYH7B, MYH8, MYLPF, MYO3A, MYOM2, MYOT, NDE1, NEFH, NOG, OFD1, ORC4, ORC6, PANK2, PDHA1, PEX1, PEX10, PEX12, PEX13, PEX14, PEX2, PEX26, PEX3, PEX6, PFKM, PI4KA, PIP5K1C, PITX1, PLEC, PLP1, POLR3D, POMGNT1, POMGNT2, PRG4, PRICKLE1, PRPH, PSD3, PTH1R, PTHLH, PTLAH, RELN, RFT1, RIPK4, RMND1, RYR3, SCARF2, SCN8A, SCYL2, SEMA3A, SEPSECS, SHPK, SIM1, SLC25A19, SLC3A1, SLC6A5, SLC6A9, SLC9A6, SLCO5A1, SMARCAD1, SMN1, SMN2, SMPD4, SOD1, SOX10, SRD5A3, STRA6, SULF1, SZT2, TAF8, TARP, TBX15, TBX22, TBX3, TNN, TNNT1, TOE1, TRPV5, TSC1, TSC2, TSEN2, TSEN34, TSEN54, TSPAN7, TUBA1A, TUBB2A, TUBB2B, TWIST2, TYMP, UNC50, UPK3A, USP16, UTRN, VIPAS39, VPS16, VPS33B, VPS8, VRK1, WNT7A, ZBTB42, ZIC3, ZNF335

AMC associated genes

Figure 2. Scoliosis-associated genes are indicated in the green box, AMC genes are indicated in the red box, and scoliosis and AMC genes are in the orange region.

This set of 227 genes was then analyzed using IPA. Overall, this group of 227 genes encompasses a diverse group of cellular functions including transcription regulation, transmembrane receptors, growth factor-related genes, and ion channels. (Table 1).

Table 1. The 227 Genes Associated with AMC-SC Stratified by Function.

Homeostasis: mechanisms include genes associated with early-onset nuclear DNA excision/repair disorders (*ERCC1*, *ERCC2*, and *ERCC6*) [13]. Monoallelic mutations in ERCC1 are associated with cerebro-oculo-facio-skeletal syndrome. *ERLIN1* encodes for a lipid raft-associated protein localized to the mitochondrion and nuclear envelope, and is a component of the *ERLIN1/ERLIN2* complex. The complex mediates the endoplasmic reticulum-associated degradation of inositol 1,4,5-trisphosphate receptors (ITPRs) which are important in calcium homeostasis [14].

BAG3, BIN1, ERCC1, ERCC2, ERCC6, ERLIN2, SELENON

Table 1. *Cont.*

Cytoskeleton: matrix proteins involved in the sarcomere such as nebulin, a giant protein of thick and thin filaments of striated muscle, encoded by *NEB*. Mutations in NEB are responsible for the majority of cases of nemaline myopathy [15] which can be diagnosed by Gomori trichrome staining on a muscle biopsy or by electron microscopic preparation. *ACTA1*, a member of the cytoskeletal grouping, encodes the principal skeletal muscle isoform of adult skeletal muscle, alpha-actin. Residing in the core of the thin filament of the sarcomere, it assists in the generation of muscle contraction [16]

ACTA1, ACTB, ACTG1, AP1S2, COL12A1, COL13A1, COL1A1, COL1A2, COL2A1, COL3A1, COL6A1, COL6A2, COL6A3, COLEC11, DCX, DES, DYNC1H1, EMD, FBN1, FBN2, FLNA, FLNB, HSPG2, LMNA, NEB, SPTBN4, SYNE1, TBCD

Extra Cellular Matrix: Extracellular matrix (ECM) protein-associated genes include *ADAMTS10* and *DCHS1*. *ADAMTS10* is a zinc-dependent protease composed of one cysteine-rich domain, and five thrombospondin type 1 (THBS1) repeats and plays an important role in the formation of the extracellular matrix [17]. *DHCS1* is a member of the protocadherin superfamily and encodes a transmembrane cell adhesion molecule responsible for apical anchoring in the brain [18].

ADAMTS10, CDON, DCHS1, MMP2, RAPSN

Signal Transduction: Promotes signaling within a cell via enzyme network cascades to generate precise and appropriate physiologic responses, particularly in skeletal development. *FGFR3* codes for an important tyrosine kinase signal transducer in chondrocytes, functioning to attenuate cartilage growth. FGFR 1–4 transmit at least 18 different fibroblast growth factor (FGF) ligands, therefore, exhibiting a variety of physiological functions [19]. *GDF5* fulfills important functions with respect to bone and muscle [20]. Through its high affinity for BMPR1B, GDF5 positively regulates chondrogenesis, leading to SMAD signal transduction [21]. Through NOG mediated interaction, GDF5 paradoxically also negatively regulates chondrogenesis.

ADGRG6, CAVIN1, CCDC22, CD96, CFL2, CRLF1, CRTAP, DOK7, EBP, FGFR1, FGFR2, FGFR3, GDF5, IFIH1, KIAA0586, MAGEL2, NF1, PEX5, PEX7, PMP22, RAB3GAP1, RAB3GAP2, STAC3, WNT5A, KBTBD13

Proto-oncogenes: Proto-oncogenes act to facilitate dysregulated cell growth and differentiation. Mutations in *HRAS* are associated with Costello syndrome, characterized by distinct facial features, papilloma of the face, cardiac anomalies, growth restriction, developmental delays, and tumor predisposition. An *HRAS* mutation was identified in an infant with features of Costello syndrome and distal arthrogryposis [22].

AKT1, CBL, HRAS, RAB18, RET, SKI

Enzyme: Account for the largest category of genes identified through IPA analysis. 7-dehydrocholesterol reductase (DHCR7) encodes the penultimate step in the cholesterol biosynthetic pathway. Smith-Lemli-Opitz Syndrome is an autosomal recessive disorder caused by an inherited deficiency of DHCR7 which is associated with a variety of birth defects, joint contractures, and intellectual disability [23]. *UBE3* which encodes E3 ubiquitin-protein ligase, a maternally expressed imprinted E3 ubiquitin-protein ligase expressed mainly in the brain, is an integral part of the ubiquitin protein degradation system. Angelman syndrome, characterized by severe cognitive impairment, seizures, an ataxic puppet-like gait, and paroxysms of laughter, is caused by an absence of expression of maternal *UBE3A* [24].

ALG2, ASAH1, B3GAT3, CANT1, CHAT, CHST14, CHST3, DHCR7, DPAGT1, DSE, ECEL1, EXTL3, EZH2, FBXL4, FKRP, FUCA1, GAD1, GBA, GFPT1, GUSB, HSD17B4, INPP5K, LARGE1, MASP1, MTM1, NAA10, NEU1, OCRL, P3H1, PAFAH1B1, PHGDH, PLOD1, PLOD2, PLOD3, PMM2, POLR3A, POMT1, POMT2, POR, PPIB, PPP3CA, PSAT1, PTDSS1, RNASEH2A, RNASEH2B, RNASEH2C, SAMHD1, TOR1A, TREX1, UBA1, UBE3A, ZMPSTE24

Transcription Factor/regulation: Transcription factors have a pivotal role in the regulation of genes associated with limb and muscle development. T-Box Transcription Factor 5 (*TBX5*) mutations are associated with Holt Oram syndrome characterized by upper limb defects and cardiac malformations [25,26]. *TRIP4* encodes ASC-1, a transcription co-activator. Infants with *TRIP4* mutations present with a congenital muscular dystrophy and respiratory failure. Muscle biopsy shows decreased mitochondria and sarcomere disorganization [27].

ARX, ASXL3, ATN1, ATRX, AUTS2, EGR2, FGD1, GZF1, IGHMBP2, IRF6, LMX1B, MED12, NSD1, PAX3, PLEKHG5, PQBP1, RBM10, SETBP1, SETX, SHOX, SOX9, TBX5, TRIP4, ZC4H2, ZEB2, ZIC2

Mitochondria: Mitochondria are depended upon highly by the brain and skeletal muscle tissues for energy. Ganglioside Differentiation Associated Protein 1 (*GDAP1*) encodes a mitochondrial protein postulated to play a role in signal transduction in the brain. Mutations in *GDAP1* are associated with various subtypes of the hereditary and sensory-motor neuropathy disease Charcot Marie Tooth (CMT), including an autosomal recessive intermediate type [28–32]. *RMRP* codes for non-coding RNA involved in mitochondrial DNA replication through the encoding of a mitochondrial RNA processing endonuclease which cleaves mitochondrial RNA at a priming site necessary for mitochondrial DNA replication. Mutations in *RMRP* are associated with cartilage-hair hypoplasia [33]. *RMRP* is essential for early murine development [34].

ATAD3A, C12orf65, GDAP1, GFM2, MFN2, RMRP, SPAR

Table 1. *Cont.*

Membrane Receptor/Ion Channel: Membrane receptor and ion channels is the second largest group of affected genes leading to AMC-SC. *CHRNA1* (cholinergic receptor nicotinic receptor alpha 1 subunit 1) is one of 5 subunits of the acetylcholine receptor (*AChR*). This gene encodes an alpha subunit and functions as part of acetylcholine binding and channel. Mutations in *CHRNA1* are associated with lethal multiple pterygium syndrome, characterized by the presence of multiple pterygia, intrauterine growth retardation, and flexion contractures resulting in severe arthrogryposis and fetal akinesia [35]. *PIEZO2* is postulated to function as an integral part of mechanically activated cation channel in somatosensory neurons through establishing connections between mechanical forces and biological signals. Mutations in *PIEZO2* are associated with distal arthrogryposis type 5, Gordon syndrome, and Marden–Walker syndrome [36].

ATP7A, CHRNA1, CHRNB1, CHRND, CHRNE, CHRNG, GPC3, GRIN1, KCNA1, KCNH1, MEGF10, NALCN, NRXN1, NUP88, PIEZO2, PIGS, PIGT, ROR2, RYR1, SCN4A, SGCG, SLC12A6, SLC18A3, SLC26A2, SLC2A10, SLC35A3, SLC39A13, SLC5A7, SNAP25, SYT2, TGFBR1, TGFBR2, TRPV4, VAMP1, WASHC5

Kinase: Kinases phosphorylate target molecules for activation or inactivation. ATR encodes a serine/threonine kinase and halts cell cycling entry upon DNA stress to enable DNA repair [37]. Compound heterozygous mutations in ATR are associated with Seckel syndrome characterized by dwarfism, microcephaly, and cognitive impairment [38]. MAP3K7 mediates cellular transduction in response to environmental changes through association with interleukin receptor (ILR1). Through the cytokine IL-1 mediated interaction with the hypothalamic IL-1 receptor, the hypothalamo-pituitary-adrenocortical axis and sympathetic nervous system pathways suppressing bone formation are activated [39]. Fronto-metaphyseal dysplasia, a progressive sclerosing skeletal dysplasia characterized by small bone undermodeling, supraorbital hyperostosis, large and small joint contractures as well as developmental abnormalities, of the cardiorespiratory system and the genitourinary tract is associated with *MAP3K7* mutations [40].

ATR, CASK, MAP3K7, MUSK, NEK9, PRKAR1A

Intracellular transport: Intracellular transport proteins are structural proteins that facilitate the movement of vesicles and substances within a cell. *BICD2* codes for a structural protein functioning as an intracellular adaptor for the dynein motor complex, linking it to various cargos. Through the stabilization of the interaction between dynein and dynactin, the movement of dynein is facilitated along the microtubule [41]. Mono-allelic mutations in *BICD2* cause congenital spinal muscular atrophy [42]. *GLE1* is postulated to act as a terminal step in the transport of mature messenger RNA messages from the nucleus to the cytoplasm. Bi-allelic mutations in *GLE1* are associated with a lethal congenital contracture syndrome characterized by fetal hydrops, degeneration of anterior horn cells, and congenital contractures [43].

BICD2, DYM, FKBP10, GLE1, KIF1A, VPS53

Structural: Structural proteins provide the framework for a cell or complex of cells. The *LAMA2* gene encodes laminin-2 or merosin, a major component of the extrasynaptic membrane of muscle cell basement membrane. Laminin-211 binds to the glycosylated residues of alpha-dystroglycan (*DAG1*) in skeletal muscle fibers [44]. Bi-allelic mutations in *LAMA2* are associated merosin-deficient congenital muscular dystrophy. Affected patients have hypotonia, joint contractures and may develop scoliosis. Myosin, the major contractile protein in muscle, is composed of two heavy chains and two light chains. *MYH3* encodes the embryonic myosin heavy chain 3. *MYH3* mutations appear to reside near a groove that is part of the myosin head and are associated with distal arthrogryposis type 1 in which contractures are limited to distal joints, Freeman –Sheldon, Sheldon -Hall syndromes [45]. Affected patients with Freeman Sheldon and Sheldon Hall syndromes have distal joint contractures, characteristic facial features and may develop scoliosis. *MYH3* mutations are postulated to cause structural changes in myosin that potentially alter myosin domain-domain interactions during ATP catalysis or affect nucleotide-binding site conformation.

FHL1, FKTN, KLHL41, LAMA2, LMOD3, MYBPC1, MYH2, MYH3, MYMK, MYO18B, MYO9A, MYOD1, MYPN, PRX, TNNI2, TNNT3, TPM2, TPM3, TTN, VMA21

Figure 3A shows examples of the 227 genes common to scoliosis and AMC analyzed in IPA. On the canonical pathway panel (A, left) describing the actin cytoskeletal pathway, major disruption points representing inactivating variants in some key genes such as F-actin, Myosin, and Filamin A (FLNA) that crosslinks actin filaments to membrane glycoproteins can be seen.

As of the right panel (B), the complex intricate interaction network clearly shows the close functional relationship and involvement of key genes such as AKT Serine/Threonine Kinase 1 (Akt), cholinergic receptor family (CHRN), and Neuregulin gene family (NRG) involved in neuromuscular junctions.

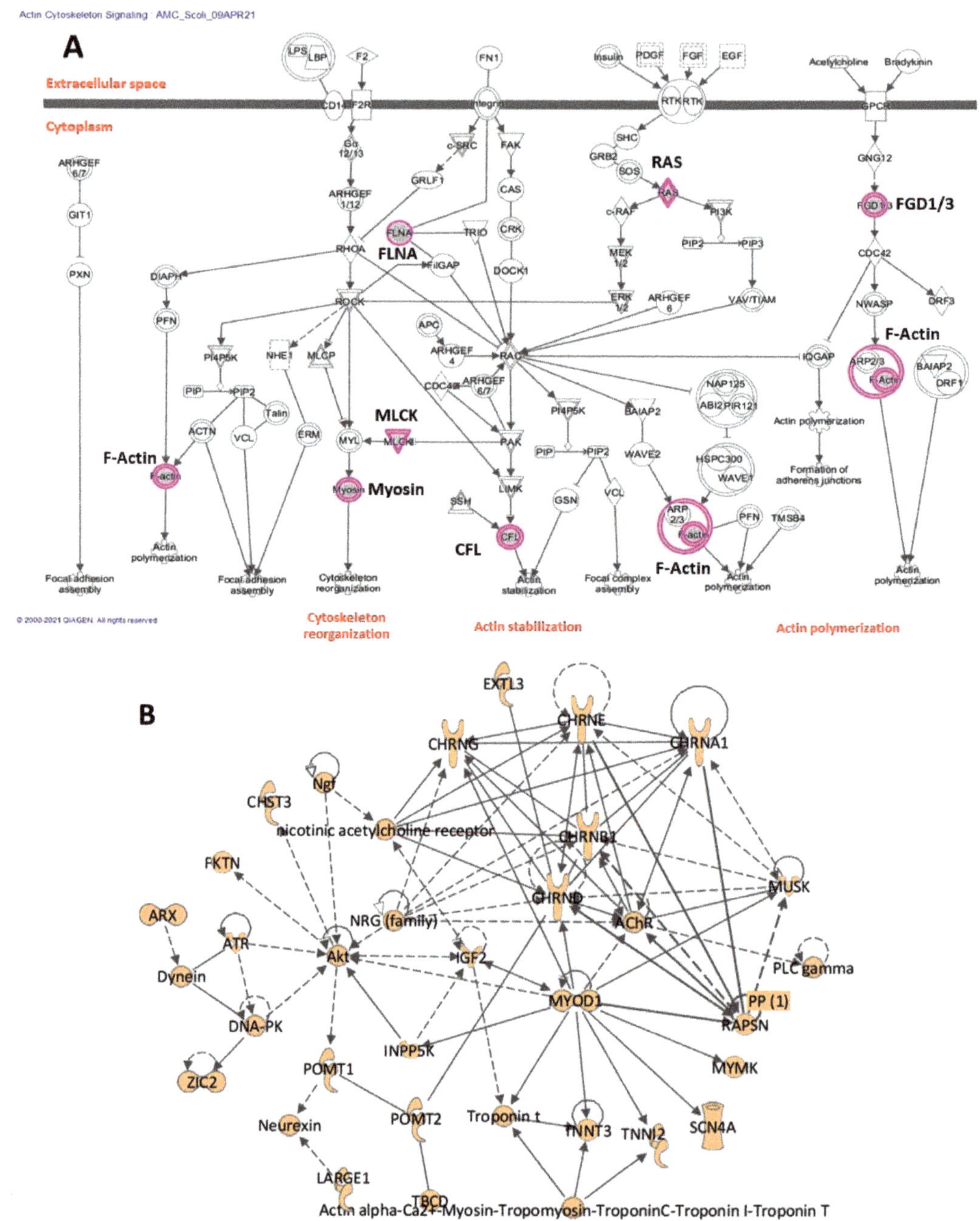

Figure 3. Two examples representative of pathway analysis of the final gene set by Ingenuity Pathway Analysis (QIAGEN). (**A**) Actin cytoskeleton signaling canonical pathway. Genes that are part of the AMC-SC list are shown in purple. (**B**) Highest score pathway predicted by IPA connecting the largest number of gene list members. Solid lines indicate direct interaction between genes while dotted lines symbolize indirect connection. Molecule shapes indicate gene functions, legends can be found here: https://qiagen.secure.force.com/KnowledgeBase/KnowledgeIPAPage?id=kA41i000000L5rTCAS (accessed on Wednesday 13 January 2021).

The DECIPHER database search (outlined in Figure 4) identified only two patients harboring CNVs associated with scoliosis and arthrogryposis, for which details are summarized in Table 2.

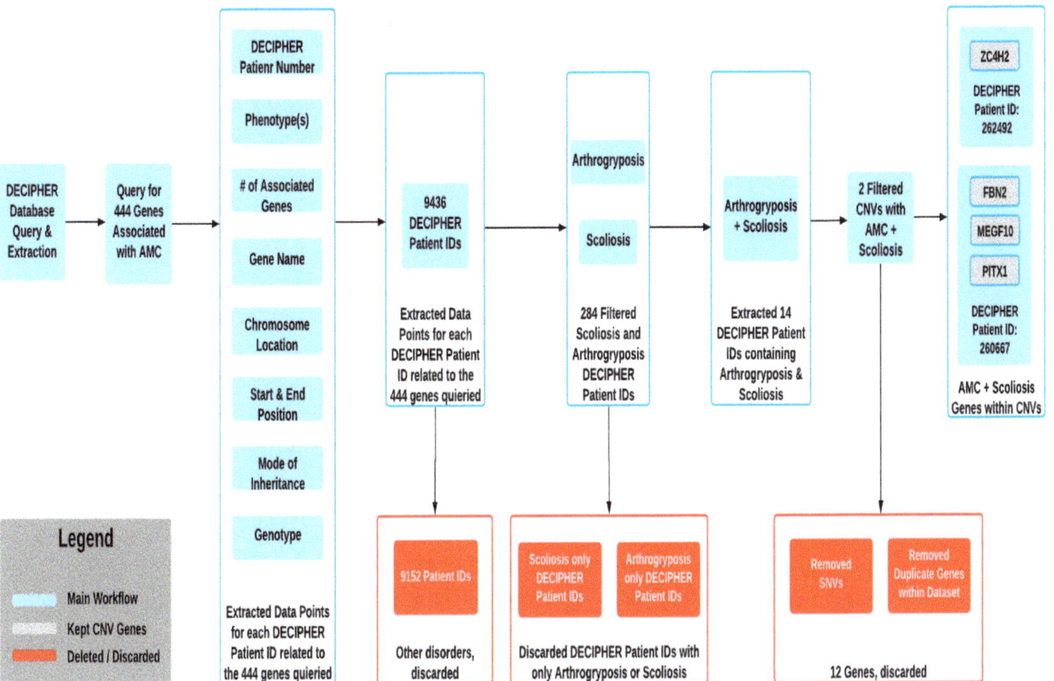

Figure 4. DECIPHER Gene Extraction Workflow.

Table 2. CNV Associated with Scoliosis and Arthrogryposis.

Title	Title	Altered Genes		
	Phenotype	Gene 1	Gene 2	Gene 3
Patient 1	Cleft palate, crumpled ear, distal arthrogryposis, intellectual disability, micrognathia, scoliosis, syringomyelia; mild pulmonary stenosis	*FBN2* (Chr5, de novo, loss, heterozygous) *	*MEGF10* (Ch5, de novo, loss, heterozygous) *	*PITX1* (Ch5, de novo, loss, heterozygous) *
Patient 2	AMC, dysphagia, dystonia, global developmental delay, laryngomalacia, thoracolumbar scoliosis	*ZC4H2* (ChX, de novo, loss, het) *	N/A	N/A

Patient 1 (DECIPHER #260667): https://decipher.sanger.ac.uk/patient/260667/overview/general (accessed on Monday 28 June 2021). Deletion chr 5: Start position 125286403, length: 10815843, contains 133 genes. Patient 2 (DECIPHER #262492): https://decipher.sanger.ac.uk/patient/262492/overview/general (accessed on Monday 28 June 2021). Deletion chr X: Start position: 64954439, length: 233145, contains 1 gene. * For each gene the following information is provided between parentheses: chromosome number (chr), CNV inheritance (de novo, heterozygous) and category (gain/loss).

The first patient (#260667) had a chromosome 5 deletion encompassing 133 genes. Three relevant genes contained within the CNV which potentially impacted the phenotype i.e., *FBN2*, *MEGF10,* and *PITX1* [46]. The CNV is a 10.82 Mb heterozygous deletion containing 133 genes resulting in a contiguous gene deletion syndrome. This deletion has been documented with a haploinsufficiency score of 50.51, i.e., a high likelihood of causing a loss of function [47].

Mono-allelic *FBN2* mutations are associated with Beals congenital contractural arachnodactyly [48]. Bi-allelic mutations in *MGF10* are associated with myopathy, areflexia, respiratory distress, and dysphagia. Mono-allelic mutations in *PITX1* are associated with congenital clubfoot, with or without deficiency of long bones and/or mirror-image polydactyly in addition to Liebenberg syndrome, defined by the presence of carpal synostosis, dysplastic elbow joints, and brachydactyly [49].

Regarding patient #2, the de novo heterozygous CNV is a fragment of 233.15 Kb located on the X chromosome, containing only the 2C4H2 gene and reported as "likely pathogenic" according to the ClinVar classification.

The second patient (#262492) had a heterozygous or hemizygous (on the X chromosome) deletion encompassing the *ZC4H2* part of the CNV. *ZC4H2* is associated with Wieacker–Wolff syndrome, characterized by the presence of foot contractures, muscle atrophy, and oculomotor apraxia [50].

4. Discussion

Despite the significant prevalence of scoliosis in the AMC patient population, there is little information regarding genetic contributions to scoliosis development in AMC in the literature. In one study of 46 patients with AMC, 32 patients (65.6%) developed scoliosis between the ages of 5–16 years [10]. Five of 32 patients (15.7%) presented with scoliosis at birth, reflecting the position of the immobile fetus in the uterus, and therefore referred to as "prenatal scoliosis". Several patterns of scoliosis have been noted to occur in AMC and include a "paralytic curve" which is more common in the hypotonic types of AMC and tends to progress; it is typically observed before the age of 2 years but can arise at any age. These curves tend to be thoracolumbar in local, often with pelvic obliquity and severe hip contractures. The second curve pattern is the less prevalent "idiopathic-like", with more balanced double curves, localized to the thoracic and thoracolumbar regions, and often manifesting in later childhood or adolescence.

The aims of this review were to identify genes that are associated with both AMC and scoliosis, and describe the functional pathways and CNVs associated with both conditions. While additional analysis of comparison between pathways associated with arthrogryposis without scoliosis, scoliosis without arthrogryposis and pathways that are associated with both arthrogryposis and scoliosis may be potentially complementary this was not the ultimate focus of our investigation. To our knowledge, this is the first study that has utilized a genomic approach to identify genes that are associated with both AMC and scoliosis. We identified a list of 227 genes that were associated with AMC-SC. The collection of genes encompasses a diverse group of cellular functions, which, once impaired, contribute to AMC-SC: cytoskeletal elements, neurotransmitter enzyme function, molecular chaperone function, ion channel regulation, extracellular matrix, DNA repair, growth factor, transmembrane receptor, transcription factor/regulator, messenger RNA regulation and cellular transport (see Table 1).

IPA analysis of the 227 AMC-SC genes suggest some common causal mechanisms and pathways such as critical "housekeeping" functions (cellular ion balance, DNA excision/repair, mRNA trafficking and post-translational modification), embryologic development, and structural families of genes expressed in bones and/or muscles (Figure 3A,B). The affected genes can heavily impair the naturally occurring or canonical pathways at crucial points, degrading the normal progression of embryologic development and/or after birth differentiation. This process depends on the chronological expression of involved genes and their transcriptional factors (TFs). Several guiding principles were demonstrated by the intricate relationships of the studied genes, as visualized by the pathway analysis: (1) It is likely that the majority of these genes are related to the pathological processes involved with the development of arthrogryposis and scoliosis, (2) IPA analysis facilitates a birds-eye view of potentially impaired key processes, (3) In the central nervous system, dysfunction can be related to mutations of genes surrounding the AKT1/2 kinases or of the growth factors regulating their activities, neurotransmitter receptors, or intracellular ion balance impairing the transmission of electrical impulses, (4) Mutations in structural genes such as actin, myosin, titin, and dynein in bone and muscle related pathways may cause impaired cytoskeletal function and/or decreased contractile ability, (5) Mutations in regulatory genes such as *TBX5*, *TRIP4* or *NFkB* act at the level of transcription to regulate activity of these genes, (6) There are inflammatory mediated effects on cellular differentiation in these organs. As an example of how the IPA analysis can reflect actual findings, the analysis attests an interaction between MYH3 and actin (Figure 3A). Mutations in *MYH3* are associated with Freeman-Sheldon syndrome (FSS), a form of DA characterized by a small mouth and joint contractures. Drosophila modeling of FSS provided molecular evidence for MYH3 and actin interaction as *MYH3* mutations are associated with myofibrillary disarray and result in decreased catalytic efficiency of actin-activated ATP hydrolysis [51]. IPA analysis did identify other potential disorders and conditions that may be attributed to alterations in genes which are members of pathways in which the 227 genes identified with AMC-SC. These include skeletal, muscular, limb defects and cognitive disability. Further validation would require a more in depth analysis, which is not the focus of this paper.

Figure 3B highlights among other interactions, interplay between *MYOD1* and *IGF2*. Literature review supports this interaction. Recently, two siblings presenting with a lethal form of fetal akinesia deformation sequence (FADS) including deficient pectoralis and proximal limb musculature, distal joint contractures and neonatal respiratory have been described. Watson et al. [52] found a homozygous probably pathogenic loss of function variant, c.188C>A/ p.Ser63*, in *MYOD1*. *MYOD1* encodes MyoD. MyoD is a key player in cell proliferation and differentiation of myoblasts and its expression fine tunes the balance between myoblast proliferation and differentiation [53]. MyoD directly activates the expression of a long non coding mRNA, called LncMYOD, encoded next to the *MYOD1* gene [54]. LncMyoD then interacts directly with IMP2 (insulin-like growth factor 2 mRNA-binging protein 2). LncMyoD downregulates IMP2-mediated mRNA translation of genes involved in cell proliferation, such as *N-RAS* and c-myc and *IGF2*. Interestingly *IGF2* is part of the imprinting control region 1 (ICR1) at chromosome 11p15.5. *IGF2*, as well as the *H19* gene, when hypomethylated at the ICR1 locus, are associated with Silver-Russell syndrome [55]. Patients with Silver-Russell syndrome have major clinical features consistent with pre- and postnatal growth restriction, frontal bossing with relative macrocephaly, feeding difficulties and low body mass index. In some individuals musculoskeletal features have also been mentioned with muscle hypoplasia and congenital joint contractures [56,57].

Querying the DECIPHER database yielded only two patients who had CNV associated with AMC-SC (Figure 4 and Table 2). Patient 1 had a much larger region of CNV, containing three relevant genes: *FBN2*, *MEGF10* and *PITX1* [46]. *FBN2* (Fibrillin 2) codes for cytoskeletal element, and mutations are associated with Beals congenital contractural arachnodactyly [48]. *FBN2* intragenic deletions or splice site mutations have been published on some occasions associated with contractural congenital arachnodactyly [58]. Rare nonsense mutations are present in the ClinVar database and reported as pathogenic. These latter are not known to be associated with an AMC phenotype. Therefore we cannot either exclude or confirm a link between *FBN2* haploinsufficiency and AMC *MEGF10* (multiple epidermal growth factor-like domains protein 1) codes for a membrane receptor involved in the phagocytosis of apoptotic cells by macrophages and astrocytes, and biallelic mutations are associated with myopathy, areflexia, respiratory distress and dysphagia. *PITX1* (Paired Like Homeodomain 1) was the only gene of the CNV analysis that had not also been identified as a gene associated with AMC-SC, and in the literature, it has not yet been associated with AMC. It is associated with congenital clubfoot, occasionally with bony malformations of the foot, and Liebenberg syndrome, defined by the presence of carpal synostosis, dysplastic elbow joints and brachydactyly [49,59]. Other nonsense variants (2) have only been mentioned in ClinVar.

It is unclear if the monoallelic *MEGF10* was responsible for any part of the patient's phenotype of either AMC or scoliosis. There is a possibility *FBN2*, a gene known to cause Beals syndrome (a form of distal arthrogryposis with scoliosis), was the only gene responsible for the AMC-SC phenotype; and there is a possibility of some type of additive effect between monoallelic *FBN2*, *MEGF10*, and *PITX1* resulting in an arthgrogrypotic phenotype. We suspect this patient has a contiguous gene syndrome. Additional literature reports describing patients with similar phenotypic features and deletions would provide support for this hypothesis.

One microdeletion involving only *PITX1* has been associated in one family with clubfoot over three generations (Alvarado et al., 2011 [59]). Other nonsense variants (2) have only been mentioned in ClinVar.

Patient 2 had a CNV for *ZC4H2* (Zinc Finger C4H2-Type Containing), a gene causing an X-linked arthrogryposis, usually only in females, known as Wieacker-Wolff syndrome or *ZC4H2*-Associated Rare Disease (ZARD). This condition is characterized by hypotonia, moderate to severe developmental delay, and early and progressive onset of scoliosis. Loss of function mutations have been described to be pathogenic on several occasions in *ZC4H2* [50,60–62].

It is of interest to notice the small number of mapped CNVs in patients as opposed to coding region variants in the literature. This can be at least partially explained by a historic bias of research toward the exome. With a current increase in whole genome sequencing clinical projects, we should see an increase in the non-coding variants for all disease-related literature in the next few years. Furthermore, geneticists do not necessarily need to submit CNVs to DECIPHER that can be easily linked to clinical symptoms.

Future directions related to this research could be exploration of possible treatments, either to ameliorate the effects of AMC and possibly avoid the scoliosis, or to prevent the fetal akinesia altogether. IPA analysis has in some instances the ability to suggest possible drug targets for specific gene pathways. For example, many of the genes implicated in fetal akinesia also are associated with later life cancers. A number of medications have been used or are being developed to treat these cancers. Viewing these drugs as therapeutic for AMC must be done with extreme caution. Mechanistically, many of the genetic AMCs are due to lack of function or altered function of the gene product, whereas many cancers are often an uncontrolled overexpression of the proliferative genes (also shut off of differentiation genes) IPA analysis is limited with respect to drugs' interactions and genes or gene products. While IPA may suggest a certain drug for its interaction with a particular gene or gene product, the nature of the interaction could be unclear. For instance, curarizing agents are listed with *CHRNG* which codes for a subunit of the acetylcholine receptor (AChR), and a mutation of which is associated with multiple pterygium syndrome (MPS). But mutations of *CHRNG* that cause MPS are loss of function mutations which result in a failure of export of the subunit to the cell surface or no protein expression [63]. The mutation already has a disconnecting effect on the neuromuscular junction, which curarizing agent would only exacerbate. Additionally, the *CHRNG* encoded protein is only expressed up to the 33rd week of pregnancy, and is replaced on the AChR with an "adult" subunit, which is presumably functional. In fact, it should be assumed that any chemical treatment for underlying causes of AMC-SC would need to be given early in pregnancy in order to prohibit development of deformities related to fetal akinesia.

Research using animal models such as zebrafish has shown some promise in the identification of pathologic mechanisms which may be amenable to targeted therapies. *MYBPC1* appears to be a novel gene responsible for DA1, though the mechanism of disease may differ from how some cardiac *MYBPC3* mutations cause hypertrophic cardiomyopathy [64]. *MYBPC1* is necessary in slow skeletal muscle development and can be used in established zebrafish models as a tractable model of human distal arthrogryposis [65]. Mutations in *MYH3*, which encodes embryonic heavy chain (MyHC) expressed initially during slow skeletal muscle development are also associated with multiple pterygium syndrome (MPS) and spondylocarpotarsal synostosis syndrome. The latter condition is characterized by joint contractures in addition to vertebral, carpal and tarsal fusions, and could present a mechanistic link between vertebral fusions and joint contractures, with hypercontraction of the surrounding muscle leading to excessive notochord tension [66,67]. Zebrafish homozygous for the smyhc (slow myosin heavy chain) are analogous to the most common distal arthrogryposis caused by MYH3 mutations. The zebrafish develop notochord kinks characterized by vertebral fusions, progression to scoliosis in addition to motor deficits accompanied the disorganized and shortened slow-twitch skeletal muscle myofibers. Slow twitch muscle fibers rely on aerobic metabolism and are recruited for smaller range of activities as compared to fast twitch fibers which rely on anaerobic metabolism and are utilized for larger bursts of activity. Treatment of the zebrafish embryos with the myosin ATPase inhibitor, para-aminoblebbistatin, which decreases actin-myosin affinity, normalized the vertebral fusions and notochord phenotype [68]. These findings hold tremendous promise for the treatment of AMC-SC.

TNNI2 is also associated with distal arthrogryposis, types DA1 and DA2B, encoding a subunit of the troponin complex. Tnni2^{K175del} transgenic mice with a heterozygous gain of function mutation in *TNNI2*, encoding a subunit of the troponin complex have small body size and joint contractures. Hypoxia-inducible factor3a (Hif3a) was found to be increased with decreased Vegf levels in bone in these mice resulting in decreased angiogenesis, delays in endochondral ossification, decreased chondrocyte differentiation and osteoblast proliferation [69]. Interestingly, both *HIF3A* silencing using Hif3A/Hif-3α siRNA and HIF-prolyl hydroxylase inhibition effectively increased the cell viability during anoxia/reoxygenation injury in cardiomyocytes and led to changes in mRNA expression of HIF1-target genes, and reduced the loss of mitochondrial membrane potential ($\Delta\psi_m$) [70]. These results show promise towards applications for AMC bone related targeted treatments.

We noted a number of methodological barriers in our research. Although similar database and literature searches were implemented to identify scoliosis- and AMC-associated genes (see Figure 1), it is noteworthy that HPO scoliosis-associated terminology accounted for 98.5% of scoliosis-associated genes, with literature review adding only another 1.5%, whereas using HPO terminology for AMC identified only 21.7% of AMC-associated genes, lagging significantly compared to the literature search. There are multiple reasons to use the HPO terminology to identify genes associated with AMC-SC. HPO is freely available, provides standardized vocabulary for phenotypic manifestations of genetic disorders and can aid in specific diagnoses. HPO also provides linkages with different disease coding systems. The lack of AMC-associated terminology in HPO stems from the scarcity of codes associated with rare AMC-associated disorders, particularly those conditions categorized under Bamshad's syndromic category [71]. Since its founding in 2008, HPO continues to expand its coverage of disease-associated phenotypes [72]. However, some disorders such as epilepsy have very deep phenotypic characterization, whereas other disorders such as respiratory diseases are less well represented in HPO. In searching the HPO database with AMC-associated terms, 18 genes were singled out as not having been identified in the literature search (see Figure 1); subsequently 6 were discarded as on further review as they did not have an associated AMC phenotype.

We strived to be comprehensive in this study, with the identification of genes associated with AMC-SC using PubMed, HPO, and DECIPHER. Despite those efforts, genes may have been missed or their phenotypic spectrum not completely realized, particularly those in research or other databases that are not public/accessible and have not been published. For instance, *MYH3* is the only gene associated with Freeman-Sheldon syndrome (and the only gene associated with DA8–autosomal dominant MPS). We would suggest further investigation into the specific gene mutations of *MYH3* that leads to the occurrence of scoliosis. *MYH3* mutations can also lead to DA1, the "classic" distal arthrogryposis, as well as DA2B, Sheldon-Hall syndrome, both of which rarely have associated scoliosis. Understanding the differences in the specific mutations of *MYH3* between these three conditions may shed light on the origins of scoliosis.

Ascertainment bias or failure to include patients with AMC or scoliosis could lead to a misrepresentation of the number of the genes associated with AMC-SC. We used a systematic HPO and literature searching approach to identify genes associated with AMC, scoliosis so we could subsequently identify genes associated with both conditions.

Syndromes/genes associated with AMC-SC are relatively rare, which highlights the need to share findings and contribute to more easily accessible platforms such as HPO. An ongoing registry project for children with AMC, funded by the Shriners Hospitals for Children and implemented in seven regions in North America, will make future contributions to genotype/phenotype associations in AMC. This should lead to a better understanding of mechanisms that lead to AMC, and possibly to better care and outcomes. International collaborations to expand this registry have started and will necessitate the identification of common data elements and terms. There will also be opportunities for the findings from this registry to contribute to such platforms as HPO.

5. Conclusions

Using a combination of HPO analysis and literature review, we identified 908 genes associated with scoliosis and 444 genes associated with AMC resulting in 227 genes associated with AMC-SC. These genes act through a variety of cellular mechanisms including transcription regulation, transmembrane receptor, growth factor, and ion channels. Through query of the DECIPHER database, we identified two patients each with one CNV associated with AMC-SC. The first case had a CNV involving three genes (*FBN2*, *MEGF10*, and *PITX1*), while the second case had a CNV involving *ZC4H2*. As we continue to learn more about genetic mechanisms responsible for AMC we anticipate the ability to better provide prognostic information and targeted therapies for affected patients.

Supplementary Materials: The following are available online at https://www.mdpi.com/article/10.3390/genes12071052/s1, Table S1: Additional genes (*n* = 30) to the list of 402 list from Kiefer & Hall, 2019 [6,73–98].

Author Contributions: Writing-Original Draft: X.L. Data Curation: S.G.C. Formal Analysis: N.D.-O. Investigation: A.G.G. Methodology: S.W.-H.H. Project administration: T.B. Resources: K.S. Software: H.J.P.v.B. Conceptualization: P.F.G., K.D. All authors have read and agreed to the published version of the manuscript.

Funding: Dahan-Oliel holds a clinical research scholar award from the Fonds de la Recherche en Santé du Québec. Research reported in this publication was also supported in part by the Eunice Kennedy Shriver National Institute of Child Health & Human Development of the National Institutes of Health under Award Number R03HD099516 to P.F.G.

Acknowledgments: This study makes use of data generated by the DECIPHER community. A full list of centers that contributed to the generation of the data is available from https://deciphergenomics.org/about/stats and via email from contact@deciphergenomics.org. Funding for the DECIPHER project was provided by Wellcome. "Those who carried out the original analysis and collection of the data bear no responsibility for the further analysis or interpretation of the data." We gratefully acknowledge the support of the Malika Ray, Asok K. Ray, FRCS/(Edin) Initiative for Child Health. We gratefully acknowledge the technical assistance of Lourdes Richardson RN, MSN, FNP-BC.

Conflicts of Interest: The authors declare no conflict of interest.

References

1. Hall, J.G. Arthrogryposis (multiple congenital contractures): Diagnostic approach to etiology, classification, genetics, and general principles. *Eur. J. Med. Genet.* **2014**, *57*, 464–472. [CrossRef] [PubMed]
2. Hall, J.G.; Kiefer, J. Arthrogryposis as a Syndrome: Gene Ontology Analysis. *Mol. Syndr.* **2016**, *7*, 101–109. [CrossRef] [PubMed]
3. Kiefer, J.; Hall, J.G. Gene ontology analysis of arthrogryposis (multiple congenital contractures). *Am. J. Med. Genet. Part C Semin. Med. Genet.* **2019**, *181*, 310–326. [CrossRef]
4. Bamshad, M.; Van Heest, A.; Pleasure, D. Arthrogryposis: A Review and Update. *J. Bone Jt. Surg. Am.* **2009**, *91*, 40–46. [CrossRef]
5. Hall, J.G.; Kimber, E.; Dieterich, K. Classification of arthrogryposis. *Am. J. Med. Genet. Part C Semin. Med. Genet.* **2019**, *181*, 300–303. [CrossRef]
6. Hall, J.G.; Aldinger, K.A.; Tanaka, K.I. Amyoplasia revisited. *Am. J. Med. Genet. A* **2014**, *164*, 700–730. [CrossRef]
7. Drummond, D.S.; A Mackenzie, D. Scoliosis in Arthrogryposis Multiplex Congenita. *Spine* **1978**, *3*, 146–151. [CrossRef]

8.   Herron, L.D.; Westin, G.W.; Dawson, E.G. Scoliosis in arthrogryposis multiplex congenita. *J. Bone Jt. Surg. Am.* **1978**, *60*, 293–299. [CrossRef]

9.   Komolkin, I.; Ulrich, E.V.; Agranovich, O.E.; van Bosse, H.J. Treatment of Scoliosis Associated with Arthrogryposis Multiplex Congenita. *J. Pediatr. Orthop.* **2017**, *37*, S24–S26. [CrossRef]

10.   Yingsakmongkol, W.; Kumar, S.J. Scoliosis in arthrogryposis multiplex congenita: Results after nonsurgical and surgical treatment. *J. Pediatr. Orthop.* **2000**, *20*, 656–661. [CrossRef]

11.   Cheng, J.C.; Castelein, R.M.; Chu, W.C.; Danielsson, A.J.; Dobbs, M.B.; Grivas, T.B.; Gurnett, C.A.; Luk, K.D.; Moreau, A.; Newton, P.O.; et al. Adolescent idiopathic scoliosis. *Nat. Rev. Dis. Primers* **2015**, *1*, 15030. [CrossRef] [PubMed]

12.   Pérez-Machado, G.; Berenguer-Pascual, E.; Bovea-Marco, M.; Rubio-Belmar, P.A.; García-López, E.; Garzón, M.J.G.; Mena-Mollá, S.; Pallardó, F.V.; Bas, T.; Viña, J.R.; et al. From genetics to epigenetics to unravel the etiology of adolescent idiopathic scoliosis. *Bone* **2020**, *140*, 115563. [CrossRef] [PubMed]

13.   Baer, S.; Obringer, C.; Julia, S.; Chelly, J.; Capri, Y.; Gras, D.; Baujat, G.; Felix, T.M.; Doray, B.; Del Pozo, J.S.; et al. Early-onset nucleotide excision repair disorders with neurological impairment: Clues for early diagnosis and prognostic counseling. *Clin. Genet.* **2020**, *98*, 251–260. [CrossRef] [PubMed]

14.   Wang, Y.; Pearce, M.P.; Sliter, D.A.; Olzmann, J.A.; Christianson, J.C.; Kopito, R.R.; Boeckmann, S.; Gagen, C.; Leichner, G.S.; Roitelman, J.; et al. *SPFH1* and SPFH2 mediate the ubiquitination and degradation of inositol 1,4,5-trisphosphate receptors in muscarinic receptor-expressing HeLa cells. *Biochim. Biophys. Acta BBA Bioenerg.* **2009**, *1793*, 1710–1718. [CrossRef]

15.   Pelin, K.; Hilpelä, P.; Donner, K.; Sewry, C.; Akkari, P.A.; Wilton, S.D.; Wattanasirichaigoon, D.; Bang, M.-L.; Centner, T.; Hanefeld, F.; et al. Mutations in the nebulin gene associated with autosomal recessive nemaline myopathy. *Proc. Natl. Acad. Sci. USA* **1999**, *96*, 2305–2310. [CrossRef]

16.   Laing, N.G.; Dye, D.E.; Wallgren-Pettersson, C.; Richard, G.; Monnier, N.; Lillis, S.; Winder, T.L.; Lochmuller, H.; Graziano, C.; Mitrani-Rosenbaum, S.; et al. Mutations and polymorphisms of the skeletal muscle alpha-actin gene (*ACTA1*). *Hum. Mutat.* **2009**, *30*, 1267–1277. [CrossRef]

17.   Somerville, R.P.; Jungers, K.A.; Apte, S.S. Discovery and characterization of a novel, widely expressed metalloprotease, *ADAMTS10*, and its proteolytic activation. *J. Biol. Chem.* **2004**, *279*, 51208–51217. [CrossRef]

18.   Cappello, S.; Gray, M.J.; Badouel, C.; Lange, S.; Einsiedler, M.; Srour, M.; Chitayat, D.; Hamdan, F.F.; Jenkins, Z.A.; Morgan, T.R.; et al. Mutations in genes encoding the cadherin receptor-ligand pair *DCHS1* and *FAT4* disrupt cerebral cortical development. *Nat. Genet.* **2013**, *45*, 1300–1308. [CrossRef] [PubMed]

19.   Foldynova-Trantirkova, S.; Wilcox, W.R.; Krejci, P. Sixteen years and counting: The current understanding of fibroblast growth factor receptor 3 (*FGFR3*) signaling in skeletal dysplasias. *Hum. Mutat.* **2012**, *33*, 29–41. [CrossRef]

20.   Hellmann, T.V.; Nickel, J.; Mueller, T.D. Missense Mutations in *GDF-5* Signaling: Molecular Mechanisms behind Skeletal Malformation. In *Mutations in Human Genetic Disease*; David, D., Chen, J.-M., Eds.; IntechOpen: Wuerzburg, Germany, 2012; Volume 1, pp. 1–45.

21.   Bai, X.; Xiao, Z.; Pan, Y.; Hu, J.; Pohl, J.; Wen, J.; Li, L. Cartilage-derived morphogenetic protein-1 promotes the differentiation of mesenchymal stem cells into chondrocytes. *Biochem. Biophys. Res. Commun.* **2004**, *325*, 453–460. [CrossRef]

22.   Quélin, C.; Loget, P.; Rozel, C.; D'Hervé, D.; Fradin, M.; Demurger, F.; Odent, S.; Pasquier, L.; Cavé, H.; Marcorelles, P. Fetal costello syndrome with neuromuscular spindles excess and p.Gly12Val *HRAS* mutation. *Eur. J. Med. Genet.* **2017**, *60*, 395–398. [CrossRef] [PubMed]

23.   Wassif, C.A.; Maslen, C.; Kachilele-Linjewile, S.; Lin, D.; Linck, L.M.; Connor, W.E.; Steiner, R.D.; Porter, F.D. Mutations in the human sterol delta7-reductase gene at 11q12-13 cause Smith-Lemli-Opitz syndrome. *Am. J. Hum. Genet.* **1998**, *63*, 55–62. [CrossRef] [PubMed]

24.   Kishino, T.; Lalande, M.; Wagstaff, J. *UBE3A/E6-AP* mutations cause Angelman syndrome. *Nat. Genet.* **1997**, *15*, 70–73. [CrossRef] [PubMed]

25.   Basson, C.T.; Huang, T.; Lin, R.C.; Bachinsky, D.R.; Weremowicz, S.; Vaglio, A.; Bruzzone, R.; Quadrelli, R.; Lerone, M.; Romeo, G.; et al. Different *TBX5* interactions in heart and limb defined by Holt-Oram syndrome mutations. *Proc. Natl. Acad. Sci. USA* **1999**, *96*, 2919–2924. [CrossRef] [PubMed]

26.   Hasson, P.; DeLaurier, A.; Bennett, M.; Grigorieva, E.; Naiche, L.; Papaioannou, V.; Mohun, T.J.; Logan, M.P. Tbx4 and Tbx5 Acting in Connective Tissue Are Required for Limb Muscle and Tendon Patterning. *Dev. Cell* **2010**, *18*, 148–156. [CrossRef]

27.   Davignon, L.; Chauveau, C.; Julien, C.; Dill, C.; Duband-Goulet, I.; Cabet, E.; Buendia, B.; Lilienbaum, A.; Rendu, J.; Minot, M.C.; et al. The transcription coactivator ASC-1 is a regulator of skeletal myogenesis, and its deficiency causes a novel form of congenital muscle disease. *Hum. Mol. Genet.* **2016**, *25*, 1559–1573. [CrossRef]

28.   Birouk, N.; Azzedine, H.; Dubourg, O.; Muriel, M.-P.; Benomar, A.; Hamadouche, T.; Maisonobe, T.; Ouazzani, R.; Brice, A.; Yahyaoui, M.; et al. Phenotypical Features of a Moroccan Family With Autosomal Recessive Charcot-Marie-Tooth Disease Associated With the S194X Mutation in the *GDAP1* Gene. *Arch. Neurol.* **2003**, *60*, 598–604. [CrossRef]

29.   Chung, K.W.; Kim, S.M.; Sunwoo, I.N.; Cho, S.Y.; Hwang, S.J.; Kim, J.; Kang, S.H.; Park, K.-D.; Choi, K.-G.; Choi, I.S.; et al. A novel *GDAP1* Q218E mutation in autosomal dominant Charcot-Marie-Tooth disease. *J. Hum. Genet.* **2008**, *53*, 360–364. [CrossRef]

30.   Nelis, E.; Erdem, S.; Bergh, P.Y.V.D.; Belpaire-Dethiou, M.-C.; Ceuterick, C.; Van Gerwen, V.; Cuesta, A.; Pedrola, L.; Palau, F.; Gabreels-Festen, A.A.; et al. Mutations in *GDAP1*: Autosomal recessive CMT with demyelination and axonopathy. *Neurology* **2002**, *59*, 1865–1872. [CrossRef]

31. Baxter, R.V.; Ben Othmane, K.; Rochelle, J.M.; Stajich, J.; Hulette, C.; Dew-Knight, S.; Hentati, F.; Ben Hamida, M.; Bel, S.; Stenger, J.E.; et al. Ganglioside-induced differentiation-associated protein-1 is mutant in Charcot-Marie-Tooth disease type 4A/8q21. *Nat. Genet.* **2001**, *30*, 21–22. [CrossRef]

32. Cuesta, A.; Pedrola, L.; Sevilla, T.; García-Planells, J.; Chumillas, M.J.; Mayordomo, F.; LeGuern, E.; Marín, I.; Vílchez, J.J.; Palau, F. The gene encoding ganglioside-induced differentiation-associated protein 1 is mutated in axonal Charcot-Marie-Tooth type 4A disease. *Nat. Genet.* **2001**, *30*, 22–25. [CrossRef]

33. Ridanpää, M.; van Eenennaam, H.; Pelin, K.; Chadwick, R.; Johnson, C.; Yuan, B.; Vanvenrooij, W.; Pruijn, G.; Salmela, R.; Rockas, S.; et al. Mutations in the RNA Component of RNase MRP Cause a Pleiotropic Human Disease, Cartilage-Hair Hypoplasia. *Cell* **2001**, *104*, 195–203. [CrossRef]

34. Rosenbluh, J.; Nijhawan, D.; Chen, Z.; Wong, K.-K.; Masutomi, K.; Hahn, W.C. RMRP Is a Non-Coding RNA Essential for Early Murine Development. *PLoS ONE* **2011**, *6*, e26270. [CrossRef]

35. Morgan, N.; Brueton, L.A.; Cox, P.; Greally, M.T.; Tolmie, J.; Pasha, S.; Aligianis, I.A.; van Bokhoven, J.; Marton, T.; Al-Gazali, L.; et al. Mutations in the Embryonal Subunit of the Acetylcholine Receptor (*CHRNG*) Cause Lethal and Escobar Variants of Multiple Pterygium Syndrome. *Am. J. Hum. Genet.* **2006**, *79*, 390–395. [CrossRef]

36. McMillin, M.J.; Beck, A.E.; Chong, J.X.; Shively, K.M.; Buckingham, K.J.; Gildersleeve, H.I.; Aracena, M.I.; Aylsworth, A.S.; Bitoun, P.; Carey, J.C.; et al. Mutations in *PIEZO2* cause Gordon syndrome, Marden-Walker syndrome, and distal arthrogryposis type 5. *Am. J. Hum. Genet.* **2014**, *94*, 734–744. [CrossRef] [PubMed]

37. Cortez, D.; Guntuku, S.; Qin, J.; Elledge, S.J. ATR and ATRIP: Partners in Checkpoint Signaling. *Science* **2001**, *294*, 1713–1716. [CrossRef]

38. Mokrani-Benhelli, H.; Gaillard, L.; Biasutto, P.; Le Guen, T.; Touzot, F.; Vasquez, N.; Komatsu, J.; Conseiller, E.; Picard, C.; Gluckman, E.; et al. Primary microcephaly, impaired DNA replication, and genomic instability caused by compound heterozygous *ATR* mutations. *Hum. Mutat.* **2013**, *34*, 374–384. [CrossRef] [PubMed]

39. Bajayo, A.; Goshen, I.; Feldman, S.; Csernus, V.; Iverfeldt, K.; Shohami, E.; Yirmiya, R.; Bab, I. Central IL-1 receptor signaling regulates bone growth and mass. *Proc. Natl. Acad. Sci. USA* **2005**, *102*, 12956–12961. [CrossRef]

40. Wade, E.M.; Daniel, P.B.; Jenkins, Z.A.; McInerney-Leo, A.; Leo, P.; Morgan, T.; Addor, M.C.; Adès, L.C.; Bertola, D.; Bohring, A.; et al. Mutations in *MAP3K7* that Alter the Activity of the TAK1 Signaling Complex Cause Frontometaphyseal Dysplasia. *Am. J. Hum. Genet.* **2016**, *99*, 392–406. [CrossRef]

41. Matanis, T.; Akhmanova, A.; Wulf, P.S.; Del Nery, E.; Weide, T.; Stepanova, T.; Galjart, N.; Grosveld, F.; Goud, B.; De Zeeuw, C.I.; et al. Bicaudal-D regulates COPI-independent Golgi–ER transport by recruiting the dynein–dynactin motor complex. *Nat. Cell Biol.* **2002**, *4*, 986–992. [CrossRef] [PubMed]

42. Oates, E.C.; Rossor, A.M.; Hafezparast, M.; Gonzalez, M.; Speziani, F.; MacArthur, D.G.; Lek, M.; Cottenie, E.; Scoto, M.; Foley, A.R.; et al. Mutations in *BICD2* Cause Dominant Congenital Spinal Muscular Atrophy and Hereditary Spastic Paraplegia. *Am. J. Hum. Genet.* **2013**, *92*, 965–973. [CrossRef] [PubMed]

43. Nousiainen, H.; Kestilä, M.; Pakkasjärvi, N.; Honkala, H.; Kuure, S.; Tallila, J.; Vuopala, K.; Ignatius, J.; Herva, R.; Peltonen, L. Mutations in mRNA export mediator *GLE1* result in a fetal motoneuron disease. *Nat. Genet.* **2008**, *40*, 155–157. [CrossRef] [PubMed]

44. Oliveira, J.; Gruber, A.; Cardoso, M.; Taipa, R.; Fineza, I.; Gonçalves, A.; Laner, A.; Winder, T.L.; Schroeder, J.; Rath, J.; et al. *LAMA2*gene mutation update: Toward a more comprehensive picture of the laminin-α2 variome and its related phenotypes. *Hum. Mutat.* **2018**, *39*, 1314–1337. [CrossRef] [PubMed]

45. Toydemir, R.M.; Bamshad, M.J. Sheldon-Hall syndrome. *Orphanet J. Rare Dis.* **2009**, *4*, 11–15. [CrossRef]

46. Ansari, M.; Rainger, J.K.; Murray, J.E.; Hanson, I.; Firth, H.V.; Mehendale, F.; Amiel, J.; Gordon, C.; Percesepe, A.; Mazzanti, L.; et al. A syndromic form of Pierre Robin sequence is caused by 5q23 deletions encompassing *FBN2* and *PHAX*. *Eur. J. Med. Genet.* **2014**, *57*, 587–595. [CrossRef]

47. Huang, N.; Lee, I.; Marcotte, E.M.; Hurles, M.E. Characterising and Predicting Haploinsufficiency in the Human Genome. *PLoS Genet.* **2010**, *6*, e1001154. [CrossRef] [PubMed]

48. Callewaert, B.; Loeys, B.; Ficcadenti, A.; Vermeer, S.; Landgren, M.; Kroes, H.Y.; Yaron, Y.; Pope, M.; Foulds, N.; Boute, O.; et al. Comprehensive clinical and molecular assessment of 32 probands with congenital contractural arachnodactyly: Report of 14 novel mutations and review of the literature. *Hum. Mutat.* **2009**, *30*, 334–341. [CrossRef]

49. Kragesteen, B.K.; Brancati, F.; Digilio, M.C.; Mundlos, S.; Spielmann, M. H2AFY promoter deletion causes *PITX1* endoactivation and Liebenberg syndrome. *J. Med. Genet.* **2019**, *56*, 246–251. [CrossRef]

50. Frints, S.G.; Hennig, F.; Colombo, R.; Jacquemont, S.; Terhal, P.; Zimmerman, H.H.; Hunt, D.; Mendelsohn, B.A.; Kordaß, U.; Webster, R.; et al. Deleterious de novo variants of X-linked *ZC4H2* in females cause a variable phenotype with neurogenic arthrogryposis multiplex congenita. *Hum. Mutat.* **2019**, *40*, 2270–2285. [CrossRef]

51. Rao, D.; Kronert, W.A.; Guo, Y.; Hsu, K.H.; Sarsoza, F.; Bernstein, S.I. Reductions in ATPase activity, actin sliding velocity, and myofibril stability yield muscle dysfunction in Drosophila models of myosin-based Freeman–Sheldon syndrome. *Mol. Biol. Cell* **2019**, *30*, 30–41. [CrossRef]

52. Watson, C.M.; A Crinnion, L.; Murphy, H.; Newbould, M.; Harrison, S.M.; Lascelles, C.; Antanaviciute, A.; Carr, I.M.; Sheridan, E.; Bonthron, D.; et al. Deficiency of the myogenic factor MyoD causes a perinatally lethal fetal akinesia. *J. Med. Genet.* **2016**, *53*, 264–269. [CrossRef] [PubMed]

53. Buckingham, M.; Rigby, P.W. Gene Regulatory Networks and Transcriptional Mechanisms that Control Myogenesis. *Dev. Cell* **2014**, *28*, 225–238. [CrossRef] [PubMed]

54. Gong, C.; Li, Z.; Ramanujan, K.; Clay, I.; Zhang, Y.; Lemire-Brachat, S.; Glass, D.J. A Long Non-coding RNA, LncMyoD, Regulates Skeletal Muscle Differentiation by Blocking IMP2-Mediated mRNA Translation. *Dev. Cell* **2015**, *34*, 181–191. [CrossRef] [PubMed]

55. Gicquel, C.; Rossignol, S.; Cabrol, S.; Houang, M.; Steunou, V.; Barbu, V.; Danton, F.; Thibaud, N.; Le Merrer, M.; Burglen, L.; et al. Epimutation of the telomeric imprinting center region on chromosome 11p15 in Silver-Russell syndrome. *Nat. Genet.* **2005**, *37*, 1003–1007. [CrossRef] [PubMed]

56. Bruce, S.; Hannula-Jouppi, K.; Peltonen, J.; Kere, J.; Lipsanen-Nyman, M. Clinically Distinct Epigenetic Subgroups in Silver-Russell Syndrome: The Degree of*H19*Hypomethylation Associates with Phenotype Severity and Genital and Skeletal Anomalies. *J. Clin. Endocrinol. Metab.* **2009**, *94*, 579–587. [CrossRef]

57. Saal, H.M.; Harbison, M.D.; Netchine, I. Silver-Russell Syndrome. In *GeneReviews®*; Adam, M.P., Ardinger, H.H., Pagon, R.A., Wallace, S.E., Bean, L.J.H., Mirzaa, G., Amemiya, A., Eds.; University of Washington: Seattle, WA, USA, 1993.

58. Meerschaut, I.; De Coninck, S.; Steyaert, W.; Barnicoat, A.; Bayat, A.; Benedicenti, F.; Berland, S.; Blair, E.M.; Breckpot, J.; De Burca, A.; et al. A clinical scoring system for congenital contractural arachnodactyly. *Genet. Med.* **2020**, *22*, 124–131. [CrossRef]

59. Alvarado, D.M.; McCall, K.; Aferol, H.; Silva, M.J.; Garbow, J.R.; Spees, W.M.; Patel, T.; Siegel, M.; Dobbs, M.B.; Gurnett, C.A. Pitx1 haploinsufficiency causes clubfoot in humans and a clubfoot-like phenotype in mice. *Hum. Mol. Genet.* **2011**, *20*, 3943–3952. [CrossRef]

60. Hirata, H.; Nanda, I.; van Riesen, A.; McMichael, G.; Hu, H.; Hambrock, M.; Papon, M.-A.; Fischer, U.; Marouillat, S.; Ding, C.; et al. ZC4H2 Mutations Are Associated with Arthrogryposis Multiplex Congenita and Intellectual Disability through Impairment of Central and Peripheral Synaptic Plasticity. *Am. J. Hum. Genet.* **2013**, *92*, 681–695. [CrossRef] [PubMed]

61. May, M.; Hwang, K.-S.; Miles, J.; Williams, C.; Niranjan, T.; Kahler, S.G.; Chiurazzi, P.; Steindl, K.; Van Der Spek, P.J.; Swagemakers, S.; et al. ZC4H2, an XLID gene, is required for the generation of a specific subset of CNS interneurons. *Hum. Mol. Genet.* **2015**, *24*, 4848–4861. [CrossRef] [PubMed]

62. Zanzottera, C.; Milani, D.; Alfei, E.; Rizzo, A.; D'Arrigo, S.; Esposito, S.; Pantaleoni, C. ZC4H2 deletions can cause severe phenotype in female carriers. *Am. J. Med. Genet. Part A* **2017**, *173*, 1358–1363. [CrossRef]

63. Vogt, J.; Morgan, N.V.; Rehal, P.; Faivre, L.; Brueton, L.A.; Becker, K.; Fryns, J.P.; Holder, S.; Islam, L.; Kivuva, E.; et al. *CHRNG* genotype-phenotype correlations in the multiple pterygium syndromes. *J. Med. Genet.* **2012**, *49*, 21–26. [CrossRef]

64. Gurnett, C.A.; Desruisseau, D.M.; McCall, K.; Choi, R.; Meyer, Z.I.; Talerico, M.; Miller, S.E.; Ju, J.-S.; Pestronk, A.; Connolly, A.M.; et al. Myosin binding protein C1: A novel gene for autosomal dominant distal arthrogryposis type 1. *Hum. Mol. Genet.* **2010**, *19*, 1165–1173. [CrossRef] [PubMed]

65. Ha, K.; Buchan, J.G.; Alvarado, D.M.; McCall, K.; Vydyanath, A.; Luther, P.; Goldsmith, M.I.; Dobbs, M.B.; Gurnett, C.A. MYBPC1 mutations impair skeletal muscle function in zebrafish models of arthrogryposis. *Hum. Mol. Genet.* **2013**, *22*, 4967–4977. [CrossRef]

66. Cameron-Christie, S.R.; Wells, C.F.; Simon, M.; Wessels, M.; Tang, C.Z.; Wei, W.; Takei, R.; Aarts-Tesselaar, C.; Sandaradura, S.; Sillence, D.O.; et al. Recessive Spondylocarpotarsal Synostosis Syndrome Due to Compound Heterozygosity for Variants in *MYH3. Am. J. Hum. Genet.* **2018**, *102*, 1115–1125. [CrossRef] [PubMed]

67. Chong, J.X.; Burrage, L.C.; Beck, A.E.; Marvin, C.T.; McMillin, M.J.; Shively, K.M.; Harrell, T.M.; Buckingham, K.J.; Bacino, C.A.; Jain, M.; et al. Autosomal-Dominant Multiple Pterygium Syndrome Is Caused by Mutations in *MYH3. Am. J. Hum. Genet.* **2015**, *96*, 841–849. [CrossRef] [PubMed]

68. Whittle, J.; Antunes, L.; Harris, M.; Upshaw, Z.; Sepich, D.S.; Johnson, A.N.; Mokalled, M.; Solnica-Krezel, L.; Dobbs, M.B.; A Gurnett, C. *MYH 3*-associated distal arthrogryposis zebrafish model is normalized with para-aminoblebbistatin. *EMBO Mol. Med.* **2020**, *12*, e12356. [CrossRef]

69. Zhu, X.; Wang, F.; Zhao, Y.; Yang, P.; Chen, J.; Sun, H.; Liu, L.; Li, W.; Pan, L.; Guo, Y.; et al. A Gain-of-Function Mutation in Tnni2 Impeded Bone Development through Increasing Hif3a Expression in DA2B Mice. *PLoS Genet.* **2014**, *10*, e1004589. [CrossRef]

70. Drevytska, T.; Gonchar, E.; Okhai, I.; Lynnyk, O.; Mankovska, I.; Klionsky, D.; Dosenko, V.; Linnyk, O. The protective effect of Hif3a RNA interference and HIF-prolyl hydroxylase inhibition on cardiomyocytes under anoxia-reoxygenation. *Life Sci.* **2018**, *202*, 131–139. [CrossRef]

71. Bedard, T.; Lowry, R.B. Disease coding systems for arthrogryposis multiplex congenita. *Am. J. Med. Genet. Part C Semin. Med. Genet.* **2019**, *181*, 304–309. [CrossRef] [PubMed]

72. Köhler, S.; Gargano, M.; Matentzoglu, N.; Carmody, L.C.; Lewis-Smith, D.; A Vasilevsky, N.; Danis, D.; Balagura, G.; Baynam, G.; Brower, A.M.; et al. The Human Phenotype Ontology in 2021. *Nucleic Acids Res.* **2021**, *49*, D1207–D1217. [CrossRef]

73. Pergande, M.; Motameny, S.; Özdemir, Ö.; Kreutzer, M.; Wang, H.; Msc, H.-S.D.; Becker, K.; Karakaya, M.; Ehrhardt, H.; Elcioglu, N.; et al. The genomic and clinical landscape of fetal akinesia. *Genet. Med.* **2020**, *22*, 511–523. [CrossRef]

74. Böhm, J.; Malfatti, E.; Oates, E.; Jones, K.; Brochier, G.; Boland, A.; Deleuze, J.-F.; Romero, N.B.; Laporte, J. Novel ASCC1 mutations causing prenatal-onset muscle weakness with arthrogryposis and congenital bone fractures. *J. Med. Genet.* **2018**, *56*, 617–621. [CrossRef] [PubMed]

75. Chi, B.; O'Connell, J.D.; Iocolano, A.D.; Coady, J.A.; Yu, Y.; Gangopadhyay, J.; Gygi, S.P.; Reed, R. The neurodegenerative diseases ALS and SMA are linked at the molecular level via the ASC-1 complex. *Nucleic Acids Res.* **2018**, *46*, 11939–11951. [CrossRef] [PubMed]

76. Churchill, L.E.; Delk, P.R.; Wilson, T.E.; Torres-Martinez, W.; Rouse, C.E.; Marine, M.B.; Piechan, J.L. Fetal MRI and ultrasound findings of a confirmed asparagine synthetase deficiency case. *Prenat. Diagn.* **2020**, *40*, 1343–1347. [CrossRef] [PubMed]

77. Monteiro, F.P.; Curry, C.J.; Hevner, R.; Elliott, S.; Fisher, J.H.; Turocy, J.; Dobyns, W.B.; Costa, L.A.; Freitas, E.; Kitajima, J.P.; et al. Biallelic loss of function variants in ATP1A2 cause hydrops fetalis, microcephaly, arthrogryposis and extensive cortical malformations. *Eur. J. Med. Genet.* **2020**, *63*, 103624. [CrossRef]

78. Mohassel, P.; Liewluck, T.; Hu, Y.; Ezzo, D.; Ogata, T.; Saade, D.; Neuhaus, S.; Bolduc, V.; Zou, Y.; Donkervoort, S.; et al. Dominant collagen XII mutations cause a distal myopathy. *Ann. Clin. Transl. Neurol.* **2019**, *6*, 1980–1988. [CrossRef] [PubMed]

79. Pérez-Cerdá, C.; Girós, M.L.; Serrano, M.; Ecay, M.J.; Gort, L.; Dueñas, B.P.; Medrano, C.; García-Alix, A.; Artuch, R.; Briones, P.; et al. A Population-Based Study on Congenital Disorders of Protein N- and Combined with O-Glycosylation Experience in Clinical and Genetic Diagnosis. *J. Pediatr.* **2017**, *183*, 170–177. [CrossRef]

80. Lieu, M.T.; Ng, B.G.; Rush, J.S.; Wood, T.; Basehore, M.J.; Hegde, M.; Chang, R.C.; Abdenur, J.E.; Freeze, H.H.; Wang, R.Y. Severe, fatal multisystem manifestations in a patient with dolichol kinase-congenital disorder of glycosylation. *Mol. Genet. Metab.* **2013**, *110*, 484–489. [CrossRef]

81. Nishimura, N.; Kumaki, T.; Murakami, H.; Enomoto, Y.; Katsumata, K.; Toyoshima, K.; Kurosawa, K. Arthrogryposis multiplex congenita with polymicrogyria and infantile encephalopathy caused by a novel GRIN1 variant. *Hum. Genome Var.* **2020**, *7*, 1–4. [CrossRef]

82. Chen, K.; Yang, K.; Luo, S.; Chen, C.; Wang, Y.; Wang, Y.; Li, D.; Yang, Y.; Tang, Y.; Liu, F.; et al. A homozygous missense variant in HSD17B4 identified in a consanguineous Chinese Han family with type II Perrault syndrome. *BMC Med. Genet.* **2017**, *18*, 1–9. [CrossRef]

83. Nagata, K.; Itaka, K.; Baba, M.; Uchida, S.; Ishii, T.; Kataoka, K. Muscle-targeted hydrodynamic gene introduction of insulin-like growth factor-1 using polyplex nanomicelle to treat peripheral nerve injury. *J. Control. Release* **2014**, *183*, 27–34. [CrossRef] [PubMed]

84. Jeffries, L.; Olivieri, J.E.; Ji, W.; Spencer-Manzon, M.; Bale, A.; Konstantino, M.; Lakhani, S.A. Two siblings with a novel nonsense variant provide further delineation of the spectrum of recessive KLHL7 diseases. *Eur. J. Med. Genet.* **2019**, *62*, 103551. [CrossRef] [PubMed]

85. Chong, J.X.; Talbot, J.C.; Teets, E.M.; Previs, S.; Martin, B.L.; Shively, K.M.; Marvin, C.T.; Aylsworth, A.S.; Saadeh-Haddad, R.; Schatz, U.A.; et al. Mutations in MYLPF cause a novel segmental amyoplasia that manifests as distal arthrogryposis. *Am. J. Hum. Genet.* **2020**, *107*, 293–310. [CrossRef] [PubMed]

86. Tan, L.; Bi, B.; Zhao, P.; Cai, X.; Wan, C.; Shao, J.; He, X. Severe congenital microcephaly with 16p13.11 microdeletion combined with NDE1 mutation, a case report and literature review. *BMC Med. Genet.* **2017**, *18*, 141. [CrossRef]

87. Bonnin, E.; Cabochette, P.; Filosa, A.; Jühlen, R.; Komatsuzaki, S.; Hezwani, M.; Dickmanns, A.; Martinelli, V.; Vermeersch, M.; Supply, L.; et al. Biallelic mutations in nucleoporin NUP88 cause lethal fetal akinesia deformation sequence. *PLoS Genet.* **2018**, *14*, e1007845. [CrossRef]

88. Garcia, A.M.G.; Tutmaher, M.S.; Upadhyayula, S.R.; Russo, R.S.; Verma, S. Novel PLEC gene variants causing congenital myasthenic syndrome. *Muscle Nerve* **2019**, *60*, E40–E43. [CrossRef]

89. Abdel-Salam, G.M.H.; Miyake, N.; Abdel-Hamid, M.S.; Sayed, I.S.M.; Gadelhak, M.I.; Ismail, S.I.; Aglan, M.S.; Afifi, H.H.; Temtamy, S.A.; Matsumoto, N. Phenotypic and molecular insights into PQBP1 -related intellectual disability. *Am. J. Med. Genet. Part A* **2018**, *176*, 2446–2450. [CrossRef]

90. Genini, S.; Nguyen, T.T.; Malek, M.; Talbot, R.; Gebert, S.; Rohrer, G.; Nonneman, D.; Stranzinger, G.; Vögeli, P. Radiation hybrid mapping of 18 positional and physiological candidate genes for arthrogryposis multiplex congenita on porcine chromosome 5. *Anim. Genet.* **2006**, *37*, 239–244. [CrossRef]

91. Gardella, E.; Møller, R. Phenotypic and genetic spectrum of SCN 8A -related disorders, treatment options, and outcomes. *Epilepsia* **2019**, *60*, S77–S85. [CrossRef]

92. Seidahmed, M.Z.; Al-Kindi, A.; Alsaif, H.S.; Miqdad, A.; Alabbad, N.; Alfifi, A.; Abdelbasit, O.B.; Alhussein, K.; Alsamadi, A.; Ibrahim, N.; et al. Recessive mutations in SCYL2 cause a novel syndromic form of arthrogryposis in humans. *Qual. Life Res.* **2020**, *139*, 513–519. [CrossRef]

93. Hakonen, A.H.; Polvi, A.; Saloranta, C.; Paetau, A.; Heikkilä, P.; Almusa, H.; Ellonen, P.; Jakkula, E.; Saarela, J.; Aittomäki, K. SLC18A3 variants lead to fetal akinesia deformation sequence early in pregnancy. *Am. J. Med. Genet. Part A* **2019**, *179*, 1362–1365. [CrossRef] [PubMed]

94. Magini, P.; Smits, D.J.; Vandervore, L.; Schot, R.; Columbaro, M.; Kasteleijn, E.; van der Ent, M.; Palombo, F.; Lequin, M.H.; Dremmen, M.; et al. Loss of SMPD4 Causes a Developmental Disorder Characterized by Microcephaly and Congenital Arthrogryposis. *Am. J. Hum. Genet.* **2019**, *105*, 689–705. [CrossRef] [PubMed]

95. Ravenscroft, G.; Clayton, J.S.; Faiz, F.; Sivadorai, P.; Milnes, D.; Cincotta, R.; Moon, P.; Kamien, B.; Edwards, M.; Delatycki, M.; et al. Neurogenetic fetal akinesia and arthrogryposis: Genetics, expanding genotype-phenotypes and functional genomics. *J. Med. Genet.* **2020**, *15*, 106901. [CrossRef]

96. Montes-Chinea, N.I.; Guan, Z.; Coutts, M.; Vidal, C.; Courel, S.; Rebelo, A.P.; Abreu, L.; Zuchner, S.; Littleton, J.T.; Saporta, M.A. Identification of a new SYT2 variant validates an unusual distal motor neuropathy phenotype. *Neurol. Genet.* **2018**, *4*, e282. [CrossRef]

97. Loeys, B.; Chen, J.; Neptune, E.R.; Judge, D.; Podowski, M.; Holm, T.; Meyers, J.; Leitch, C.C.; Katsanis, N.; Sharifi, N.; et al. A syndrome of altered cardiovascular, craniofacial, neurocognitive and skeletal development caused by mutations in TGFBR1 or TGFBR2. *Nat. Genet.* **2005**, *37*, 275–281. [CrossRef] [PubMed]
98. Woolnough, R.; Dhawan, A.; Dow, K.; Walia, J.S. Are Patients with Loeys-Dietz Syndrome Misdiagnosed with Beals Syndrome? *Pediatrics* **2017**, *139*, 139–e20161281. [CrossRef]

genes

Article

Identification of Copy Number Variants in a Southern Chinese Cohort of Patients with Congenital Scoliosis

Wenjing Lai [1,†], Xin Feng [1,†], Ming Yue [1], Prudence W. H. Cheung [2], Vanessa N. T. Choi [1], You-Qiang Song [1], Keith D. K. Luk [2], Jason Pui Yin Cheung [2,*] and Bo Gao [1,*]

1   School of Biomedical Sciences, Li Ka Shing Faculty of Medicine, The University of Hong Kong, Hong Kong, China; floralai@connect.hku.hk (W.L.); FengxinCassie@connect.hku.hk (X.F.); nervym7@connect.hku.hk (M.Y.); vntchoi@hku.hk (V.N.T.C.); songy@hku.hk (Y.-Q.S.)
2   Department of Orthopaedics and Traumatology, The University of Hong Kong, Hong Kong, China; gnuehcp6@hotmail.com (P.W.H.C.); hrmoldk@hku.hk (K.D.K.L.)
*   Correspondence: cheungjp@hku.hk (J.P.Y.C.); gaobo@hku.hk (B.G.); Tel.: +852-2255-4581 (J.P.Y.C.); +852-3917-6809 (B.G.)
†   These authors contribute equally to this work.

Abstract: Congenital scoliosis (CS) is a lateral curvature of the spine resulting from congenital vertebral malformations (CVMs) and affects 0.5–1/1000 live births. The copy number variant (CNV) at chromosome 16p11.2 has been implicated in CVMs and recent studies identified a compound heterozygosity of 16p11.2 microdeletion and *TBX6* variant/haplotype causing CS in multiple cohorts, which explains about 5–10% of the affected cases. Here, we studied the genetic etiology of CS by analyzing CNVs in a cohort of 67 patients with congenital hemivertebrae and 125 family controls. We employed both candidate gene and family-based approaches to filter CNVs called from whole exome sequencing data. This identified 12 CNVs in four scoliosis-associated genes (*TBX6*, *NOTCH2*, *DSCAM*, and *SNTG1*) as well as eight recessive and 64 novel rare CNVs in 15 additional genes. Some candidates, such as *DHX40*, *NBPF20*, *RASA2*, and *MYSM1*, have been found to be associated with syndromes with scoliosis or implicated in bone/spine development. In particular, the *MYSM1* mutant mouse showed spinal deformities. Our findings suggest that, in addition to the 16p11.2 microdeletion, other CNVs are potentially important in predisposing to CS.

Keywords: congenital scoliosis; congenital vertebral malformation; copy number variant; CNV

Citation: Lai, W.; Feng, X.; Yue, M.; Cheung, P.W.H.; Choi, V.N.T.; Song, Y.-Q.; Luk, K.D.K.; Cheung, J.P.Y.; Gao, B. Identification of Copy Number Variants in a Southern Chinese Cohort of Patients with Congenital Scoliosis. *Genes* **2021**, *12*, 1213. https://doi.org/10.3390/genes12081213

Academic Editor: Luisa Politano

Received: 31 May 2021
Accepted: 3 August 2021
Published: 5 August 2021

Publisher's Note: MDPI stays neutral with regard to jurisdictional claims in published maps and institutional affiliations.

1. Introduction

Among all musculoskeletal disorders, scoliosis is one of the most common diseases, affecting around 3% of the world population, which can occur as an isolated defect or as a concomitant symptom in other diseases or syndromes [1]. Scoliosis is categorized into several main groups, including congenital scoliosis (CS), idiopathic scoliosis (IS), neuromuscular scoliosis, and degenerative scoliosis. CS, which usually has first onset at birth or shortly after birth, affects approximately 0.5–1 in 1000 live births [2–5]. Compared with IS, CS is generally more severe due to the high risk of progressive deformity and associated problems such as pulmonary compromise [6]. One of the most significant differences between CS and IS is that IS does not have an association with congenital vertebral malformation (CVM), whereas CVM is the major cause leading to CS. CVM can be classified into several subclasses, including failure of vertebral formation (e.g., hemivertebrae, wedged vertebrae), failure of vertebral segmentation (e.g., unilateral bar, block vertebrae), and mixed type. Of all CVMs, congenital hemivertebrae is the most common anomaly that causes CS [4,5].

During vertebral development, the paraxial mesoderm forms bilaterally paired blocks, named somites, along the anterior–posterior axis. The vertebral bodies are derived from somites formed in the presomitic mesoderm. This fundamental process is called somitogenesis. Once somitogenesis is disturbed, the resulting CVM may lead to spinal deformities.

The most commonly accepted mechanism governing somitogenesis is the clock and wavefront model, which is controlled and coordinated by several key signaling pathways, such as Notch, Wnt, Fgf and retinoic acid signaling pathways [7,8]. Genetic studies of human patients with CVM have identified a variety of mutations in components of Notch signaling pathway (e.g., *NOTCH2*, *DLL3*, *MESP2*, *LFNG*, *HES7*, and *RIPPLY2*) and also in several key transcription factors essential for somitogenesis (e.g., *TBX6*, *TBXT*, and *SOX9*). Nevertheless, the genetic basis for majority of patients with CS still remains unclear [1,9].

Copy number variation (CNV) is a type of structural variation of genome. With the advancement of genome-wide analysis tools, it has been revealed that CNVs are widespread in the human genome and account for a large fraction of human genetic diversity [10]. CNVs have been, so far, implicated in many disease states including scoliosis. Although a number of CNVs were found to be associated with adolescent idiopathic scoliosis (AIS) [11,12], there have not been many reports about CS-associated CNVs. The 16p11.2 microdeletion was found to be associated with CS [13], and recent studies demonstrated that a compound inheritance of a *TBX6*-containing 16p11.2 microdeletion and a *TBX6* mutation or hypomorphic haplotype accounted for 5–10% of patients with CS in different populations [14–17]. Additional CNVs, including 10q24.31, 17p11.2, 20p11, 22q11.2, and a few other regions, were respectively reported in individual patients with CVMs [18,19]. Besides 16p11.2 microdeletion, it is unknown whether other CNVs are prevalent in CS.

Here, we analyzed CNVs in a Southern Chinese cohort of patients with congenital hemivertebrae. CNVs were called from whole-exome sequencing (WES) data of 67 cases and 125 family members (controls). We identified 12 rare CNVs in 4 known scoliosis-associated genes and eight recessive CNVs in three genes. We also found 64 novel, rare CNVs in 14 genes that occurred in multiple patients but are very rare in our control group and the general population, suggesting a potential role for genetic susceptibility in the development of CS.

2. Materials and Methods

2.1. Patient Recruitment

The patients studied in this project were recruited from the Duchess of Kent Children's Hospital (DKCH), a tertiary scoliosis referral center in Hong Kong. The patients with CS were diagnosed by imaging such as plain standing whole-spine radiographs and computed tomography. A total of 67 patients with hemivertebrae were chosen for this study, of which 31 had single congenital hemivertebrae while 36 had multiple congenital hemivertebrae. Patients' personal data and medical records were collected under ethical privacy guidelines and approval. Ethics was approved by the Institutional Review Board of the University of Hong Kong/Hospital Authority Hong Kong West Cluster (HKU/HA HKW IRB Ref # UW 15-216), and written informed consent was obtained from all participants and/or parents/siblings.

2.2. Control Cohort

The control cohort studied in this project consisted of 125 participating family members of the recruited patients with CS. Only unaffected parents and siblings (without CS) were included. Accordingly, 58 out of 67 patients had family member(s) participating in this study, including 2 quintets, 14 quartets, 33 trios, and 9 duos.

2.3. Genomic DNA Extraction

Genomic DNAs were extracted from peripheral blood samples of 67 patients and 125 of their family members using Invitrogen™ ChargeSwitch gDNA Serum Kit. The purified genomic DNA was quantified by NanoDrop.

2.4. Whole-Exome Sequencing (WES) and Copy Number Variations (CNVs) Calling

WES was performed for all recruited patients with congenital hemivertebrae and participating family members by Novogene Co, Ltd. (Hong Kong, China), using the

Agilent SureSelect Human All Exon Kit on the Illumina sequencing platform. The WES data were processed as described previously [20]. The raw sequence data were first analyzed by fastp for quality control and filtering [21]. After filtering, the Q20 base of most samples was greater than 95%, and the Q30 base was greater than 90%. The sequence reads were mapped to the reference genome (GRCh37/hg19) by Burrows-Wheller Aligner v0.7.17 (BWA-MEM) [22] and further processed using SAMtools v1.10 to sort and index aligned reads [23]. The sam format files generated by BWA was converted to bam format files by SAMtools. CNVs were called from bam files with ExomeDepth v1.1.15, which is an R package based on a read depth algorithm [24]. ExomeDepth uses a robust statistical model to build an optimized reference set in maximizing the CNVs detection power. In this study, four healthy control family members (CS59A, CS71A, CS71B, and CS81A) were selected to generate the reference set.

2.5. CNVs Filtering

Several criteria were used to filter CNVs: (i) Bayes factor (BF) values were calculated for each variant. BF equals to the log10 likelihood ratio of the alternative hypothesis (i.e., there is a CNV) over the null hypothesis (i.e., there is no CNV). BF = log10 (alternative hypothesis/null hypothesis). BF value greater than 1 was regarded as a strong supporting evidence of CNV. CNVs with BF values smaller than 1 were excluded. (ii) As ExomeDepth cannot detect small size CNVs accurately, CNVs with size smaller than 100 bp were excluded. (iii) Because CNVs with high allele frequency in the general population are likely benign and less susceptible, the CNVs with the allele frequency greater than 0.01 were excluded (a minimum sample size of 100 is required). Database of Genomic Variants (DGV) and Genome Aggregation Database (gnomAD) were used. If available, the allele frequency in East Asian population was also checked. As different CNVs often overlap and have no clear boundaries, this filtration was conducted in a gene-based manner. If there were multiple CNVs covering the same gene, the maximum allele frequency was used for filtering. (iv) In a gene-based manner, the number of CNV recurrence was counted in patients and controls.

2.6. Real-Time Quantitative PCR (qPCR)

Real-time qPCR was performed to validate some of the detected CNVs. Briefly, ROX Reference Dye (0.4 μL, 50X), forward and reverse primers (0.4 μL each, 10 μM), TB Green Premix Ex Taq (10 μL, 2X, Tli RNaseH Plus, Takara), patients' genomic DNA (0.5 μL, 10 ng/μL), and sterile ddH$_2$O (8.3 μL) were mixed for qPCR, which was performed using Applied Biosystem^{TM} StepOnePlus^{TM} Real-Time PCR System. A locus outside of the detected CNV region of *NOTCH2*, *DSCAM* and *SNTG1* was used as reference locus (P1). P1 is near the region of chromosome 16p11.2 and previously used as a reference site to detect 16p11.2/TBX6 deletion [14,17]. Each sample was analyzed in triplicate. Quantities of the copy numbers of specific locus were determined by the delta Ct method. The $2^{-\Delta\Delta CT}$ method was used to analyze the relative changes. The qPCR primer sequences: *NOTCH2*-F: 5′- AGGAGGCGACCGAGAAGATG-3′; *NOTCH2*-R: 5′-CGATACTCACCATGCGCG-GG-3′; *DSCAM*-F: 5′-AGCGAACGTTCCTATCGCTT-3′; *DSCAM*-R: 5′-TTTCACTTATGCGCCCTGGG-3′; *SNTG1*-F: 5′-GTCTACATGGGCTGGTG-TGA-3′; *SNTG1*-R: 5′-CTGGAGGTGCCAGAAACTTG-3′; P1-F: 5′-GGGGAAGGAACTTA-CATGAC-3′; P1-R: 5′-TCGTGTTTCCCTGTTGTACC-3′.

3. Results

3.1. CS Cohort and WES

In our cohort, we recruited a total of 92 patients with CS, in which vertebral malformations, such as hemivertebrae, unilateral bar, or block vertebrae, were identified. This operational definition thus excluded other types of scoliosis such as AIS. Because hemivertebrae is the most common type of vertebral malformation in CS and has the greatest potential for rapid progression (5–10 degrees/year) [25], 67 patients with congenital hemivertebrae were first selected. Further, 125 healthy family members of 58 patients were enrolled for this study, including parents and siblings from two quintets, 14 quartets, 33 trios, and nine duos. WES was performed for all 67 patients and 125 participating family members (controls). The contaminating sequencing adaptors and low-quality reads were first removed and the filtered reads were then aligned to the reference human genome (GRCh37/hg19), sorted and indexed.

3.2. CNV Calling

CNVs were called from the sequence reads with the read-depth analysis tool ExomeDepth, which has high sensitivity and specificity at the exon level [24,26]. Four healthy parents who were not carriers of 16p11.2 microdeletion but whose children have been previously diagnosed with TBX6 compound heterozygosity [17] were selected to generate the reference set for ExomeDepth analysis. After CNV calling of the 67 patients with CS, a total of 15,671 CNVs were detected. On average, each patient carries around 234 CNVs. By counting repeatedly occurring CNVs among different cases, there were 6084 distinct CNVs. This strategy successfully identified *TBX6*-containing 16p11.2 microdeletion in four patients as previously reported [17]. For the control group, a total of 27,116 CNVs were detected from 125 family control members. On average, each control carried around 217 CNVs. By counting repeatedly occurring CNVs among different controls, there were 7171 distinct CNVs. Although more CNVs were detected in a few individuals (six patients and four controls), there was no significant difference between the patient group and the control group (Supplementary Figure S1). The average CNV numbers in patients and controls were similar to the previous report [24]. Afterwards, we analyzed all CNVs by employing both a candidate gene approach and family-based filtering and prioritization strategies. A workflow is shown in Figure 1.

3.3. CNVs in Candidate Genes

To identify CNVs associated with CS, we firstly used the candidate gene approach, and focused on CNVs that contained genes known to be involved in scoliosis or somitogenesis. After checking allele frequencies of CNVs in the Database of Genomic Variants (DGV) and the Genome Aggregation Database (gnomAD), a total of 12 rare CNVs that influence four candidate genes were found in 12 patients, including known *TBX6*-containing 16p11.2 heterozygous deletion in four cases [17]. We also identified two rare CNVs that contained *NOTCH2*, a key component in the Notch signaling pathway, in two patients, and six rare CNVs in AIS-associated genes, *DSCAM* [27] and *SNTG1* [28,29], in six patients (Table 1). We then checked these CNVs in their available family members and found that they are either *novel* mutations (*NOTCH2* in CS043, *DSCAM* in CS018 and CS036, and *SNTG1* in CS048) or paternally inherited (*DSCAM* in CS050) (Table 1). We were unable to determine the inheritance patterns of other patients (*NOTCH2* in CS033, *DSCAM* in CS053 and CS064) due to the lack of family members.

Among the identified rare CNVs, the TBX6-containing chromosome 16p11.2 microdeletion had been previously validated [17]. Here, we further examined CNVs that contained *NOTCH2*, *DSCAM* or *SNTG1* genes. Indeed, qPCR analysis detected heterozygous deletions within the *NOTCH2*, *DSCAM* and *SNTG1* loci (Supplementary Figure S2), indicating the reliability of CNVs called from WES data by ExomeDepth.

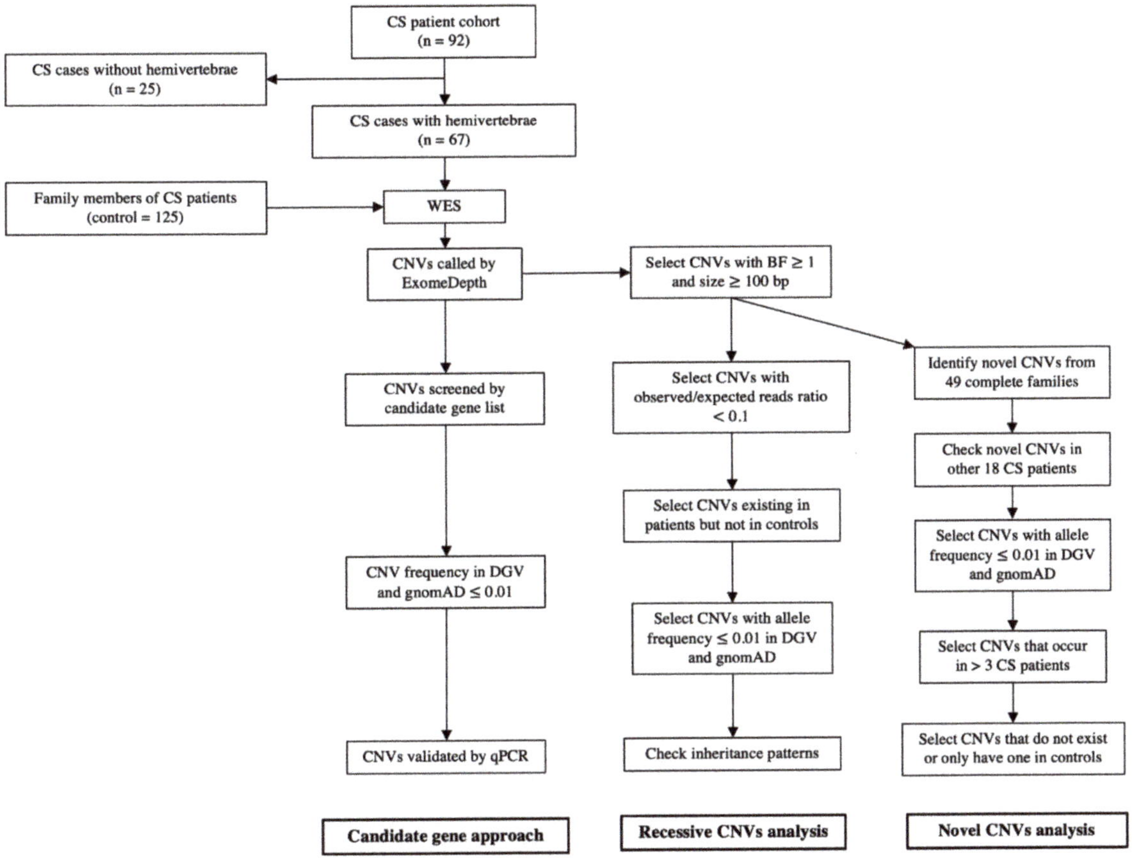

Figure 1. The workflow of CNV analysis. This strategy detected CNVs in several candidate genes and identified recessive and novel rare CNVs enriched in patients with CS.

3.4. Recessive CNVs in Patients with CS

We then searched for homozygous CNVs (observed/expected reads ratio < 0.1) in 67 patients and 125 controls. After excluding the homozygous CNVs that existed in both patients and controls, we identified unique homozygous CNVs in eight patients with CS. The heterozygous deletions of these loci are rare in DGV or gnomAD database (Table 2). Considering that homozygous CNVs might be inherited from parents, we further checked their inheritance pattern and found that they were either *novel* mutations or unknown due to lack of parents' data. These recessive CNVs contained three genes, *NBPF20* (Neuroblastoma Breakpoint Family Member 20), *FAM138C* (Family with Sequence Similarity 138 Member C), and *DHX40* (DEAH-Box Helicase 40). Interestingly, the *DHX40*-containing homozygous CNVs were detected in six patients but was not reported in DGV or gnomAD. *DHX40*-containing heterozygous CNVs are also very rare (Table 2). *FMA138C* is an RNA gene and *NBPF20* is a member of NBPF family characterized by tandemly repeats of DUF1220 domain, but their functions are unclear. *DHX40* encodes a member of the DExD/H-box RNA helicase superfamily that catalyzes the unwinding of double-stranded RNA and has an essential role in RNA metabolism [30].

Table 1. CNVs found with candidate gene approach (N.D., not determined; N.A., not applied).

Gene	Patient	Type	Chr	Start	End	Size (bp)	Bayes Factor	Reads Ratio (Observed/ Expected)	Exons Annotation (hg19) (Gene_exon)	Inheritance Pattern	Highest Frequency in DGV (Sample Size >100)	gnomAD_ Structural_Variants Frequency (Heterozygous Loss)
NOTCH2	CS033	deletion	1	120,611,949	120,612,020	71	4.62	0.657	NOTCH2_1	N.D.	0.0037	0.00037 (0 in East Asia)
	CS043	deletion	1	120,539,621	120,612,020	72,399	8.46	0.722	NOTCH2_1-4	De novo	0.0037	0.00037 (0 in East Aisa)
DSCAM	CS018	deletion	21	41,452,080	41,452,267	187	4.65	0.429	DSCAM_25	De novo	0.00049	N.A.
	CS036	deletion	21	41,452,080	41,452,267	187	6.02	0.415	DSCAM_25	De novo	0.00049	N.A.
	CS050	deletion	21	41,452,080	41,452,267	187	5.8	0.468	DSCAM_25	Paternal	0.00049	N.A.
	CS053	deletion	21	41,452,080	41,452,267	187	5.45	0.494	DSCAM_25	N.D.	0.00049	N.A.
	CS064	deletion	21	41,452,080	41,452,267	187	6.1	0.463	DSCAM_25	N.D.	0.00049	N.A.
SNTG1	CS048	deletion	8	51,503,440	51,571,223	67,783	6.37	0.43	SNTG1_13-15	De novo	0.0002	0.000046 (0 in East Asia)
TBX6	CS059	deletion	16	29,674,601	30,199,897	525,296	644	0.555	SPN_2,AC009133.19_2-3,QPRT_1-4,C16orf54_2,ZG16_2-4,KIF22_1-13,MAZ_1-5,PRRT2_2-3,PAGR1_1-3,CTD-2574D22.6_1-2,MVP_2-15,CDIPT_6-2,SEZ6L2_16-1,ASPHD1_1-3,KCTD13_6-1,TMEM219_1-4,TAOK2_2-16,HIRIP3_7-1,INO80E_1-7,DOC2A_11-2,C16orf92_2-3,FAM57B_5-1,ALDOA_8-16,PPP4C_2-9,**TBX6_9-2**,YPEL3_4-1,GDPD3_10-1,MAPK3_8-1,CORO1A_2-3,CORO1A_4-10	N.D.	0.0005	0.0001462 (0 in East Asia)
	CS071	deletion	16	29,495,011	30,218,221	723,210	754	0.572	NPIPL3_3-1,SPN_2,AC009133.19_2-3,QPRT_1-4,C16orf54_2,ZG16_3-4,KIF22_2-12,MAZ_1-5,PRRT2_2-3,PAGR1_1-3,CTD-2574D22.6_1-2,MVP_2-15,CDIPT_6-2,SEZ6L2_16-1,ASPHD1_1-3,KCTD13_6-1,TMEM219_1-4,TAOK2_2-16,HIRIP3_7-2,INO80E_1-7,DOC2A_11-2,C16orf92_2-3,FAM57B_5-1,ALDOA_8-16,PPP4C_2-9,**TBX6_9-2**,YPEL3_4-1,GDPD3_10-1,MAPK3_8-1,CORO1A_2-11,BOLA2B_3-1,SLX1A_1-5,SULT1A3_3-9,RP11-347C12.3_5-2	De novo	0.0005	0.0001462 (0 in East Asia)
	CS078	deletion	16	29,498,516	30,199,897	701,381	690	0.578	NPIPL3_1,SPN_2,AC009133.19_2-3,QPRT_1-4,C16orf54_2,ZG16_3-4,KIF22_2-12,MAZ_1-5,PRRT2_2-3,PAGR1_1-3,CTD-2574D22.6_1-2,MVP_2-15,CDIPT_6-2,SEZ6L2_16-1,ASPHD1_1-3,KCTD13_6-1,TMEM219_1-4,TAOK2_2-16,HIRIP3_7-2,INO80E_1-7,DOC2A_11-2,C16orf92_2-3,FAM57B_5-1,ALDOA_8-16,PPP4C_2-9,**TBX6_9-2**,YPEL3_4-1,GDPD3_10-1,MAPK3_8-1,CORO1A_2-10	N.D.	0.0005	0.0001462 (0 in East Asia)
	CS081	deletion	16	29,498,516	30,199,897	701,381	645	0.558	NPIPL3_1,SPN_2,AC009133.19_2-3,QPRT_1-4,C16orf54_2,ZG16_3-4,KIF22_2-12,MAZ_1-5,PRRT2_2-3,PAGR1_1-3,CTD-2574D22.6_1-2,MVP_2-15,CDIPT_6-2,SEZ6L2_16-1,ASPHD1_1-3,KCTD13_6-1,TMEM219_1-4,TAOK2_2-16,HIRIP3_7-2,INO80E_1-7,DOC2A_11-2,C16orf92_2-3,FAM57B_5-1,ALDOA_8-16,PPP4C_2-9,**TBX6_9-2**,YPEL3_4-1,GDPD3_10-1,MAPK3_8-1,CORO1A_2-10	N.D.	0.0005	0.0001462 (0 in East Asia)

Table 2. Recessive CNVs unique in patients with CS (N.D., not determined; N.A., not applied).

Gene	Patient	Type	Chr	Start	End	Size (bp)	Bayes Factor	Reads Ratio (Observed/Expected)	Exons Annotation (hg19) (Gene_exon)	Inheritance Pattern	Highest Frequency in DGV (Sample Size > 100)	gnomAD_ Structural Variants Frequency (Heterozygous Loss)
NBPF20	CS047	deletion	1	148,261,458	148,262,366	908	5.27	0.04	NBPF20_98-99	De novo	0	0 (0 in East Asia)
FAM138C	CS048	deletion	9	35,061	35,519	458	6.51	0	FAM138C_1-2	* De novo	0.0074	N.A.
DHX40	CS004 CS035 CS043 CS050 CS053 CS057	deletion	17	57,656,834	57,657,240	406	5.38 4.91 6 7.23 7.02 6.29	0	DHX40_9-10	N.D. N.D. * De novo * De novo N.D. De novo	0.0000922	0.0025 (0.008152 in East Asia)

* The CNV is also not present in the healthy siblings.

3.5. Novel CNVs in Patients with CS

We also sought to identify CS-associated novel CNVs and first analyzed the data from 49 complete families (two quintets, 14 quartets or 33 trios). The detected novel CNVs were then checked in the other 18 patients (nine singlets and nine duos). Eventually, we identified 64 CNVs in 14 genes that occurred in more than three patients but did not exist or was very rare (<1%) in family control group. Those with high CNV allele frequency (>1%) in the general population were also filtered out. This strategy successfully identified the known *TBX6*-containing CNVs in four patients [17] and *DHX40*-containing homozygous CNVs in six patients. Interestingly, we also found there are four additional heterozygous *DHX40* CNVs (Table 3 and Supplementary Table S1). Most of the identified novel CNVs were heterozygous loss, and one was gain of one copy. Our CNV shortlist includes genes involved in ubiquitination (*NAE1*, *MYSM1*), enzymatic activities (*MME*, *PHKB*), ion/small molecule transportation (*SCN7A*, *ABCA6*), meiosis (*MNS1*, *SPO11*), spermatogenesis (*GMCL1*), GTPase activity (*RASA2*), TNF signaling (*NSMAF*), or with unknown function (*LRRC40*).

Table 3. Novel CNVs enriched in patients with CS (N.A., not applied).

Gene	Type	Chr	Size (bp)	Count in 67 Patients	Count in 125 Controls	Highest Frequency in DGV (Sample Size > 100)	gnomad_ Structural_Variants Frequency (Heterozygous Loss)	gnomad_East Asia_Structural_Variants Frequency (Heterozygous Loss)
LRRC40	deletion	1	30,080–383,938	4	0	0.009556907	0.0000461	0.000
SCN7A	deletion	2	544–11,914	4	0	0.002257336	0.0000461	0.000
MME	deletion	3	279–55,232	4	0	0.000798722	0.0000462	0.000
NAE1	deletion	16	6724–71,803	4	0	N.A.[b]	0	0.000
TBX6	deletion	16	525,296–723,210	4	0	0.0005	0.0001462	0.000
DHX40	deletion	17	107–2354	10 [a]	1	0.0000922	0.0025	0.008152
GMCL1	deletion	2	11,694–24,032	5	1	0.001303781	0.0000479	0.000
MYSM1	deletion	1	190–10,053	4	1	0.0000922	0.0000461	0.0004139
RASA2	deletion	3	2892–100,149	4	1	0.00086881	0	0.000
NSMAF	deletion	8	203–12,325	4	1	0.001145475	0	0.000
MNS1	deletion	15	52,610–323,156	4	1	0.00518807	0.0000922	0.000
PHKB	deletion	16	7697–99,254	4	1	0.001198083	0.0000481	0.000
SPO11	deletion	20	730–33,608	4	1	0.00064226	0	0.000
ABCA6	duplication	17	560–13,580	4	1	0.0009219	0	0.000

[a] DHX40 has 6 homozygous (listed in Table 2) and 4 heterozygous CNVs. [b] No CNV with sample size more than 100 is found within the *NAE1* locus. Note: Detailed information of these novel CNVs is shown in Table S1.

4. Discussion

CS is a genetically heterogeneous disorder with evidence for multiple causative genes. However, the genetic causes of the majority of patients still remain unknown. As most cases of CS are of sporadic etiology, CNVs may have greater influence than single nucleotide variations (SNVs) [31]. This was well exemplified by the *TBX6*-containing 16p11.2 microdeletion in previous CS studies [14–17]. Here, we systematically investigated CNVs in a cohort of patients with congenital hemivertebrae and their family controls. We identified the well-known CNVs at chromosome 16p11.2, as well as a number of new CNVs that are potentially associated with CS. Haploinsufficiency of Notch signaling pathway has been demonstrated to cause CS [32] and mutations in *NOTCH2* caused Alagille syndrome and Hajdu–Cheney syndrome, both of which showed abnormal curvature of the spine [33,34]. In our study, we found one short and one long CNVs at *NOTCH2* locus in two patients, spanning one and four exons of *NOTCH2*, respectively. As no significant coding SNV could be detected in *NOTCH2* of these two patients, it is unclear whether heterozygous loss of *NOTCH2* is sufficient to cause CS or other non-coding *NOTCH2* SNVs or environmental factors [32] may contribute. Interestingly, CNVs in two AIS-associated genes, *DSCAM* [27] and *SNTG1* [28,29], were found in six patients, suggesting CS and AIS may be genetically related to each other.

An intriguing finding in our analysis is the identification of CNVs spanning various exons of *DHX40* in ten patients, including six homozygous and four heterozygous CNVs. Most of them are novel mutations. *DHX40* belongs to the conserved DExD/H-box RNA helicase family, which facilitates the ATP-dependent unwinding of RNA secondary structures [30]. However, the biological functions of each member remained poorly understood. Interestingly, the *DHX40* mutant mice were described to exhibit abnormal bone structure and bone mineralization (Mouse Genome Informatics, MGI: 1914737), indicating a role of *DHX40* in bone development. The mutant of its family member *DHX35* was described to have abnormal vertebrae morphology and scoliosis in mice (MGI: 1918965). Patients carrying *DHX37* mutations showed developmental delay and intellectual disability as well as vertebral anomalies [30]. These observations in the mouse and human might suggest a potential link between DHX family members and CVM.

Although the function of *NBPF20* is unknown, it locates at chromosome 1q21.1, which microdeletion is associated with a variety of phenotypes including skeletal malformations such as scoliosis [35,36]. This region also contains other NBPF family members, such as *NBPF10*, whose genetic variants were implicated in Mayer-Rokitansky-Küster-Hauser (MRKH) syndrome (OMIM # 277000) [37], a disease associated with CS [38].

Among the candidate genes of the identified novel CNVs, *RASA2* (RAS P21 Protein Activator 2) and *MYSM1* (Myb-Like, SWIRM, and MPN domains 1) are of particular interest. *RASA2* encodes a GAP (GTPase-activating protein) protein and functions as a suppressor of RAS by promoting its intrinsic GTPase activity. Rare variants in *RASA2* have been found associated with Noonan syndrome [39]. As scoliosis occurs frequently in Noonan syndrome [40], *RASA2* is a potential candidate gene for CS. It would be interesting to investigate the *RASA2* mutant mouse phenotype. MYSM1 is a deubiquitinase reported to be essential for bone formation [41] and its mutant mice have truncated and kinky tails [42–44], which are often associated with vertebral malformations [45]. Indeed, an X-ray from the International Mouse Phenotyping Consortium (MGI: 2444584) exhibited a spinal deformity in the *MYSM1* mutant mouse (Supplementary Figure S3), indicating a potential role of *MYSM1* in spinal development and predisposition to CS. Further detailed phenotypic analysis of mutant animals is needed to validate its pathogenicity in CS.

Although there are a few reports of heterozygous point mutations in CS patients [17,46–48], the dominant negative effect was only demonstrated by a novel *TBXT* mutation [17]. Considering the low familial recurrence rate in CS, recessive or compound heterozygous mutations are more likely to be the major cause of CS. In this regard, heterozygous CNVs are not sufficient to induce CS. Their pathogenicity may be explained by the following genetic models. First, in our cases, the patients carrying heterozygous CNVs may have additional risk variant or haplotype on the other allele. This possibility has been well exemplified by the 16p11.2/TBX6 mutations and haplotype. However, further analysis of risk variant/haplotype in our study is severely limited by our dataset from WES as they may reside in non-coding regions that regulate gene transcription. We did not detect significant deleterious mutations in the coding regions of these genes. Second, additional mutations in other relevant genes may increase the risk of CS (polygenic model). Other possibilities include environmental contributions and novel mutations in somatic tissues. Environmental factors, such as short-term gestational hypoxia, have been found to cause CS in combination with haploinsufficiency of Notch signaling pathway genes [32]. On the other hand, somatic mutations may serve as the "second hit" in addition to the heterozygous germline CNV mutations (first hit). This genetic model has been well demonstrated in other diseases as well as dystrophic scoliosis caused by *NF1* [49–52]. Testing the above models will require whole genome sequencing, more comprehensive data analysis, and the isolation of malformed vertebral tissues in future studies.

5. Conclusions

In this study, we investigated the genetic basis of CS by analyzing CNVs in a cohort of CS families. Based on the candidate gene approach and family-based filtering of CNVs, we identified both known CS-associated genes and a set of new susceptibility genes, some of which (e.g., *DHX40*, *RASA2*, and *MYSM1*) warrant further investigations in larger cohorts as well as functional characterization. Given the well-defined example of the *TBX6* compound inheritance and the complex genetic nature of CS, future studies examining the combined effects of SNVs and CNVs and somatic tissues may help better decipher the genetic etiology and heterogeneity of CS.

Supplementary Materials: The following are available online at https://www.mdpi.com/article/10.3390/genes12081213/s1, Figure S1: Total number CNVs in each patient and family control, Figure S2: Quantitative PCR analysis for validating heterozygous deletion, Figure S3: The X-ray of *MYSM1* mutant mouse from IMPC. Table S1: Novel CNVs enriched in CS patients.

Author Contributions: Conceptualization, B.G., J.P.Y.C. and K.D.K.L.; methodology, W.L., X.F., M.Y., V.N.T.C. and P.W.H.C.; software, W.L., M.Y. and X.F.; validation, W.L., X.F. and B.G.; resources, B.G., J.P.Y.C., Y.-Q.S. and K.D.K.L.; data curation, J.P.Y.C. and B.G.; manuscript preparation, W.L., X.F., J.P.Y.C. and B.G.; supervision, B.G., J.P.Y.C. and Y.-Q.S.; project administration, B.G. and J.P.Y.C.; funding acquisition, B.G., J.P.Y.C. and K.D.K.L. All authors have read and agreed to the published version of the manuscript.

Funding: This research was supported by the Hong Kong Health and Medical Research Fund (06171406) to B.G., and the Tam Sai Kit Endowment Fund to K.D.K.L.

Institutional Review Board Statement: The study was conducted according to the guidelines of the Declaration of Helsinki and approved by the Institutional Review Board of the University of Hong Kong/Hospital Authority Hong Kong West Cluster (HKU/HAHKW, IRB Reference Number UW 15-216).

Informed Consent Statement: Informed consents was obtained from all subjects involved in the study.

Data Availability Statement: Data available on request due to restrictions.

Acknowledgments: Nil.

Conflicts of Interest: The authors declare no conflict of interest.

References

1. Giampietro, P.F. Genetic Aspects of Congenital and Idiopathic Scoliosis. *Scientifica* **2012**, *2012*, 1–15. [CrossRef]
2. Shands, A.R.; Eisberg, H.B. The incidence of scoliosis in the state of Delaware; a study of 50,000 minifilms of the chest made during a survey for tuberculosis. *J. Bone Jt. Surg. Am. Vol.* **1955**, *37*, 1243–1249. [CrossRef]
3. Wynne-Davies, R. Congenital vertebral anomalies: Aetiology and relationship to spina bifida cystica. *J. Med. Genet.* **1975**, *12*, 280–288. [CrossRef]
4. McMaster, M.J.; Ohtsuka, K. The natural history of congenital scoliosis. A study of two hundred and fifty-one patients. *J. Bone Jt. Surg. Am. Vol.* **1982**, *64*, 1128–1147. [CrossRef]
5. McMaster, M.J.; Singh, H. Natural History of Congenital Kyphosis and Kyphoscoliosis. A Study of One Hundred and Twelve Patients. *J. Bone Jt. Surg. Am. Vol.* **1999**, *81*, 1367–1383. [CrossRef] [PubMed]
6. Cheng, J.C.; Castelein, R.M.; Chu, W.C.; Danielsson, A.J.; Dobbs, M.B.; Grivas, T.B.; Gurnett, C.A.; Luk, K.D.; Moreau, A.; Newton, P.O.; et al. Adolescent idiopathic scoliosis. *Nat. Rev. Dis. Primers* **2015**, *1*, 15030. [CrossRef] [PubMed]
7. Pourquié, O. Vertebrate Segmentation: From Cyclic Gene Networks to Scoliosis. *Cell* **2011**, *145*, 650–663. [CrossRef]
8. Maroto, M.; Bone, R.A.; Dale, J.K. Somitogenesis. *Development* **2012**, *139*, 2453–2456. [CrossRef]
9. Giampietro, P.; Raggio, C.; Blank, R.D.; McCarty, C.; Broeckel, U.; Pickart, M. Clinical, Genetic and Environmental Factors Associated with Congenital Vertebral Malformations. *Mol. Syndr.* **2012**, *4*, 94–105. [CrossRef]
10. Sebat, J.; Lakshmi, B.; Troge, J.; Alexander, J.; Young, J.; Lundin, P.; Månér, S.; Massa, H.; Walker, M.; Chi, M.; et al. Large-Scale Copy Number Polymorphism in the Human Genome. *Science* **2004**, *305*, 525–528. [CrossRef]
11. Buchan, J.G.; Alvarado, D.M.; Haller, G.; Aferol, H.; Miller, N.H.; Dobbs, M.B.; Gurnett, C.A. Are copy number variants associated with adolescent idiopathic scoliosis? *Clin. Orthop. Relat. Res.* **2014**, *472*, 3216–3225. [CrossRef] [PubMed]
12. Sadler, B.; Haller, G.; Antunes, L.; Bledsoe, X.; Morcuende, J.; Giampietro, P.; Raggio, C.; Miller, N.; Kidane, Y.; Wise, C.A.; et al. Distal chromosome 16p11.2 duplications containing SH2B1 in patients with scoliosis. *J. Med. Genet.* **2019**, *56*, 427–433. [CrossRef] [PubMed]
13. Al-Kateb, H.; Khanna, G.; Filges, I.; Hauser, N.; Grange, D.K.; Shen, J.; Smyser, C.D.; Kulkarni, S.; Shinawi, M. Scoliosis and vertebral anomalies: Additional abnormal phenotypes associated with chromosome 16p11.2 rearrangement. *Am. J. Med. Genet. Part A* **2014**, *164*, 1118–1126. [CrossRef]
14. Wu, N.; Ming, X.; Xiao, J.; Wu, Z.; Chen, X.; Shinawi, M.; Shen, Y.; Yu, G.; Liu, J.; Xie, H.; et al. TBX6 Null Variants and a Common Hypomorphic Allele in Congenital Scoliosis. *N. Engl. J. Med.* **2015**, *372*, 341–350. [CrossRef]

15. Takeda, K.; Kou, I.; Kawakami, N.; Iida, A.; Nakajima, M.; Ogura, Y.; Imagawa, E.; Miyake, N.; Matsumoto, N.; Yasuhiko, Y.; et al. Compound Heterozygosity for Null Mutations and a Common Hypomorphic Risk Haplotype in TBX6 Causes Congenital Scoliosis. *Hum. Mutat.* **2017**, *38*, 317–323. [CrossRef] [PubMed]

16. Lefebvre, M.; Duffourd, Y.; Jouan, T.; Poe, C.; Jean-Marçais, N.; Verloes, A.; St-Onge, J.; Riviere, J.-B.; Petit, F.; Pierquin, G.; et al. Autosomal recessive variations of TBX6, from congenital scoliosis to spondylocostal dysostosis. *Clin. Genet.* **2016**, *91*, 908–912. [CrossRef]

17. Feng, X.; Cheung, J.P.Y.; Je, J.S.H.; Cheung, P.W.H.; Chen, S.; Yue, M.; Wang, N.; Choi, V.N.T.; Yang, X.; Song, Y.; et al. Genetic variants of TBX6 and TBXT identified in patients with congenital scoliosis in Southern China. *J. Orthop. Res.* **2020**, *39*, 971–988. [CrossRef]

18. Liu, Z.L.; Wu, N.; Wu, Z.H.; Zuo, Y.Z.; Qiu, G.X. Advances in congenital vertebral malformation caused by genomic copy number variation. *Zhonghua Wai Ke Za Zhi* **2016**, *54*, 313–316.

19. Gao, X.; Gotway, G.; Rathjen, K.; Johnston, C.; Sparagana, S.; Wise, C.A. Genomic Analyses of Patients with Unexplained Early-Onset Scoliosis. *Spine Deform.* **2014**, *2*, 324–332. [CrossRef]

20. Li, C.; Wang, N.; Schaffer, A.A.; Liu, X.; Zhao, Z.; Elliott, G.; Garrett, L.; Choi, N.T.; Wang, Y.; Wang, Y.; et al. Mutations in COMP cause familial carpal tunnel syndrome. *Nat. Commun.* **2020**, *11*, 1–16. [CrossRef]

21. Chen, S.; Zhou, Y.; Chen, Y.; Gu, J. Fastp: An ultra-fast all-in-one FASTQ preprocessor. *Bioinformatics* **2018**, *34*, i884–i890. [CrossRef]

22. Li, H.; Durbin, R. Fast and accurate short read alignment with Burrows-Wheeler transform. *Bioinformatics* **2009**, *25*, 1754–1760. [CrossRef] [PubMed]

23. Li, H.; Handsaker, R.; Wysoker, A.; Fennell, T.; Ruan, J.; Homer, N.; Marth, G.; Abecasis, G.; Durbin, R. The Sequence Alignment/Map format and SAMtools. *Bioinformatics* **2009**, *25*, 2078–2079. [CrossRef] [PubMed]

24. Plagnol, V.; Curtis, J.; Epstein, M.; Mok, K.; Stebbings, E.; Grigoriadou, S.; Wood, N.; Hambleton, S.; Burns, S.; Thrasher, A.; et al. A robust model for read count data in exome sequencing experiments and implications for copy number variant calling. *Bioinformatics* **2012**, *28*, 2747–2754. [CrossRef]

25. Burnei, G.; Gavriliu, S.; Vlad, C.; Georgescu, I.; Ghita, R.A.; Dughilă, C.; Japie, E.M.; Onilă, A. Congenital scoliosis: An up-to-date. *J. Med. Life* **2015**, *8*, 388–397.

26. Ellingford, J.M.; Campbell, C.; Barton, S.; Bhaskar, S.; Gupta, S.; Taylor, R.L.; Sergouniotis, P.I.; Horn, B.; Lamb, J.; Michaelides, M.; et al. Validation of copy number variation analysis for next-generation sequencing diagnostics. *Eur. J. Hum. Genet.* **2017**, *25*, 719–724. [CrossRef]

27. Sharma, S.; Gao, X.; Londono, D.; Devroy, S.E.; Mauldin, K.N.; Frankel, J.T.; Brandon, J.M.; Zhang, D.; Li, Q.-Z.; Dobbs, M.B.; et al. Genome-wide association studies of adolescent idiopathic scoliosis suggest candidate susceptibility genes. *Hum. Mol. Genet.* **2011**, *20*, 1456–1466. [CrossRef] [PubMed]

28. Bashiardes, S.; Veile, R.; Allen, M.; Wise, C.A.; Dobbs, M.; Morcuende, J.A.; Szappanos, L.; Herring, J.A.; Bowcock, A.M.; Lovett, M. SNTG1, the gene encoding gamma1-syntrophin: A candidate gene for idiopathic scoliosis. *Hum. Genet.* **2004**, *115*, 81–89. [CrossRef]

29. Tassano, E.; Ronchetto, P.; Severino, M.; Divizia, M.; Lerone, M.; Uccella, S.; Nobili, L.; Tavella, E.; Morerio, C.; Coviello, D.; et al. Scoliosis with cognitive impairment in a girl with 8q11.21q11.23 microdeletion and SNTG1 disruption. *Bone* **2021**, *150*, 116022. [CrossRef]

30. Paine, I.; Posey, J.; Grochowski, C.M.; Jhangiani, S.N.; Rosenheck, S.; Kleyner, R.; Marmorale, T.; Yoon, M.; Wang, K.; Robison, R.; et al. Paralog Studies Augment Gene Discovery: DDX and DHX Genes. *Am. J. Hum. Genet.* **2019**, *105*, 302–316. [CrossRef]

31. Lupski, J.R. Genomic rearrangements and sporadic disease. *Nat. Genet.* **2007**, *39*, S43–S47. [CrossRef] [PubMed]

32. Sparrow, D.B.; Chapman, G.; Smith, A.J.; Mattar, M.Z.; Major, J.A.; O'Reilly, V.C.; Saga, Y.; Zackai, E.H.; Dormans, J.P.; Alman, B.A.; et al. A Mechanism for Gene-Environment Interaction in the Etiology of Congenital Scoliosis. *Cell* **2012**, *149*, 295–306. [CrossRef]

33. McDaniell, R.; Warthen, D.M.; Sanchez-Lara, P.A.; Pai, A.; Krantz, I.D.; Piccoli, D.A.; Spinner, N.B. NOTCH2 Mutations Cause Alagille Syndrome, a Heterogeneous Disorder of the Notch Signaling Pathway. *Am. J. Hum. Genet.* **2006**, *79*, 169–173. [CrossRef]

34. Gray, M.J.; Kim, C.A.; Bertola, D.R.; Arantes, P.R.; Stewart, H.; Simpson, M.A.; Irving, M.D.; Robertson, S.P. Serpentine fibula polycystic kidney syndrome is part of the phenotypic spectrum of Hajdu–Cheney syndrome. *Eur. J. Hum. Genet.* **2011**, *20*, 122–124. [CrossRef] [PubMed]

35. Haldeman-Englert, C.R.; Jewett, T. 1q21.1 Recurrent Microdeletion. In *Gene Reviews(R)*; Adam, M.P., Ardinger, H.H., Eds.; University of Washington: Seattle, WA, USA, 1993.

36. Gamba, B.F.; Zechi-Ceide, R.M.; Kokitsu-Nakata, N.M.; Vendramini-Pittoli, S.; Rosenberg, C.; Santos, A.C.K.; Ribeiro-Bicudo, L.; Richieri-Costa, A. Interstitial 1q21.1 Microdeletion Is Associated with Severe Skeletal Anomalies, Dysmorphic Face and Moderate Intellectual Disability. *Mol. Syndr.* **2016**, *7*, 344–348. [CrossRef]

37. Pan, H.X.; Luo, G.N.; Wan, S.Q.; Qin, C.L.; Tang, J.; Zhang, M.; Du, M.; Xu, K.K.; Shi, J.Q. Detection of de novo genetic variants in Mayer-Rokitansky-Kuster-Hauser syndrome by whole genome sequencing. *Eur. J. Obstet. Gynecol. Reprod. Biol. X* **2019**, *4*, 100089. [CrossRef]

38. Fisher, K.; Esham, R.H.; Thorneycroft, I. Scoliosis associated with typical Mayer-Rokitansky-Küster-Hauser syndrome. *South. Med. J.* **2000**, *93*, 243–246. [CrossRef]

39. Chen, P.-C.J.; Yin, J.; Yu, H.-W.; Yuan, T.; Fernandez, M.; Yung, C.; Trinh, Q.M.; Peltekova, V.D.; Reid, J.G.; Tworog-Dube, E.; et al. Next-generation sequencing identifies rare variants associated with Noonan syndrome. *Proc. Natl. Acad. Sci. USA* **2014**, *111*, 11473–11478. [CrossRef]

40. Lee, C.-K.; Chang, B.S.; Hong, Y.M.; Yang, S.W.; Seo, J.B. Spinal deformities in Noonan syndrome: A clinical review of sixty cases. *J. Bone Jt. Surg. Am. Vol.* **2001**, *83*, 1495–1502. [CrossRef]

41. Li, P.; Yang, Y.M.; Sanchez, S.; Cui, D.C.; Dang, R.J.; Wang, X.Y.; Lin, Q.X.; Wang, Y.; Wang, C.; Chen, D.F.; et al. Deubiquitinase MYSM1 Is Essential for Normal Bone Formation and Mesenchymal Stem Cell Differentiation. *Sci. Rep.* **2016**, *6*, 1–11. [CrossRef]

42. Jiang, X.-X.; Nguyen, Q.; Chou, Y.; Wang, T.; Nandakumar, V.; Yates, P.; Jones, L.; Wang, L.; Won, H.; Lee, H.-R.; et al. Control of B Cell Development by the Histone H2A Deubiquitinase MYSM1. *Immunity* **2011**, *35*, 883–896. [CrossRef]

43. Liakath-Ali, K.; Vancollie, V.; Heath, E.H.; Smedley, D.P.; Estabel, J.; Sunter, D.; DiTommaso, T.; White, J.K.; Ramirez-Solis, R.; Smyth, I.; et al. Novel skin phenotypes revealed by a genome-wide mouse reverse genetic screen. *Nat. Commun.* **2014**, *5*, 1–11. [CrossRef] [PubMed]

44. DiTommaso, T.; Jones, L.K.; Cottle, D.L.; Gerdin, A.-K.; Vancollie, V.E.; Watt, F.M.; Ramirez-Solis, R.; Bradley, A.; Steel, K.P.; Sundberg, J.P.; et al. Identification of Genes Important for Cutaneous Function Revealed by a Large Scale Reverse Genetic Screen in the Mouse. *PLoS Genet.* **2014**, *10*, e1004705. [CrossRef] [PubMed]

45. Giampietro, P.F.; Blank, R.D.; Raggio, C.L.; Merchant, S.; Jacobsen, F.S.; Faciszewski, T.; Shukla, S.K.; Greenlee, A.R.; Reynolds, C.; Schowalter, D.B. Congenital and Idiopathic Scoliosis: Clinical and Genetic Aspects. *Clin. Med. Res.* **2003**, *1*, 125–136. [CrossRef]

46. Wu, N.; Wang, L.; Hu, J.; Zhao, S.; Liu, B.; Li, Y.; Du, H.; Zhang, Y.; Li, X.; Yan, Z.; et al. A Recurrent Rare SOX9 Variant (M469V) is Associated with Congenital Vertebral Malformations. *Curr. Gene Ther.* **2019**, *19*, 242–247. [CrossRef]

47. Lin, M.; Zhao, S.; Liu, G.; Huang, Y.; Yu, C.; Zhao, Y.; Wang, L.; Zhang, Y.; Yan, Z.; Wang, S.; et al. Identification of novel FBN1 variations implicated in congenital scoliosis. *J. Hum. Genet.* **2019**, *65*, 221–230. [CrossRef]

48. Al Dhaheri, N.; Wu, N.; Zhao, S.; Wu, Z.; Blank, R.D.; Zhang, J.; Raggio, C.; Halanski, M.; Shen, J.; Noonan, K.; et al. KIAA1217: A novel candidate gene associated with isolated and syndromic vertebral malformations. *Am. J. Med. Genet. Part A* **2020**, *182*, 1664–1672. [CrossRef]

49. Qian, F.; Germino, G.G. Mistakes happen: Somatic mutation and disease. *Am. J. Hum. Genet.* **1997**, *61*, 1000–1005. [CrossRef]

50. Watnick, T.J.; Torres, V.E.; Gandolph, M.A.; Qian, F.; Onuchic, L.F.; Klinger, K.W.; Landes, G.; Germino, G.G. Somatic Mutation in Individual Liver Cysts Supports a Two-Hit Model of Cystogenesis in Autosomal Dominant Polycystic Kidney Disease. *Mol. Cell* **1998**, *2*, 247–251. [CrossRef]

51. Lamlum, H.; Ilyas, M.; Rowan, A.; Clark, S.; Johnson, V.; Bell, J.; Frayling, I.; Efstathiou, J.; Pack, K.; Payne, S.; et al. The type of somatic mutation at APC in familial adenomatous polyposis is determined by the site of the germline mutation: A new facet to Knudson's 'two-hit' hypothesis. *Nat. Med.* **1999**, *5*, 1071–1075. [CrossRef] [PubMed]

52. Margraf, R.L.; VanSant-Webb, C.; Mao, R.; Viskochil, D.H.; Carey, J.; Hanson, H.; D'Astous, J.; Grossmann, A.; Stevenson, D.A. NF1 Somatic Mutation in Dystrophic Scoliosis. *J. Mol. Neurosci.* **2019**, *68*, 11–18. [CrossRef]

genes

Article

Severity of Idiopathic Scoliosis Is Associated with Differential Methylation: An Epigenome-Wide Association Study of Monozygotic Twins with Idiopathic Scoliosis

Patrick M. Carry [1,2], Elizabeth A. Terhune [2], George D. Trahan [3], Lauren A. Vanderlinden [4], Cambria I. Wethey [2], Parvaneh Ebrahimi [5], Fiona McGuigan [5], Kristina Åkesson [5,6] and Nancy Hadley-Miller [1,2,*]

[1] Musculoskeletal Research Center, Children's Hospital Colorado, Aurora, CO 80045, USA; patrick.carry@cuanschutz.edu
[2] Department of Orthopedics, University of Colorado Anschutz Medical Campus, Aurora, CO 80045, USA; elizabeth.a.terhune@cuanschutz.edu (E.A.T.); cambria.wethey@cuanschutz.edu (C.I.W.)
[3] Department of Pediatrics, University of Colorado Anschutz Medical Campus, Aurora, CO 80045, USA; devon.trahan@cuanschutz.edu
[4] Department of Biostatistics and Informatics, Colorado School of Public Health, Aurora, CO 80045, USA; lauren.vanderlinden@cuanschutz.edu
[5] Clinical Sciences Malmo, Clinical and Molecular Osteoporosis Research Unit, Lund University, S-205 02 Malmö, Sweden; parvaneh.ebrahimi@med.lu.se (P.E.); fiona.mcguigan@med.lu.se (F.M.); kristina.akesson@med.lu.se (K.Å.)
[6] Department of Orthopedics, Skane University Hospital, S-205 02 Malmö, Sweden
[*] Correspondence: nancy.hadley-miller@cuanschutz.edu; Tel.: +1-303-724-0357

Citation: Carry, P.M.; Terhune, E.A.; Trahan, G.D.; Vanderlinden, L.A.; Wethey, C.I.; Ebrahimi, P.; McGuigan, F.; Åkesson, K.; Hadley-Miller, N. Severity of Idiopathic Scoliosis Is Associated with Differential Methylation: An Epigenome-Wide Association Study of Monozygotic Twins with Idiopathic Scoliosis. *Genes* **2021**, *12*, 1191. https://doi.org/10.3390/genes12081191

Academic Editors: Philip Giampietro and Cathy L. Raggio

Received: 16 July 2021
Accepted: 29 July 2021
Published: 30 July 2021

Publisher's Note: MDPI stays neutral with regard to jurisdictional claims in published maps and institutional affiliations.

Abstract: Epigenetic mechanisms may contribute to idiopathic scoliosis (IS). We identified 8 monozygotic twin pairs with IS, 6 discordant (Cobb angle difference > 10°) and 2 concordant (Cobb angle difference ≤ 2°). Genome-wide methylation in blood was measured with the Infinium HumanMethylation EPIC Beadchip. We tested for differences in methylation and methylation variability between discordant twins and tested the association between methylation and curve severity in all twins. Differentially methylated region (DMR) analyses identified gene promoter regions. Methylation at cg12959265 (chr. 7 *DPY19L1*) was less variable in cases (false discovery rate (FDR) = 0.0791). We identified four probes (false discovery rate, FDR < 0.10); cg02477677 (chr. 17, *RARA* gene), cg12922161 (chr. 2 *LOC150622* gene), cg08826461 (chr. 2), and cg16382077 (chr. 7) associated with curve severity. We identified 57 DMRs where hyper- or hypo-methylation was consistent across the region and 28 DMRs with a consistent association with curve severity. Among DMRs, 21 were correlated with bone methylation. Prioritization of regions based on methylation concordance in bone identified promoter regions for *WNT10A* (WNT signaling), *NPY* (regulator of bone and energy homeostasis), and others predicted to be relevant for bone formation/remodeling. These regions may aid in understanding the complex interplay between genetics, environment, and IS.

Keywords: idiopathic scoliosis; monozygotic twin; epigenome-wide association study; DNA methylation; bone; discordant; curve severity; differentially methylated region

1. Introduction

Adolescent idiopathic scoliosis (IS) is a three-dimensional spinal deformity affecting 1–3% of otherwise normal prepubescent and adolescent individuals [1,2]. Screening programs, conservative treatment, and surgical care in the case of progressive curvatures impose significant personal, familial, financial, and societal costs across the lifetime of affected individuals. The etiology of IS remains unknown. However, it has been shown to have a strong familial component [3] with a sibling recurrence risk of 18%, and heritability estimates of approximately 87.5% [4–6].

Traditional genetic association methods including familial linkage studies [7–17], exome sequencing [18–27], and genome wide association studies (GWAS) [28–36] have resulted in a number of positive associations with IS, of which only a few loci, notably those in or near *ADGRG6* [31,37–43] and *LBX1* [21,29,30,44–57], have been replicated across multiple independent study populations [58,59]. While familial and case–control designs have added to our understanding of IS, the complex heterogenic nature of IS [58–62] has limited our understanding of the genetic underpinnings of this particular disorder. The combination of our inability to relate specific genetic variants to the biology of IS, the low reproducibility of results, increased prevalence of more severe curves among females [63], and the wide variation in phenotype has increased interest in the potential role of environmental and/or epigenetic factors in the etiology of IS [59,64,65].

One frequently studied mechanism of epigenetic regulation is DNA methylation in which a methyl group is added to the cytosine nucleotide within a DNA sequence. Typically occurring in a cytosine phosphate guanine (CpG) dinucleotide pair, methylation has the capacity to change chromatin structure and alter transcription factor binding [66]. This is a reversible event that may provide a link between genetic variation, environment, and disease [67,68]. Although tissue-specific, up to 80% of the variation in the epigenome may be due to genotype [69], therefore a challenge among epigenome wide association studies (EWAS) is determining whether the observed epigenetic phenotype association is due to environmental or genetic effects. Studying monozygotic (MZ) twins, is one way to minimize this concern. MZ twins discordant for the phenotype of interest are near perfect genetic matches, therefore their DNA methylation levels can be compared to shed light on the phenotypic expression of the disease.

Previous epigenome-wide association studies have provided evidence supporting the role of DNA methylation in numerous complex musculoskeletal diseases including osteoarthritis [70], osteoporosis [70], cerebral palsy [71], and Paget's disease of bone [72]. The role of DNA methylation in IS has not been well studied. Targeted studies have reported associations between methylation and IS near the *COMP* [73] and *PITX1* [74] genes. Two recent studies [75,76] of *ESR1* and *ESR2* methylation from paravertebral muscle tissue in females with IS supports potential interrelationship between sex hormone levels, methylation, and the clinical manifestation of IS. *ESR1* methylation levels from paravertebral muscle tissue on the concave side of curve were associated with curve severity [75], and furthermore, *ESR2* promoter methylation levels differed between concave and convex sides of the curves [76]. Epigenome-wide discovery analyses in MZ twins are limited to studies including only one [77] and two [78] MZ twin pairs discordant for IS. There is a strong need for additional epigenome wide analyses to understand the potential role of DNA methylation in IS. Therefore, the aim of this EWAS was to identify differences in DNA methylation levels between monozygotic twin pairs discordant for IS. Within twin pairs, we also aimed to determine if differences in methylation were associated with differences in curve severity.

2. Materials and Methods

2.1. Study Population

Peripheral whole blood samples were obtained from 8 female monozygotic twin pairs (n = 16 individuals) diagnosed with idiopathic scoliosis (IS). Participants were identified from an existing registry, the Genetics of Idiopathic Scoliosis project (the GenesIS project), that has been described by Baschal et al. [19]. A diagnosis of IS required that subjects had no congenital deformities or other co-existing genetic disorders and a standing anteroposterior radiograph showing a curvature of at least 10° by the Cobb method [79]. There were 6 twin discordant twin pairs (difference in primary curve Cobb angle >10°) and 2 concordant twin pairs in our study population. The difference in the primary curvature among the two concordant twin pairs was <1 and 2°, respectively (Table 1). Written informed consent and assent, when appropriate, was obtained from all study participants and/or their legal guardians in accordance with protocols approved through the Johns Hopkins School of Medicine Institutional Review Board and the University of Colorado Anschutz Medical Campus Institutional Review Board.

Table 1. Demographics and Clinical Characteristics.

Twin Pair	ID	Case Status	Curve Degree [†]	Age [*]
		Discordant		
1	15505	Case	48	44.8
	15501	Control	41/37	44.8
2	15643	Case	75	81.4
	15642	Control	35	81.9
3	16012	Case	50	16.3
	16009	Control	22	16.3
4	16615	Case	52/48	33.4
	16611	Control	32/28	33.3
5	18453	Case	34	5.6
	18454	Control	23	5.6
6	19294	Case	56/43	48.7
	19292	Control	12	48.7
		Concordant		
7	16037	NA	45	42.8
	16038	NA	45	42.8
8	18721	NA	26/33	25.3
	18722	NA	29/31	25.5

[†] Major curve(s), [*] Age at sample collection.

2.2. DNA Methylation Processing

Genomic DNA was extracted from fresh whole blood using a standard phenol-chloroform purification procedure [80]. DNA was further purified using the Zymo DNA Clean & Concentrator kit, followed by Nanodrop quantification. Approximately 1 μg DNA was bisulfite converted using the Zymo EZ DNA Methylation kit (Zymo Research, Irvine, CA, USA). The precipitated DNA was dispensed onto the Infinium MethylationEPIC 850 K BeadChip (Illumina). The EPIC chip provides methylation measurements across the genome. All sequencing was performed at the University of Colorado Genomics and Microarray Shared Resource. Twin pairs were consistently arranged in sequential order on the plate to minimize within and between batch effects.

The 850 K Infinium platform includes 866,836 annotated probes representing individual cytosine phosphate guanine (CpG) probe sites. Data were normalized using the SWAN normalization method implemented within the minfi R package [81]. Standard quality control checks were performed at both the sample and probe level using the minfi R package [81]. Probes or samples that failed any of the following standard filtering procedures were treated as missing data: probe was not detectable above background noise ($n = 776$ [<0.1%]), probes with a low bead count (>5% of samples with a beadcount <3) ($n = 17,272$ [2%]), cross reactive probes ($n = 17,028$ [2%]), dropped during initial quality control processing ($n = 528$ [<0.1%]). Probes located on sex chromosomes ($n = 15,648$ [1.8%]) and probes that included known SNPs ($n = 165,678$ [19.1%]) were also excluded from the analysis. Consistent with recommendations of Logue et al. [82], we filtered low variability probes known to be associated with poor reproducibility. We filtered out all probes with β range value < 0.05%, $n = 145,163$ (16.7%). In total, 504,743 probes met the inclusion criteria and were included in subsequent steps.

The microarray methylation measurements were performed in two batches. The ComBat function implemented in the sva R package [83] was used to adjust the SWAN normalized M values for potential batch effects. The batch adjusted M-values were used in all statistical analyses. The current Illumina annotation for the 850 K platform is on hg19. Only probes which match 100% to a single location were used for further analyses.

We used three analytical approaches to identify individual CpG sites (Figure 1): (A) we performed a discordant differentially methylated position (DMP) analysis, testing for differences in methylation (batch adjusted M-values) between six twin pairs where the difference in the primary spinal curvature Cobb angle was >10°, (B) Using all twins, we tested the association between within twin pair differences in methylation ($\Delta_{methylation}$ = methylation levels in affected/more severe twin–methylation levels unaffected/less severe twin) and differences in curve severity within twin pairs (Δ_{curve} = primary curve magnitude in affected/more severe twin–primary curve magnitude is unaffected/less severe twin), (C) We tested for differences in methylation variability between the discordant twin pairs (Differentially Variable Position [DVP] analysis). The methods workflow is outlined in Figure 1.

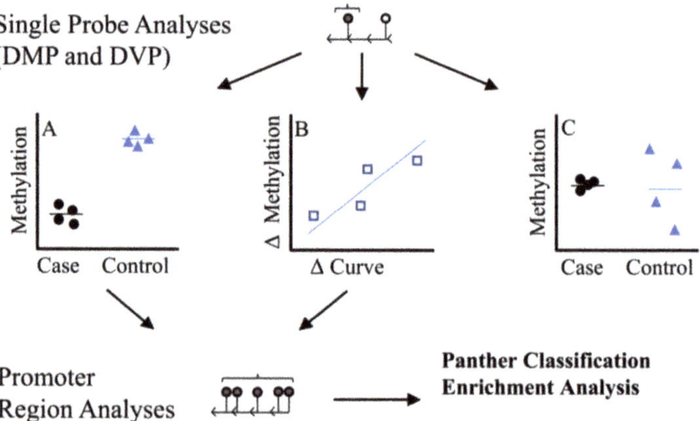

Figure 1. Study Methodologic Workflow. Differentially methylated position (DMP) analyses were used to identify single probes. We used two DMP strategies (1) Discordant DMP Analysis (**A**), differences in methylation between cases (twin with more severe IS) relative to controls (twin with less sever IS) (2) Severity DMP Analysis (**B**) looked at association between difference in curve severity and methylation Δ_{curve} **vs** $\Delta_{methylation}$, where $\Delta_{methylation}$ = methylation levels in affected/more severe twin–methylation levels unaffected/less severe twin and Δ_{curve} = primary curve magnitude in affected/more severe twin–primary curve magnitude is unaffected/less severe twin. We also looked at differences in variability at single probes (differentially variable position analysis, DVP) among cases compared to controls (**C**). Region analyses based on single probes from (**A,B**) were used to identify regions of consistent methylation effects within promoter regions. We only considered regions with 5 or more probes where direction of effect was consistent across all probes. DVP probes (**C**) were not considered in the region analysis due to challenges interpreting a 'consistent' direction of effect based on variability.

Methylation is tissue specific. Confounding due to differences in cell composition between cases and controls is a potential concern in epigenome-wide association studies [84]. Cell proportions were estimated from methylation values using the minifi [81] R package. The distribution of CD8T, CD4T, B Cells, natural killer Cells, monocytes, and neutrophils was similar in cases compared to controls (Appendix A, Table A1). Due to the small sample size and similar distribution of cell proportions across all individuals, we did not adjust for cell type in subsequent analyses.

2.3. Statistical Analysis

We used descriptive statistics to summarize the demographic and clinical characteristics of all subjects included in the study. Wilcoxon signed rank tests were used to compare the distribution of the five cell types between case and control twins (Table 1 and Appendix A, Table A1). Normalized M-values were used in all analyses. β-values and percent methylation were also reported to facilitate biological interpretation. In the discordant DMP analysis, paired t-tests were used to test for differences in M values between discordant twins (n = 6 pairs, n = 12 individuals). The individual with the more severe curvature was designated as the "case", and the corresponding twin with the less severe curvature was designated as the "control." For the DVP discordant analysis, a regularized version of Bartlett's test was used to identify differentially variable probes between discordant twin pairs. For the curve severity DMP analysis involving all twin pairs, linear regression models were used to test the association between differences in methylation ($\Delta_{methylation}$) and differences in curve severity within twin pairs (Δ_{curve}). To account for multiple testing, false-discovery rate (FDR) adjusted p values were estimated using the algorithm described by Benjamini and Hochberg [85].

An exploratory analysis was used to identify differentially methylated regions (DMR) using the mCSEA [86] R package among promoter regions with a minimum of 5 probes. Significance was assessed based on 100,000 permutations. The DMR analysis was implemented for the discordant DMP and the curve severity DMP analyses. Based on the exploratory nature of the DMR analysis, only regions where 100% of probes were in the same direction of effect and the FDR adjusted p value was < 0.05 were considered significant.

Functional gene overrepresentation analysis was performed based on the results of the discordant DMP and curves severity DMP analyses. The top DMRs, nominal p value <0.001, were included in the DMR overrepresentation analysis. Overrepresentation analyses of the severity and discordant gene lists were conducted using PANTHER v16.0, test release 20,210,224 (PantherDB.org) [87–89]. Custom gene background inputs were used in accordance with gene promoter region DMRs analyzed within the Infinium Human-Methylation EPIC Beadchip platform. The Annotation Data Sets Gene Ontology (GO) Cellular Component Complete, GO Molecular Function Complete, and GO Biological Process Complete were analyzed separately, each using a Fisher's Exact Test and Bonferroni correction. We report overrepresented GO terms with a Bonferroni-adjusted $p < 0.05$. Significant terms were reviewed; parent terms were manually removed and REVIGO [90] was used to eliminate redundant terms.

2.4. Prioritization of Candidates: Blood Methylation Correlation

While the tissue of origin for IS has not been determined, the primary clinical manifestation is that of the bony spinal column. Ebrahimi et al. [91] previously reported on methylation levels in whole blood versus trabecular bone. We utilized these data to prioritize our methylation candidates as ones that may have a functional role in bone. We reviewed the distribution of correlation coefficients, representing the strength of association between methylation levels in blood versus bone, among all probes evaluated by Ebrahimi et al. [91] Probes were considered strongly correlated if the correlation coefficients exceeded the 75th percentile among all probes evaluated by Ebrahimi et al. [91] ($\rho = 0.49$). We then reviewed the correlation coefficients for probes identified as candidates in our DMP and DMR analyses. For the DMR analysis, we reported the maximum correlation coefficient among all probes in the DMR as well as the percentage of strongly, positively correlated probes across the entire region.

3. Results

3.1. Demographics

The study population included 6 discordant and 2 concordant female monozygotic twin pairs with idiopathic scoliosis (IS, see Table 1). The average age among all individuals at the time of blood acquisition was 37.3 years ($\pm$22.5). The average Cobb angle of the primary curve was 39.6° ($\pm$15.3). The average difference in age between twin pairs at the time of sample acquisition was 1.2 months (range: 0 to 6 months). The average difference in curve severity among all twins was 19° (range: 0 to 44°). Among discordant twins, the average difference in curve severity was 25° (range: 11 to 44°).

3.2. Discordant Curvature Analysis

In the discordant analysis, none of the individual CpG probes were significant at the FDR adjusted $p = 0.10$. Differentially methylated region (DMR) analyses identified 200 promoter regions (containing 5–14 CpG sites) that were significant at the FDR adjusted p value of 0.05. Among these regions, 58 DMRs included probes/sites where the direction of effect (hypermethylation or hypomethylation) was consistent across 100% of the probes (Appendix A, Table A2). The most significant DMR represented a region on chr. 14 in the promoter region for the *BCL2L2-PABPN1* gene (FDR adjusted $p = 0.0113$, see Figure 2). Using the Panther enrichment algorithm with these 58 DMRs, we identified 1 significantly enriched gene ontology (GO) term (Appendix A, Table A3).

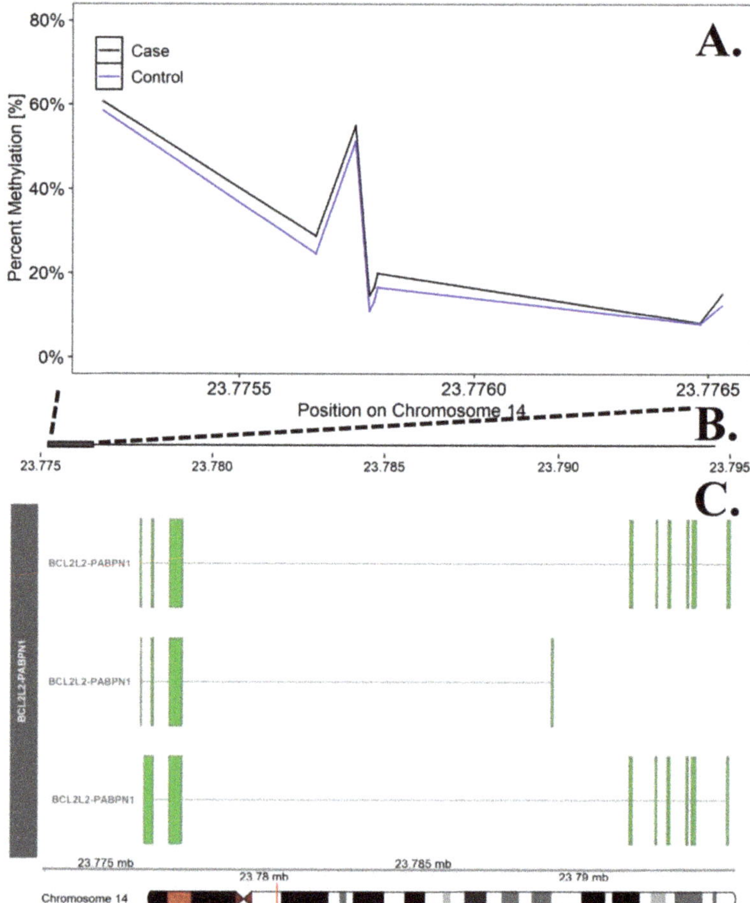

Figure 2. Differentially Methylated Region in the *BCL2L2-PABN1* Promoter Region on Chromosome 14. The top panel (**A**) describes differences in percent methylation between cases and controls at each probe included in the promoter region for *BCL2L2-PABN1*. This region was the most significant DMR in the discordant twin analysis. The X axis represents the position (mb) of the probes. The middle panel (**B**) represents the location of promoter region (solid square) relative to the entire gene, represented in the bottom panel. Multiple known isoforms of *BCL2L2-PABN1* are represented in the bottom panel (**C**), boxes represent exons and lines represent introns. The red line on the ideogram, bottom of the figure, represents location of the region within the chromosome.

In addition to differences in methylation levels, we looked for differences in methylation variability, which may provide valuable information about the heterogenous environmental exposures that contribute to disease etiology. In the differentially variable position (DVP) analysis, methylation variability at cg02477677 was significantly lower (FDR adjusted *p* value = 0.0791) in cases with a more severe curve compared to unaffected or less severely affected controls (Figure 3). The cg02477677 CpG probe is an open sea probe on chr. 7 near *DPY19L1*.

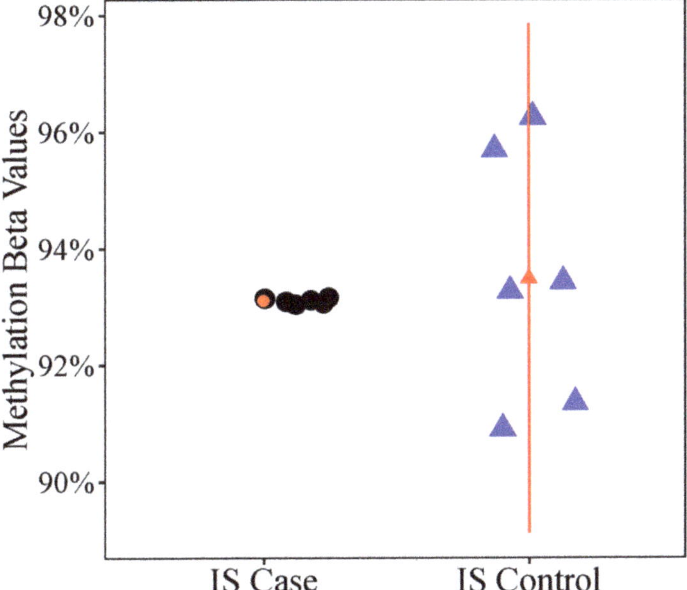

Figure 3. Scatter Plot of Differential Variability Between Case and Control Twins. Methylation levels (% methylation or β values) at the cg12959265 probe, an open sea probe near the *DPY19L1* gene on chromosome 7 in IS cases and IS controls. The triangle represents the mean % methylation and error bars represent +/− 1 standard deviation. The plot illustrates the large difference in variability at cg12959265 in IS cases vs. IS controls.

3.3. Curve Severity Analysis

We also tested whether the difference in methylation between cases and controls was associated with the difference in curve severity between cases and controls. We identified 4 CpG sites where the difference in methylation was significantly associated (FDR adjusted *p* value = 0.0753) with the difference in curve severity (Figure 4). At each of these open sea probes, increasing disparity in curve severity between cases and controls was associated with a pattern of hypomethylation.

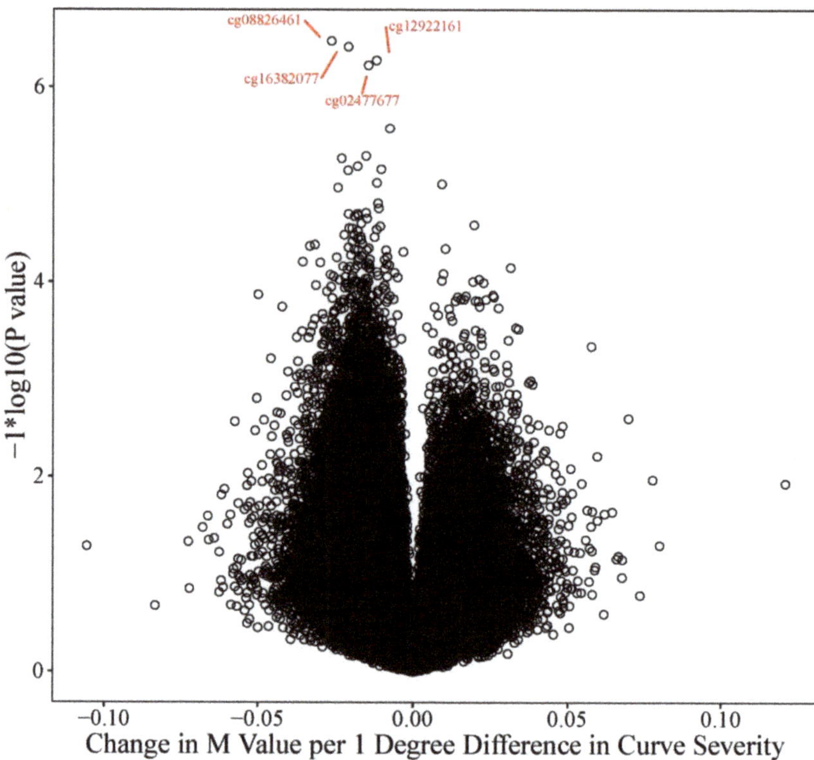

Figure 4. Volcano Plot: Curve Severity Analysis The volcano plot describes the effect size and *p* value for every probe tested in the curve severity analysis. The Y axis represents the $-\log10(p$ values) and the X axis represents the change in M value for every one-degree difference in curve severity between the twin pairs for each of the respective probes tested. Increasing curve disparity was more often associated with hypomethylation (decreased M values, left or negative side of the plot) than hypermethylation. The four FDR significant probes (FDR adj *p* = 0.0753) are highlighted in red, cg08826461 (nominal *p* value = 3.37×10^{-7}), cg16382077 (nominal *p* value = 3.85×10^{-7}), cg12922161 (nominal *p* value = 5.32×10^{-7}), and cg02477677 (nominal *p* value = 5.97×10^{-7}).

For every 1 degree increase in the difference in curve severity in cases compared to controls, batch adjusted M-values decreased by an average of between 0.012 to 0.027 units. Significant probes included cg02477677 (slope: 0.015 units, near the *RARA* gene on chr. 17, nominal *p* value = 5.97×10^{-7}); cg08826461 (slope: 0.027 units chr. 2, does not map to a known gene, nominal *p* value = 3.37×10^{-7}), cg12922161 (slope: -0.012 chr. 2, maps to *LOC150622*, nominal *p* value = 5.32×10^{-7}), and cg16382077 (slope: 0.021 units, chr. 7, does not map to a known gene, nominal *p* value = 3.85×10^{-7}).

The differentially methylated region analyses identified n = 197 promoter regions (ranging from 5 to 34 CpG sites) significant at the FDR adjusted *p* value of 0.05. Among these, 28 regions included probes where the direction of effect (difference in curve severity was either positively or negatively associated with the difference in methylation between twin pairs) was consistent across 100% of the probes (Appendix A, Table A4). The top DMR consisted of 34 probes on chr. 20 within the promoter region for the *NNAT* gene (FDR adjusted *p* value = 0.0237, Figure 5). Using Panther, we identified 15 significantly enriched ontologies (Appendix A, Table A5). The top biological process terms included pituitary gland development (GO:0021983) and anterior/posterior pattern specification (GO:0009952).

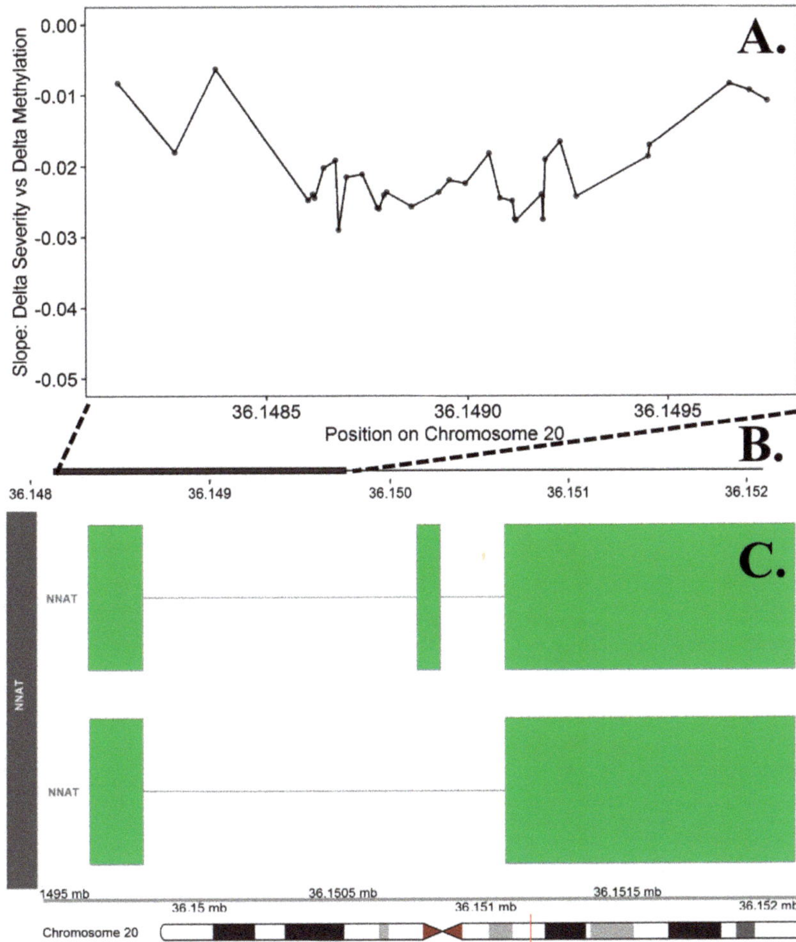

Figure 5. Differentially Methylation Region in the *NNAT* Promoter Region on Chromosome 20. The top panel (**A**) presents the slope estimates from the curve severity analysis that represent the change in methylation between cases and controls per one-degree change in curve severity at each of the 34 probes included in the promoter region for the *NNAT* gene. This region was the most significant DMR in the curve severity analysis. The X axis represents the position (mb) of the probes. The middle panel (**B**) represents the location of promoter region (solid square) relative to the entire gene, represented in the bottom panel. Multiple known isoforms of the *NNAT* gene are represented in the bottom panel (**C**), boxes represent exons and lines represent introns. The red line on the ideogram, bottom of the figure, represents location of the region within the chromosome.

3.4. Candidate Prioritization

Ebrahimi et al. [91] conducted an epigenome-wide analysis to measure the correlation between methylation levels in whole blood and trabecular bone. We used these correlation coefficients to prioritize the methylation candidates identified in our study. Among the four probes identified as candidates in our DMP analysis (cg02477677, cg12922161, cg08826461, and cg16382077), only one probe, cg08826461, was strongly correlated with bone tissue ($\rho = 0.494$, FDR adjusted *p* value = 0.41329). Among DMRs, we prioritized candidate regions where either one or more probes within the DMR was significantly correlated with bone (FDR adjusted p value of less than 0.10), or, greater than 50% of probes included in the region were strongly correlated with bone. We identified 13 priority regions based on the discordant DMR analysis and 8 priority candidate regions based on the severity DMR analysis (Table 2).

Table 2. Priority DMRs Based on High Correlation with Bone.

Nearest Gene	Chr.	Start Position	End Position	Number of Probes	DMR Nominal p Value	DMR FDR p Value	Maximum Bone Correlation	FDR Adj. p Value for Maximum Bone Correlation	Percent Strongly Positively Correlated Probes Across DMR
				Discordant DMR Analysis					
WNT10A	chr2	219,744,145	219,745,748	9	2.17×10^{-5}	0.0113	0.83	0.0307	33.3%
CRISP2	chr6	49,681,178	49,681,774	11	2.19×10^{-5}	0.0113	0.89	0.0128	100.0%
RBPJL	chr20	43,934,854	43,935,551	12	2.20×10^{-5}	0.0113	0.72	0.1048	66.7%
KDM2B	chr12	122,018,574	122,020,205	14	2.21×10^{-5}	0.0113	0.83	0.0336	50.0%
IL27	chr16	28,518,114	28,519,597	9	4.34×10^{-5}	0.0156	0.75	0.0844	33.3%
CA14	chr1	150,229,143	150,230,345	9	6.51×10^{-5}	0.0196	0.78	0.0585	33.3%
C9orf47	chr9	91,604,473	91,606,140	12	2.64×10^{-4}	0.0318	0.79	0.0518	18.2%
STAB1	chr3	52,528,714	52,529,393	8	3.04×10^{-4}	0.0329	0.79	0.0534	12.5%
ACY3	chr11	67,415,183	67,418,365	8	3.26×10^{-4}	0.0329	0.72	0.1045	62.5%
MPG	chr16	125,896	128,009	11	3.29×10^{-4}	0.0329	0.82	0.0368	10.0%
ESM1	chr5	54,281,198	54,282,459	13	3.97×10^{-4}	0.0360	0.79	0.0526	61.5%
TMEM219	chr16	29,972,752	29,974,294	6	5.57×10^{-4}	0.0431	0.77	0.0667	66.7%
CREBBP	chr16	3,930,112	3,931,489	5	6.16×10^{-4}	0.0457	0.76	0.0725	80.0%
				Severity DMR Analysis					
GANC	chr15	42,565,522	42,566,390	7	3.38×10^{-4}	0.0357	0.88	0.0153	28.6%
NME3	chr16	1,821,559	1,822,346	8	3.59×10^{-4}	0.0366	0.76	0.0729	37.5%
SLC6A5	chr11	20,619,598	20,621,109	8	3.59×10^{-4}	0.0366	0.80	0.0489	28.6%
RAB22A	chr20	56,883,532	56,885,003	8	4.38×10^{-4}	0.0391	0.78	0.0594	25.0%
ACTN4	chr19	39,137,911	39,138,334	7	4.89×10^{-4}	0.0407	0.84	0.0298	33.3%
NPY	chr7	24,322,873	24,324,570	8	5.58×10^{-4}	0.0421	0.84	0.0276	37.5%
RAB38	chr11	87,908,558	87,909,729	9	6.33×10^{-4}	0.045	0.73	0.0995	44.4%
COPB1	chr11	14,521,639	14,522,617	6	7.79×10^{-4}	0.0495	0.88	0.0149	50.0%

Maximum Bone Correlation = maximum correlation coefficient representing strength of correlation between blood and bone CpG sites across all sites included in the DMR (from Ebrahimi et al.), FDR p Value for Maximum Bone Correlation = FDR adjusted p value for maximum correlation coefficient across all sites included in the DMR (from Ebrahimi et al.)., Percent Strongly Correlated Probes Across DMR = Percentage of probes across the entire region where the correlation coefficient representing strength of correlation between blood and bone CpG sites is greater than 75th percentile among all probes tested in Ebrahimi et al.

4. Discussions

We utilized an epigenome-wide association study (EWAS) in monozygotic (MZ) twins to identify individual methylation sites and regions across the genome relevant to idiopathic scoliosis (IS). We identified a single CpG site where methylation variability was different between discordant MZ twins and identified CpG sites where increasing curve severity was more often associated with hypomethylation. Differentially methylated region (DMR) analyses identified multiple regions potentially indicative of unique methylation changes within twin pairs discordant for IS as well as unique methylation patterns associated with curve severity. Integration of a peripheral blood/bone methylation dataset allowed us to prioritize regions and sites based on their potential relevance to the IS disease process in bone. Collectively, these results highlight both new and previously reported pathways related to IS curve progression including those involved in neurogenesis and body segmentation.

Differential variability in methylation represents large shifts in methylation that may reflect differential epigenetic and/or environmental effects in cases relative to controls. Differential variability analyses in disease-discordant monozygotic twins have been used to identify DNA methylation signatures associated with Type 1 Diabetes [92] and rheumatoid arthritis [93]. In our analysis, variability at cg0247767, chr. 7 near *DPY19L1*, was significantly different between discordant twin pairs. DPY19L1 is a transmembrane protein localized in the endoplasmic reticulum that regulates neuronal migration and extension during development [94,95]. Zebrafish with mutations within this gene demonstrate spinal axial curvatures [96]; however, the phenotype has not been studied in detail.

We also identified four individual CpG sites that were associated with curve severity (cg02477677, cg12922161, cg08826461, and cg1638077). At these sites, an increase in curve severity tended to be associated with a decrease in methylation (hypomethylation) within the twin pairs. One of the probes, cg12922161, maps to a location near *LOC150622/SILC1*, a non-coding RNA gene. Although the function of this non-coding RNA is not well known, it has been shown to regulate neuron outgrowth and neuroregeneration via cis-acting activation of the transcription factor SOX11 [97]. To date, select non-coding RNAs as non-protein coding regulatory transcripts within the genome have been hypothesized to functionally participate in the initiation and progression of IS [98]. The cg02477677 probe was also associated with curve severity. This probe maps to a region near *RARA* on chr. 17, which encodes a transcription factor for the retinoic acid receptor protein During development, RA signaling plays an essential role in embryonic body axis extension, left-right somite synchronization, and limb development [99]. It is a central mechanism underlying bilateral symmetry during development of the mouse embryo [100]. Right-left asymmetries have previously been hypothesized as a potential contributing factor to IS based on the increased prevalence of IS among individuals demonstrating vestibular and posterior basicranial morphological asymmetries in MRI cross-sectional studies [101–103].

Regions of differentially methylated probes (DMRs) may have more important functional implications than methylation levels at a single CpG site, particularly in promoter regions which are areas of the genome where methylation levels tend to be negatively correlated with gene expression [67,68]. Based on the discordant analysis, we identified 58 significant DMRs in known promoter regions, the most significant region included methylation sites with the promoter region for the *BCL2L2-PABPN1* gene on chr. 14. *BCL2L2-PABPN1* is a paralog of *PABPN1*, which is associated with the development of oculopharyngeal muscular dystrophy, a disease characterized by muscular weakness in eyelids, pharyngeal musculature, and limbs [104]. In the curve severity analysis, we identified 28 significant DMRs in known promoter regions, the most significant included probes with the promoter region for the *NNAT* gene on chr. 20. This gene is important for brain development and implicated in neurodegenerative diseases including anterior horn disease [105]. The paternal copy of the *NNAT* gene is exclusively expressed due to imprinting [106,107]. This is potentially relevant to IS given the sex bias of progressive curvatures (females > males), and the higher percentage of affected offspring from paternal IS cases compared to maternal IS cases (80% vs. 56%) [63].

Enrichment analyses of the top DMRs from both the discordant and curve severity results revealed both broad, non-specific ontologies and select ontologies related to neurogenesis, axon guidance and neuron differentiation, all of which can be supported by current literature from both family-based exome sequencing and genome-wide association (GWAS) studies [49,50]. Combined with results in the current study, these data suggests a potential role of neuropathological processes underlying the development and progression of IS.

The complex genetic architecture underlying IS is further complicated by the lack of a clear tissue target. Despite the major clinical manifestation and therapeutic dilemma of the spinal curvature, IS ultimately affects multiple tissue types, one of which is bone. To understand the potential functional relevance of our methylation results in osseous tissue, we reviewed overlap between our identified CpG sites and those reported by Ebrahimi et al. [91] correlating the same methylation markers in matched blood and trabecular bone samples. We identified 21 regions where the DNA methylation in our dataset was correlated with methylation in bone tissue, and therefore, could potentially be considered biologically relevant (Table 2). Annotation of two of the regions implicated the *NPY* gene on chr. 7 and the *WNT10A* gene on chr. 2. The *NPY* gene encodes a neuropeptide expressed throughout the central and peripheral nervous systems [108] and is an essential regulator of bone homeostasis and metabolism [109]. NPY is also a local regulator of osteoblastic lineage and is responsive to mechanical stimuli with potential roles in fracture healing and osteoarthritis [109]. The *WNT10A* gene is a member of the WNT gene class and functions within the WNT10A/β-catenin signaling pathway in regulation of adult epithelial proliferation [110,111], mesenchymal stem cell regulation by stimulating osteoblastogenesis [112], coordination of vertebrate segmentation, and motile cilia function [113].

The role of DNA methylation in IS has not been well studied in current literature outside of two targeted studies and extremely small genome-wide discovery analyses. Mao et al. [73] reported IS cases were associated with increased methylation near the promotor region for the *COMP* gene on chr. 19 and more importantly, decreased COMP expression. Shi et al. [74] identified significantly higher levels of methylation in IS cases versus controls in a region near the pituitary homeobox-1 (*PITX1)* gene, a homeobox transcriptional regulator that plays a role in maintenance of side-to-side musculoskeletal symmetry during development [114]. In our study, methylation levels in the promoter regions for the *PITX1* and/or the *COMP* gene were not differentially methylated in the discordant and/or curve severity analysis. However, pituitary gland development (GO:0021983) and the anterior/posterior pattern specification (GO:0009952) ontologies represented the top 2 most enriched terms in the curve severity DMR analysis (see Appendix A, Table A5).

Meng et al. [78] and Liu et al. [77] conducted the only other IS EWAS studies in the current literature. They used a similar strategy, testing for methylation differences in peripheral blood samples in a discovery cohort (1 and 2 MZ twin pairs, respectively) discordant for curve progression. A second cohort consisting of individuals with IS versus controls was used to confirm methylation sites or regions identified in the discovery cohort. Meng et al. [78] identified a single probe cg01374129 (near the *HSA2* gene) that was significantly hypomethylated in the progressive group compared to the non-progressive group. Liu et al. [77] identified a DMR near the promoter region for the *NDN* gene that was significantly associated with IS. These studies were limited in that the discovery EWAS was performed in a very limited number of individuals. Our study builds on these initial findings with added methylation data from MZ twins both concordant and discordant in their spinal curves. Our complementary analyses provide a list of candidate sites and regions across the genome that may assist in the development of prognostic tools capable of identifying individuals at risk for curve progression. Methylation is tissue specific, we were also able to prioritize hits identified in our analysis based on known correlation with methylation in bone tissue. Additional validation in larger cohorts is needed to confirm these methylation markers as relevant and explore their utility in the clinical setting.

Limitations

Our study includes several limitations. First, the samples were obtained after disease onset. We cannot exclude the possibility that differences in methylation were caused by changes in curve severity. Although samples within each twin pair were obtained no more than six months apart and thus age at sample acquisition was balanced across the twin pairs, there was substantial heterogeneity in age at sample acquisition across the twin pairs. This is potentially problematic if age modifies the effect of methylation on curve progression. Similarly, we used a case–control design. Cases and controls were defined based on curve pattern information available at the time of sample acquisition. Misclassification of controls is possible if spinal progression occurred over the lifetime of the individual.

5. Conclusions

A better understanding of the genetic, epigenetic, and environmental factors underlying IS onset and/or curve progression has significant clinical implications [115,116]. DNA methylation markers may provide value as a prognostic tool for predicting both the initiation and progression of this disorder and furthermore, may also aid in the identification of homogenous subgroups of individuals allowing for more personalized treatment algorithms. In the current study, we identified methylation at specific sites across the genome. Differentially methylation region (DMR) promoter enrichment analyses identified several biologically relevant ontologies related to pituitary gland development, body segmentation and neuronal differentiation. We prioritized the DMR candidates based on known correlation between methylation in blood versus bone. Priority candidates include DMRs in promoter regions related to the WNT signaling pathway (*WNT10A*), a signaling pathway that is relevant to bone formation and remodeling [117], and neuropeptide Y (*NPY*), a regulator of bone and energy homeostasis [109]. This information allows for further targeted studies aimed at understanding the functional relevance of these findings in relation to IS and axial spinal development, alignment, and side-to-side symmetry.

Author Contributions: Conceptualization: P.M.C. and N.H.-M. Methodology: P.M.C., L.A.V. and N.H.-M. Software: P.M.C., L.A.V., and G.D.T. Validation: N/A. Formal analysis: P.M.C. Investigation: N.H.-M., E.A.T. and C.I.W. Resources: N.H.-M., P.E., F.M. and K.Å. Data curation: P.M.C. and E.A.T. Writing—original draft preparation: P.M.C. Writing—review and editing: P.M.C., E.A.T., G.D.T., L.A.V., C.I.W., P.E., F.M., K.Å. and N.H.-M. Visualization: P.M.C. Supervision: N.H.-M. Project administration: E.A.T. and N.H.-M. Funding acquisition: N.H.-M. All authors have read and agreed to the published version of the manuscript.

Funding: This work was partially funded by NIH NIAMS R01 AR068292.

Institutional Review Board Statement: Written informed consent was obtained from study subjects who were enrolled in accordance with protocols approved by the Johns Hopkins School of Medicine Institutional Review Board and the University of Colorado Anschutz Medical Campus Institutional Review Board (Colorado Multiple Institutional Review Board, Studies #06-1161 and 07-0417). All procedures involving human participants were performed in accordance with the ethical standards of these institutional review boards, the 1964 Declaration of Helsinki and its later amendments, or comparable ethical standards.

Informed Consent Statement: Written informed consent was obtained for all study participants in accordance with protocols approved through the Johns Hopkins School of Medicine Institutional Review Board and the University of Colorado Anschutz Medical Campus Institutional Review Board.

Data Availability Statement: Data is available from the author upon reasonable request.

Conflicts of Interest: The authors declare no conflict of interest.

Appendix A

Table A1. Proportion of Blood Cell Populations in IS Cases and Controls.

	Cases			Control			
	Median	**Min**	**Max**	**Median**	**Min**	**Max**	***p* Value**
CD8T Cell (%)	11.5%	8.3%	33.6%	12.2%	6.9%	18.7%	0.1953
CD4T Cell (%)	18.5%	7.7%	28.3%	14.3%	9.3%	18.0%	0.1953
Natural Killer Cell Count (%)	5.3%	2.2%	8.8%	5.7%	1.9%	8.4%	0.7422
Bcell Count (%)	5.8%	3.5%	10.3%	6.2%	2.9%	10.3%	0.6406
Monocyte Cell Count (%)	7.5%	3.4%	9.5%	8.0%	5.1%	10.9%	0.6406
Neutrophil Cell Count (%)	55.3%	25.7%	67.2%	58.4%	52.0%	72.9%	0.1953

Table A2. Analysis of Discordant Curve Severity in Twin Pairs Identified 57 Differentially Methylated Regions where Hyper- or Hypo-methylation was Consistent Across all Probes in the Region.

Feature/Gene	Chr.	Start Position	End Position	Number of Probes	Nominal p Value	FDR p Value	Maximum Bone Correlation	FDR p Value for Maximum Bone Correlation	Percent Strongly Positively Correlated Probes Across DMR
WNT10A	chr2	219,744,145	219,745,748	9	2.17×10^{-5}	0.0113	0.83	0.0307	33.3%
BCL2L2 *	chr14	23,775,206	23,776,530	8	2.17×10^{-5}	0.0113	0.50	0.4069	14.3%
CRISP2	chr6	49,681,178	49,681,774	11	2.19×10^{-5}	0.0113	0.89	0.0128	100.0%
SLFN13	chr17	33,773,921	33,777,219	11	2.19×10^{-5}	0.0113	-0.27	0.7380	0.0%
RBPJL	chr20	43,934,854	43,935,551	12	2.20×10^{-5}	0.0113	0.72	0.1048	66.7%
KDM2B	chr12	122,018,574	122,020,205	14	2.21×10^{-5}	0.0113	0.83	0.0336	50.0%
MS4A3	chr11	59,822,727	59,828,426	13	2.21×10^{-5}	0.0113	-0.57	0.2917	7.7%
GPR21	chr9	125,794,756	125,797,284	14	2.21×10^{-5}	0.0113	0.38	0.5940	0.0%
IL27	chr16	28,518,114	28,519,597	9	4.34×10^{-5}	0.0156	0.75	0.0844	33.3%
AZU1	chr19	826,359	827,821	9	4.34×10^{-5}	0.0156	-0.28	0.7263	0.0%
CA14	chr1	150,229,143	150,230,345	9	6.51×10^{-5}	0.0196	0.78	0.0585	33.3%
HTRA4	chr8	38,830,814	38,831,857	11	6.58×10^{-5}	0.0196	0.29	0.7117	0.0%
ESRP2	chr16	68,269,763	68,271,177	13	6.62×10^{-5}	0.0196	0.35	0.6385	0.0%
ELANE	chr19	850,975	852,311	8	8.68×10^{-5}	0.0199	0.45	0.4788	0.0%
CLDN15	chr7	100,880,751	100,882,286	10	8.74×10^{-5}	0.0199	0.68	0.1542	20.0%
RUNX1	chr21	36,259,179	36,422,112	14	8.85×10^{-5}	0.0199	0.60	0.2518	21.4%
NFE2	chr12	54,689,278	54,696,210	14	8.85×10^{-5}	0.0199	-0.36	0.6204	0.0%
C6orf229	chr6	24,799,059	24,799,757	5	9.45×10^{-5}	0.0199	0.70	0.1293	20.0%
SLC25A16	chr10	70,287,181	70,287,493	6	1.07×10^{-5}	0.0216	-0.45	0.4813	0.0%
EPHA2	chr1	16,482,553	16,483,528	8	1.09×10^{-4}	0.0216	0.47	0.4470	0.0%
TNFSF13	chr17	7,460,690	7,462,249	12	1.32×10^{-4}	0.0223	0.65	0.1888	8.3%
UBASH3A	chr21	43,822,540	43,823,863	6	1.69×10^{-4}	0.0252	-0.64	0.1916	0.0%
C7orf49	chr7	134,852,662	134,855,381	9	1.74×10^{-4}	0.0256	0.59	0.2703	11.1%
ADAP2	chr17	29,247,612	29,248,848	9	1.95×10^{-4}	0.0266	-0.39	0.5675	0.0%
AMT	chr3	49,459,855	49,461,563	11	1.97×10^{-4}	0.0266	0.37	0.5997	0.0%
C9orf47	chr9	91,604,473	91,606,140	12	2.64×10^{-4}	0.0318	0.79	0.0518	18.2%
CD3D	chr11	118,213,272	118,214,927	8	2.78×10^{-4}	0.0318	-0.20	0.8262	0.0%
HSPB6	chr19	36,247,867	36,248,907	8	2.82×10^{-4}	0.0318	0.24	0.7811	0.0%
STAB1	chr3	52,528,714	52,529,393	8	3.04×10^{-4}	0.0329	0.79	0.0534	12.5%
ACY3	chr11	67,415,183	67,418,365	8	3.26×10^{-4}	0.0329	0.72	0.1045	62.5%
MPG	chr16	125,896	128,009	11	3.29×10^{-4}	0.0329	0.82	0.0368	10.0%
KLRD1	chr12	10,455,788	10,460,639	8	3.34×10^{-4}	0.0329	-0.73	0.1030	12.5%
FES	chr15	91,427,184	91,428,456	10	3.50×10^{-4}	0.033	0.39	0.5737	0.0%
ESM1	chr5	54,281,198	54,282,459	13	3.97×10^{-4}	0.036	0.79	0.0526	61.5%
CTSG	chr14	25,045,625	25,046,267	6	4.28×10^{-4}	0.0382	-0.36	0.6151	0.0%
HMGN2	chr1	26,797,576	267,987,40	6	4.71×10^{-4}	0.0403	-0.45	0.4827	0.0%
LRG1	chr19	4,540,003	4,540,782	6	4.71×10^{-4}	0.0403	-0.42	0.5212	0.0%
LOC100130933	chr17	73,641,809	73,642,991	10	4.81×10^{-4}	0.0405	0.44	0.5036	0.0%
HMGCR	chr5	74,632,477	74,637,028	5	5.10×10^{-4}	0.0419	0.71	0.1165	20.0%
RPSAP52	chr12	66,220,754	66,221,950	6	5.14×10^{-4}	0.0419	-0.77	0.0683	0.0%
MIR145	chr5	148,808,721	148,810,180	7	5.18×10^{-4}	0.0419	-0.38	0.5867	0.0%
PILRA	chr7	99,970,448	99,971,016	5	5.31×10^{-4}	0.0419	0.37	0.6031	0.0%
TMEM219	chr16	29,972,752	29,974,294	6	5.57×10^{-4}	0.0431	0.77	0.0667	66.7%
GIMAP1	chr7	150,412,503	150,415,143	5	5.86×10^{-4}	0.0451	0.33	0.6550	0.0%
CREBBP	chr16	3,930,112	3,931,489	5	6.16×10^{-4}	0.0457	0.76	0.0725	80.0%
MAST2	chr1	46,268,158	46,269,120	6	6.21×10^{-4}	0.0457	0.70	0.1248	16.7%
TAGLN3	chr3	111,717,534	111,718,245	8	6.29×10^{-4}	0.0457	0.68	0.1531	12.5%
S100P	chr4	6,694,923	6,695,698	9	6.30×10^{-4}	0.0457	-0.51	0.3861	0.0%
FAM53C	chr5	137,672,901	137,675,418	7	6.48×10^{-4}	0.0458	-0.25	0.7705	0.0%
GAMT	chr19	1,401,372	1,402,626	5	6.58×10^{-4}	0.0459	-0.35	0.6314	0.0%
H6PD	chr1	9,293,583	9,303,499	10	6.77×10^{-4}	0.0462	0.39	0.5714	0.0%
ITGAE	chr17	3,704,471	37,058,75	10	6.77×10^{-4}	0.0462	-0.45	0.4790	0.0%
CDH9	chr5	27,036,352	27,040,099	10	6.77×10^{-4}	0.0462	-0.42	0.5366	0.0%
ELP2	chr18	33,709,151	33,709,799	7	6.91×10^{-4}	0.0462	-0.22	0.8067	0.0%
LRP11	chr6	150,185,188	150,186,488	8	7.16×10^{-4}	0.0462	0.54	0.3423	12.5%
FIGLA	chr2	71,017,541	71,018,823	11	7.46×10^{-4}	0.0473	0.61	0.2324	18.2%
CAPN13	chr2	31,020,802	31,031,755	6	8.07×10^{-4}	0.0496	0.48	0.4419	0.0%

* BCL2L2-PABPN1, Maximum Bone Correlation = maximum correlation coefficient representing strength of correlation between blood and bone CpG sites across all sites within the DMR (from Ebrahimi et al.), FDR p Value For Maximum Bone Correlation = FDR adjusted p value for maximum correlation coefficient across all sites included in the DMR (from Ebrahimi et al.), Percent Positively Strongly Correlated Probes Across DMR = Percentage of probes across the entire region where the correlation coefficient representing strength of correlation between blood and bone CpG sites is greater than 75th percentile among all probes tested in Ebrahimi et al.

Table A3. Based on 58 DMRs for Discordant Curve Severity, one Significantly Enriched Ontology was Identified. Analysis of Discordant Curve Severity in Twin Pairs Identified n = 57 Differentially Methylated Regions where Hyper- or Hypomethylation was Consistent Across all Probes in the Region.

Category	Fold Enrichment	Bonferroni Adjusted p Value
Gene Ontology Cellular Component secretory granule lumen (GO:0034774)	3.8	7.29×10^{-3}

Table A4. Analysis of Curve Severity in Twin Pairs Identified 28 Differentially Methylated Regions where Slope Representing Association between Difference in Methylation and Difference in Curve Severity within each Twin Pair was Consistent Across all Probes in the Region.

Feature/Gen	Chr.	Start Position	End Position	Number of Probes	Nominal p Value	FDR p Value	Maximum Bone Correlation	FDR p Value for Maximum Bone Correlation	Percent Strongly Positively Correlated Probes Across DMR
NNAT	chr20	36,148,133	36,149,750	34	1.11×10^{-5}	0.0237	0.57	0.2898	12.1%
TMEM232	chr5	110,021,543	110,062,837	11	5.09×10^{-5}	0.0237	0.68	0.1441	30.0%
SLC22A20	chr11	64,979,837	64,981,596	9	8.43×10^{-5}	0.0252	0.66	0.1680	33.3%
PDE12	chr3	57,541,377	57,543,243	5	9.89×10^{-5}	0.0252	0.69	0.1347	40.0%
SUV420H2	chr19	55,850,082	55,852,507	5	1.98×10^{-4}	0.0299	0.70	0.1324	40.0%
CYR61	chr1	86,045,347	86,046,661	10	2.22×10^{-4}	0.0299	0.71	0.1148	40.0%
ZNF440	chr19	11,924,860	11,925,219	7	2.63×10^{-4}	0.0333	0.63	0.2110	14.3%
LOC150622	chr2	6,072,139	6,072,801	5	3.16×10^{-4}	0.0348	0.61	0.2383	40.0%
GANC	chr15	42,565,522	42,566,390	7	3.38×10^{-4}	0.0357	0.88	0.0153	28.6%
TMEM87A	chr15	42,565,522	42,566,390	7	3.38×10^{-4}	0.0357	NA	NA	NA
NME3	chr16	1,821,559	1,822,346	8	3.59×10^{-4}	0.0366	0.76	0.0729	37.5%
SLC6A5	chr11	20,619,598	20,621,109	8	3.59×10^{-4}	0.0366	0.80	0.0489	28.6%
HSPB6	chr19	36,247,867	36,248,907	8	3.98×10^{-4}	0.0378	0.24	0.7811	0.0%
RAB22A	chr20	56,883,532	56,885,003	8	4.38×10^{-4}	0.0391	0.78	0.0594	25.0%
SLC1A1	chr9	4,489,544	4,490,288	6	4.60×10^{-4}	0.0394	0.60	0.2551	16.7%
ACTN4	chr19	39,137,911	39,138,334	7	4.89×10^{-4}	0.0407	0.84	0.0298	33.3%
PRKD1	chr14	30,396,845	30,397,763	6	4.96×10^{-4}	0.0407	0.49	0.4133	16.7%
STL	chr6	125,284,212	125,284,659	6	4.96×10^{-4}	0.0407	0.22	0.8030	0.0%
CPXM1	chr20	2,781,122	2,782,348	9	5.06×10^{-4}	0.0411	0.43	0.5202	0.0%
INSR	chr19	7,294,087	7,295,192	5	5.27×10^{-4}	0.0413	0.61	0.2438	20.0%
NPY	chr7	24,322,873	24,324,570	8	5.58×10^{-4}	0.0421	0.84	0.0276	37.5%
RAB38	chr11	87,908,558	87,909,729	9	6.33×10^{-4}	0.045	0.73	0.0995	44.4%
CNTNAP5	chr2	124,782,117	124,783,254	9	6.33×10^{-4}	0.045	0.63	0.2146	11.1%
AKR7L	chr1	19,600,471	19,601,069	7	7.52×10^{-4}	0.0495	0.49	0.4219	0.0%
COPB1	chr11	14,521,639	14,522,617	6	7.79×10^{-4}	0.0495	0.88	0.0149	50.0%
MEI1	chr22	42,095,347	42,095,536	5	7.91×10^{-4}	0.0495	0.65	0.1891	40.0%
PPP2R1B	chr11	111,637,044	111,638,422	5	7.91×10^{-4}	0.0495	0.69	0.1381	40.0%
KCNB1	chr20	48,098,642	48,100,238	8	7.97×10^{-4}	0.0495	0.55	0.3315	12.5%

Maximum Bone Correlation = maximum correlation coefficient representing strength of correlation between blood and bone CpG sites across all sites included in the DMR (from Ebrahimi et al.), FDR p Value for Maximum Bone Correlation = FDR adjusted p value for maximum correlation coefficient across all sites included in the DMR (from Ebrahimi et al.), Percent Strongly Positively Correlated Probes Across DMR = Percentage of probes across the entire region where the correlation coefficient representing strength of correlation between blood and bone CpG sites is greater than 75th percentile among all probes tested in Ebrahimi et al., NA = probes unavailable.

Table A5. Significantly Enriched Ontologies Based on Curve Severity DMR Promoter Analysis.

Category	Fold Enrichment	Bonferroni Adjusted p Value
Gene Ontology Molecular Function		
RNA polymerase II cis-regulatory region sequence-specific DNA binding (GO:0000978)	2.4	1.27×10^{-3}
DNA-binding transcription factor activity, RNA polymerase II-specific (GO:0000981)	2.3	3.65×10^{-4}
regulatory region nucleic acid binding (GO:0001067)	2.2	1.04×10^{-3}
transcription regulator activity (GO:0140110)	1.9	2.04×10^{-2}
Gene Ontology Cellular Component		
chromatin (GO:0000785)	2.3	7.60×10^{-3}
Gene Ontology Biologic Process		
pituitary gland development (GO:0021983)	12.0	8.09×10^{-3}
anterior/posterior pattern specification (GO:0009952)	5.2	3.91×10^{-3}
mesenchyme development (GO:0060485)	4.8	8.84×10^{-3}
heart morphogenesis (GO:0003007)	4.4	2.44×10^{-3}
embryonic organ development (GO:0048568)	3.3	4.70×10^{-3}
negative regulation of cell differentiation (GO:0045596)	3.0	6.46×10^{-3}
head development (GO:0060322)	2.7	2.37×10^{-3}
tube development (GO:0035295)	2.7	3.04×10^{-3}
neuron differentiation (GO:0030182)	2.5	5.09×10^{-3}
anatomical structure morphogenesis (GO:0009653)	2.0	3.30×10^{-3}

References

1. Altaf, F.; Drinkwater, J.; Phan, K.; Cree, A.K. Systematic Review of School Scoliosis Screening. *Spine Deform.* **2017**, *5*, 303–309. [CrossRef] [PubMed]
2. Weinstein, S.L.; Dolan, L.A.; Cheng, J.C.; Danielsson, A.; Morcuende, J.A. Adolescent idiopathic scoliosis. *Lancet* **2008**, *371*, 1527–1537. [CrossRef]
3. Miller, N.H. Genetics of familial idiopathic scoliosis. *Clin. Orthop. Relat. Res.* **2007**, *462*, 6–10. [CrossRef] [PubMed]
4. Khanshour, A.M.; Wise, C.A. The Genetic Architecture of Adolescent Idiopathic Scoliosis. In *Pathogenesis of Idiopathic Scoliosis*; Springer: Berlin, Germany, 2018; pp. 51–74.
5. Tang, N.L.; Yeung, H.Y.; Hung, V.W.; Di Liao, C.; Lam, T.P.; Yeung, H.M.; Lee, K.M.; Ng, B.K.; Cheng, J.C. Genetic epidemiology and heritability of AIS: A study of 415 Chinese female patients. *J. Orthop Res.* **2012**, *30*, 1464–1469. [CrossRef]
6. Ward, K.; Ogilvie, J.; Argyle, V.; Nelson, L.; Meade, M.; Braun, J.; Chettier, R. Polygenic inheritance of adolescent idiopathic scoliosis: A study of extended families in Utah. *Am. J. Med. Genet. A* **2010**, *152*, 1178–1188. [CrossRef]
7. Miller, N.H.; Mims, B.; Child, A.; Milewicz, D.M.; Sponseller, P.; Blanton, S.H. Genetic analysis of structural elastic fiber and collagen genes in familial adolescent idiopathic scoliosis. *J. Orthop. Res.* **1996**, *14*, 994–999. [CrossRef]
8. Miller, N.H.; Schwab, D.L.; Sponseller, P.; Shugert, E.; Bell, J.; Maestri, N. Genomic Search for X-Linkage in Familial Adolescent Idiopathic Scoliosis. In *Research into Spinal Deformities 2*; Stokes, I.A., Ed.; IOS Press: Amsterdam, The Netherlands, 1998; pp. 209–213.
9. Wise, C.A.; Barnes, R.; Gillum, J.; Herring, J.A.; Bowcock, A.M.; Lovett, M. Localization of susceptibility to familial idiopathic scoliosis. *Spine (Phila Pa 1976)* **2000**, *25*, 2372–2380. [CrossRef]
10. Chan, V.; Fong, G.C.; Luk, K.D.; Yip, B.; Lee, M.K.; Wong, M.S.; Lu, D.D.; Chan, T.K. A genetic locus for adolescent idiopathic scoliosis linked to chromosome 19p13.3. *Am. J. Hum. Genet.* **2002**, *71*, 401–406. [CrossRef]
11. Justice, C.M.; Miller, N.H.; Marosy, B.; Zhang, J.; Wilson, A.F. Familial idiopathic scoliosis: Evidence of an X-linked susceptibility locus. *Spine (Phila Pa 1976)* **2003**, *28*, 589–594. [CrossRef]
12. Miller, N.H.; Justice, C.M.; Marosy, B.; Doheny, K.F.; Pugh, E.; Zhang, J.; Dietz, H.C., 3rd; Wilson, A.F. Identification of candidate regions for familial idiopathic scoliosis. *Spine (Phila Pa 1976)* **2005**, *30*, 1181–1187. [CrossRef]
13. Alden, K.J.; Marosy, B.; Nzegwu, N.; Justice, C.M.; Wilson, A.F.; Miller, N.H. Idiopathic scoliosis: Identification of candidate regions on chromosome 19p13. *Spine (Phila Pa 1976)* **2006**, *31*, 1815–1819. [CrossRef]
14. Miller, N.H.; Marosy, B.; Justice, C.M.; Novak, S.M.; Tang, E.Y.; Boyce, P.; Pettengil, J.; Doheny, K.F.; Pugh, E.W.; Wilson, A.F. Linkage analysis of genetic loci for kyphoscoliosis on chromosomes 5p13, 13q13.3, and 13q32. *Am. J. Med. Genet. A* **2006**, *140*, 1059–1068. [CrossRef]
15. Montanaro, L.; Parisini, P.; Greggi, T.; Di Silvestre, M.; Campoccia, D.; Rizzi, S.; Arciola, C.R. Evidence of a linkage between matrilin-1 gene (MATN1) and idiopathic scoliosis. *Scoliosis* **2006**, *1*, 21. [CrossRef]

16. Gurnett, C.A.; Alaee, F.; Bowcock, A.; Kruse, L.; Lenke, L.G.; Bridwell, K.H.; Kuklo, T.; Luhmann, S.J.; Dobbs, M.B. Genetic linkage localizes an adolescent idiopathic scoliosis and pectus excavatum gene to chromosome 18 q. *Spine (Phila Pa 1976)* **2009**, *34*, E94–E100. [CrossRef]

17. Marosy, B.; Justice, C.M.; Vu, C.; Zorn, A.; Nzegwu, N.; Wilson, A.F.; Miller, N.H. Identification of susceptibility loci for scoliosis in FIS families with triple curves. *Am. J. Med. Genet. A* **2010**, *152*, 846–855. [CrossRef]

18. Buchan, J.G.; Alvarado, D.M.; Haller, G.E.; Cruchaga, C.; Harms, M.B.; Zhang, T.; Willing, M.C.; Grange, D.K.; Braverman, A.C.; Miller, N.H.; et al. Rare variants in FBN1 and FBN2 are associated with severe adolescent idiopathic scoliosis. *Hum. Mol. Genet.* **2014**, *23*, 5271–5282. [CrossRef]

19. Baschal, E.E.; Wethey, C.I.; Swindle, K.; Baschal, R.M.; Gowan, K.; Tang, N.L.; Alvarado, D.M.; Haller, G.E.; Dobbs, M.B.; Taylor, M.R.; et al. Exome sequencing identifies a rare HSPG2 variant associated with familial idiopathic scoliosis. *G3* **2014**, *5*, 167–174. [CrossRef]

20. Patten, S.A.; Margaritte-Jeannin, P.; Bernard, J.C.; Alix, E.; Labalme, A.; Besson, A.; Girard, S.L.; Fendri, K.; Fraisse, N.; Biot, B.; et al. Functional variants of POC5 identified in patients with idiopathic scoliosis. *J. Clin. Investig.* **2015**, *125*, 1124–1128. [CrossRef]

21. Grauers, A.; Wang, J.; Einarsdottir, E.; Simony, A.; Danielsson, A.; Akesson, K.; Ohlin, A.; Halldin, K.; Grabowski, P.; Tenne, M.; et al. Candidate gene analysis and exome sequencing confirm LBX1 as a susceptibility gene for idiopathic scoliosis. *Spine J.* **2015**, *15*, 2239–2246. [CrossRef]

22. Li, W.; Li, Y.; Zhang, L.; Guo, H.; Tian, D.; Li, Y.; Peng, Y.; Zheng, Y.; Dai, Y.; Xia, K.; et al. AKAP2 identified as a novel gene mutated in a Chinese family with adolescent idiopathic scoliosis. *J. Med. Genet.* **2016**, *53*, 488–493. [CrossRef]

23. Haller, G.; Alvarado, D.; McCall, K.; Yang, P.; Cruchaga, C.; Harms, M.; Goate, A.; Willing, M.; Morcuende, J.A.; Baschal, E.; et al. A polygenic burden of rare variants across extracellular matrix genes among individuals with adolescent idiopathic scoliosis. *Hum. Mol. Genet.* **2016**, *25*, 202–209. [CrossRef] [PubMed]

24. Gao, W.; Chen, C.; Zhou, T.; Yang, S.; Gao, B.; Zhou, H.; Lian, C.; Wu, Z.; Qiu, X.; Yang, X.; et al. Rare coding variants in MAPK7 predispose to adolescent idiopathic scoliosis. *Hum. Mutat.* **2017**, *38*, 1500–1510. [CrossRef] [PubMed]

25. Einarsdottir, E.; Grauers, A.; Wang, J.; Jiao, H.; Escher, S.A.; Danielsson, A.; Simony, A.; Andersen, M.; Christensen, S.B.; Akesson, K.; et al. CELSR2 is a candidate susceptibility gene in idiopathic scoliosis. *PLoS ONE* **2017**, *12*, e0189591. [CrossRef] [PubMed]

26. Baschal, E.E.; Terhune, E.A.; Wethey, C.I.; Baschal, R.M.; Robinson, K.D.; Cuevas, M.T.; Pradhan, S.; Sutphin, B.S.; Taylor, M.R.G.; Gowan, K.; et al. Idiopathic Scoliosis Families Highlight Actin-Based and Microtubule-Based Cellular Projections and Extracellular Matrix in Disease Etiology. *G3* **2018**, *8*, 2663–2672. [CrossRef]

27. Terhune, E.A.; Cuevas, M.T.; Monley, A.M.; Wethey, C.I.; Chen, X.; Cattell, M.V.; Bayrak, M.N.; Bland, M.R.; Sutphin, B.; Trahan, G.D.; et al. Mutations in KIF7 implicated in idiopathic scoliosis in humans and axial curvatures in zebrafish. *Hum. Mutat.* **2021**, *42*, 392–407. [CrossRef]

28. Sharma, S.; Gao, X.; Londono, D.; Devroy, S.E.; Mauldin, K.N.; Frankel, J.T.; Brandon, J.M.; Zhang, D.; Li, Q.Z.; Dobbs, M.B.; et al. Genome-wide association studies of adolescent idiopathic scoliosis suggest candidate susceptibility genes. *Hum. Mol. Genet.* **2011**, *20*, 1456–1466. [CrossRef] [PubMed]

29. Takahashi, Y.; Kou, I.; Takahashi, A.; Johnson, T.A.; Kono, K.; Kawakami, N.; Uno, K.; Ito, M.; Minami, S.; Yanagida, H.; et al. A genome-wide association study identifies common variants near LBX1 associated with adolescent idiopathic scoliosis. *Nat. Genet.* **2011**, *43*, 1237–1240. [CrossRef]

30. Nelson, L.M.; Chettier, R.; Ogilvie, J.W.; Ward, K. Candidate Genes for Susceptibility of Adolescent Idiopathic Scoliosis Identified Through a Large Genome-Wide Association Study. In Proceedings of the Scoliosis Research Society 46th Annual Meeting & Course, Louisville, KY, USA, 14–17 September 2011.

31. Kou, I.; Takahashi, Y.; Johnson, T.A.; Takahashi, A.; Guo, L.; Dai, J.; Qiu, X.; Sharma, S.; Takimoto, A.; Ogura, Y.; et al. Genetic variants in GPR126 are associated with adolescent idiopathic scoliosis. *Nat. Genet.* **2013**, *45*, 676–679. [CrossRef]

32. Miyake, A.; Kou, I.; Takahashi, Y.; Johnson, T.A.; Ogura, Y.; Dai, J.; Qiu, X.; Takahashi, A.; Jiang, H.; Yan, H.; et al. Identification of a susceptibility locus for severe adolescent idiopathic scoliosis on chromosome 17q24.3. *PLoS ONE* **2013**, *8*, e72802. [CrossRef]

33. Zhu, Z.; Tang, N.L.; Xu, L.; Qin, X.; Mao, S.; Song, Y.; Liu, L.; Li, F.; Liu, P.; Yi, L.; et al. Genome-wide association study identifies new susceptibility loci for adolescent idiopathic scoliosis in Chinese girls. *Nat. Commun.* **2015**, *6*, 8355. [CrossRef]

34. Ogura, Y.; Kou, I.; Miura, S.; Takahashi, A.; Xu, L.; Takeda, K.; Takahashi, Y.; Kono, K.; Kawakami, N.; Uno, K.; et al. A Functional SNP in BNC2 Is Associated with Adolescent Idiopathic Scoliosis. *Am. J. Hum. Genet.* **2015**, *97*, 337–342. [CrossRef]

35. Sharma, S.; Londono, D.; Eckalbar, W.L.; Gao, X.; Zhang, D.; Mauldin, K.; Kou, I.; Takahashi, A.; Matsumoto, M.; Kamiya, N.; et al. A PAX1 enhancer locus is associated with susceptibility to idiopathic scoliosis in females. *Nat. Commun.* **2015**, *6*, 6452. [CrossRef]

36. Kou, I.; Otomo, N.; Takeda, K.; Momozawa, Y.; Lu, H.F.; Kubo, M.; Kamatani, Y.; Ogura, Y.; Takahashi, Y.; Nakajima, M.; et al. Genome-wide association study identifies 14 previously unreported susceptibility loci for adolescent idiopathic scoliosis in Japanese. *Nat. Commun.* **2019**, *10*, 3685. [CrossRef]

37. Mogha, A.; Benesh, A.E.; Patra, C.; Engel, F.B.; Schoneberg, T.; Liebscher, I.; Monk, K.R. Gpr126 functions in Schwann cells to control differentiation and myelination via G-protein activation. *J. Neurosci.* **2013**, *33*, 17976–17985. [CrossRef]

38. Xu, J.F.; Yang, G.H.; Pan, X.H.; Zhang, S.J.; Zhao, C.; Qiu, B.S.; Gu, H.F.; Hong, J.F.; Cao, L.; Chen, Y.; et al. Association of GPR126 gene polymorphism with adolescent idiopathic scoliosis in Chinese populations. *Genomics* **2015**, *105*, 101–107. [CrossRef]

39. Karner, C.M.; Long, F.; Solnica-Krezel, L.; Monk, K.R.; Gray, R.S. Gpr126/Adgrg6 deletion in cartilage models idiopathic scoliosis and pectus excavatum in mice. *Hum. Mol. Genet.* **2015**, *24*, 4365–4373. [CrossRef]

40. Qin, X.; Xu, L.; Xia, C.; Zhu, W.; Sun, W.; Liu, Z.; Qiu, Y.; Zhu, Z. Genetic Variant of GPR126 Gene is Functionally Associated With Adolescent Idiopathic Scoliosis in Chinese Population. *Spine (Phila Pa 1976)* **2017**, *42*, E1098–E1103. [CrossRef]

41. Man, G.C.; Tang, N.L.; Chan, T.F.; Lam, T.P.; Li, J.W.; Ng, B.K.; Zhu, Z.; Qiu, Y.; Cheng, J.C. Replication Study for the Association of GWAS-associated Loci with Adolescent Idiopathic Scoliosis Susceptibility and Curve Progression in a Chinese Population. *Spine (Phila Pa 1976)* **2019**, *44*, 464–471. [CrossRef]

42. Kou, I.; Watanabe, K.; Takahashi, Y.; Momozawa, Y.; Khanshour, A.; Grauers, A.; Zhou, H.; Liu, G.; Fan, Y.H.; Takeda, K.; et al. A multi-ethnic meta-analysis confirms the association of rs6570507 with adolescent idiopathic scoliosis. *Sci. Rep.* **2018**, *8*, 11575. [CrossRef]

43. Liu, G.; Liu, S.; Lin, M.; Li, X.; Chen, W.; Zuo, Y.; Liu, J.; Niu, Y.; Zhao, S.; Long, B.; et al. Genetic polymorphisms of GPR126 are functionally associated with PUMC classifications of adolescent idiopathic scoliosis in a Northern Han population. *J. Cell Mol. Med.* **2018**, *22*, 1964–1971. [CrossRef]

44. Gao, W.; Peng, Y.; Liang, G.; Liang, A.; Ye, W.; Zhang, L.; Sharma, S.; Su, P.; Huang, D. Association between common variants near LBX1 and adolescent idiopathic scoliosis replicated in the Chinese Han population. *PLoS ONE* **2013**, *8*, e53234. [CrossRef]

45. Jiang, H.; Yang, Q.; Liu, Y.; Guan, Y.; Zhan, X.; Xiao, Z.; Wei, Q. Association between ladybird homeobox 1 gene polymorphisms and adolescent idiopathic scoliosis: A MOOSE-compliant meta-analysis. *Medicine (Baltimore)* **2019**, *98*, e16314. [CrossRef]

46. Liang, J.; Xing, D.; Li, Z.; Chua, S.; Li, S. Association Between rs11190870 Polymorphism Near LBX1 and Susceptibility to Adolescent Idiopathic Scoliosis in East Asian Population: A Genetic Meta-Analysis. *Spine (Phila Pa 1976)* **2014**, *39*, 862–869. [CrossRef]

47. Chen, S.; Zhao, L.; Roffey, D.M.; Phan, P.; Wai, E.K. Association of rs11190870 near LBX1 with adolescent idiopathic scoliosis in East Asians: A systematic review and meta-analysis. *Spine J.* **2014**, *14*, 2968–2975. [CrossRef]

48. Jiang, H.; Qiu, X.; Dai, J.; Yan, H.; Zhu, Z.; Qian, B.; Qiu, Y. Association of rs11190870 near LBX1 with adolescent idiopathic scoliosis susceptibility in a Han Chinese population. *Eur. Spine J.* **2013**, *22*, 282–286. [CrossRef]

49. Li, Y.L.; Gao, S.J.; Xu, H.; Liu, Y.; Li, H.L.; Chen, X.Y.; Ning, G.Z.; Feng, S.Q. The association of rs11190870 near LBX1 with the susceptibility and severity of AIS, a meta-analysis. *Int. J. Surg.* **2018**, *54*, 193–200. [CrossRef]

50. Cao, Y.; Min, J.; Zhang, Q.; Li, H.; Li, H. Associations of LBX1 gene and adolescent idiopathic scoliosis susceptibility: A meta-analysis based on 34,626 subjects. *BMC Musculoskelet Disord* **2016**, *17*, 309. [CrossRef]

51. Guo, L.; Yamashita, H.; Kou, I.; Takimoto, A.; Meguro-Horike, M.; Horike, S.; Sakuma, T.; Miura, S.; Adachi, T.; Yamamoto, T.; et al. Functional Investigation of a Non-coding Variant Associated with Adolescent Idiopathic Scoliosis in Zebrafish: Elevated Expression of the Ladybird Homeobox Gene Causes Body Axis Deformation. *PLoS Genet.* **2016**, *12*, e1005802. [CrossRef]

52. Liu, S.; Wu, N.; Zuo, Y.; Zhou, Y.; Liu, J.; Liu, Z.; Chen, W.; Liu, G.; Chen, Y.; Chen, J.; et al. Genetic Polymorphism of LBX1 Is Associated with Adolescent Idiopathic Scoliosis in Northern Chinese Han Population. *Spine (Phila Pa 1976)* **2017**, *42*, 1125–1129. [CrossRef]

53. Xu, L.; Wu, Z.; Xia, C.; Tang, N.; Cheng, J.C.Y.; Qiu, Y.; Zhu, Z. A Genetic Predictive Model Estimating the Risk of Developing Adolescent Idiopathic Scoliosis. *Curr. Genom.* **2019**, *20*, 246–251. [CrossRef] [PubMed]

54. Chettier, R.; Nelson, L.; Ogilvie, J.W.; Albertsen, H.M.; Ward, K. Haplotypes at LBX1 have distinct inheritance patterns with opposite effects in adolescent idiopathic scoliosis. *PLoS ONE* **2015**, *10*, e0117708. [CrossRef]

55. Londono, D.; Kou, I.; Johnson, T.A.; Sharma, S.; Ogura, Y.; Tsunoda, T.; Takahashi, A.; Matsumoto, M.; Herring, J.A.; Lam, T.P.; et al. A meta-analysis identifies adolescent idiopathic scoliosis association with LBX1 locus in multiple ethnic groups. *J. Med. Genet.* **2014**, *51*, 401–406. [CrossRef] [PubMed]

56. Nada, D.; Julien, C.; Samuels, M.E.; Moreau, A. A Replication Study for Association of LBX1 Locus with Adolescent Idiopathic Scoliosis in French-Canadian Population. *Spine (Phila Pa 1976)* **2018**, *43*, 172–178. [CrossRef] [PubMed]

57. Fan, Y.H.; Song, Y.Q.; Chan, D.; Takahashi, Y.; Ikegawa, S.; Matsumoto, M.; Kou, I.; Cheah, K.S.; Sham, P.; Cheung, K.M.; et al. SNP rs11190870 near LBX1 is associated with adolescent idiopathic scoliosis in southern Chinese. *J. Hum. Genet.* **2012**, *57*, 244–246. [CrossRef]

58. Edery, P.; Margaritte-Jeannin, P.; Biot, B.; Labalme, A.; Bernard, J.C.; Chastang, J.; Kassai, B.; Plais, M.H.; Moldovan, F.; Clerget-Darpoux, F. New disease gene location and high genetic heterogeneity in idiopathic scoliosis. *Eur. J. Hum. Genet.* **2011**, *19*, 865–869. [CrossRef]

59. Gorman, K.F.; Julien, C.; Moreau, A. The genetic epidemiology of idiopathic scoliosis. *Eur. Spine J.* **2012**, *21*, 1905–1919. [CrossRef]

60. Lowe, T.G.; Edgar, M.; Margulies, J.Y.; Miller, N.H.; Raso, V.J.; Reinker, K.A.; Rivard, C.H. Etiology of idiopathic scoliosis: Current trends in research. *J. Bone Jt. Surg. Am.* **2000**, *82*, 1157–1168. [CrossRef]

61. Dunwoodie, S.L.; Kusumi, K. *The Genetics and Development of Scoliosis*, 2nd ed.; Springer International Publishing: Cham, Switzerland, 2018; p. 1. [CrossRef]

62. Heary, R.F.; Madhavan, K. Genetics of scoliosis. *Neurosurgery* **2008**, *63*, 222–227. [CrossRef]

63. Kruse, L.M.; Buchan, J.G.; Gurnett, C.A.; Dobbs, M.B. Polygenic threshold model with sex dimorphism in adolescent idiopathic scoliosis: The Carter effect. *J. Bone Jt. Surg. Am.* **2012**, *94*, 1485–1491. [CrossRef]

64. Arpon, A.; Milagro, F.I.; Ramos-Lopez, O.; Mansego, M.L.; Riezu-Boj, J.I.; Martinez, J.A.; Project, M. Methylome-Wide Association Study in Peripheral White Blood Cells Focusing on Central Obesity and Inflammation. *Genes* **2019**, *10*, 444. [CrossRef]

65. Grauers, A.; Rahman, I.; Gerdhem, P. Heritability of scoliosis. *Eur. Spine J.* **2012**, *21*, 1069–1074. [CrossRef]

66. Domcke, S.; Bardet, A.F.; Adrian Ginno, P.; Hartl, D.; Burger, L.; Schubeler, D. Competition between DNA methylation and transcription factors determines binding of NRF1. *Nature* **2015**, *528*, 575–579. [CrossRef]

67. Gutierrez-Arcelus, M.; Lappalainen, T.; Montgomery, S.B.; Buil, A.; Ongen, H.; Yurovsky, A.; Bryois, J.; Giger, T.; Romano, L.; Planchon, A.; et al. Passive and active DNA methylation and the interplay with genetic variation in gene regulation. *eLife* **2013**, *2*, e00523. [CrossRef]

68. Lappalainen, T.; Greally, J.M. Associating cellular epigenetic models with human phenotypes. *Nat. Rev. Genet.* **2017**, *18*, 441–451. [CrossRef]

69. Gertz, J.; Varley, K.E.; Reddy, T.E.; Bowling, K.M.; Pauli, F.; Parker, S.L.; Kucera, K.S.; Willard, H.F.; Myers, R.M. Analysis of DNA methylation in a three-generation family reveals widespread genetic influence on epigenetic regulation. *PLoS Genet.* **2011**, *7*, e1002228. [CrossRef]

70. Delgado-Calle, J.; Fernandez, A.F.; Sainz, J.; Zarrabeitia, M.T.; Sanudo, C.; Garcia-Renedo, R.; Perez-Nunez, M.I.; Garcia-Ibarbia, C.; Fraga, M.F.; Riancho, J.A. Genome-wide profiling of bone reveals differentially methylated regions in osteoporosis and osteoarthritis. *Arthritis Rheum.* **2013**, *65*, 197–205. [CrossRef]

71. Mohandas, N.; Bass-Stringer, S.; Maksimovic, J.; Crompton, K.; Loke, Y.J.; Walstab, J.; Reid, S.M.; Amor, D.J.; Reddihough, D.; Craig, J.M. Epigenome-wide analysis in newborn blood spots from monozygotic twins discordant for cerebral palsy reveals consistent regional differences in DNA methylation. *Clin. Epigenet.* **2018**, *10*, 25. [CrossRef]

72. Diboun, I.; Wani, S.; Ralston, S.H.; Albagha, O.M. Epigenetic analysis of Paget's disease of bone identifies differentially methylated loci that predict disease status. *eLife* **2021**, *10*, e65715. [CrossRef]

73. Mao, S.H.; Qian, B.P.; Shi, B.; Zhu, Z.Z.; Qiu, Y. Quantitative evaluation of the relationship between COMP promoter methylation and the susceptibility and curve progression of adolescent idiopathic scoliosis. *Eur. Spine J.* **2018**, *27*, 272–277. [CrossRef]

74. Shi, B.; Xu, L.; Mao, S.; Xu, L.; Liu, Z.; Sun, X.; Zhu, Z.; Qiu, Y. Abnormal PITX1 gene methylation in adolescent idiopathic scoliosis: A pilot study. *BMC Musculoskelet Disord* **2018**, *19*, 138. [CrossRef]

75. Janusz, P.; Chmielewska, M.; Andrusiewicz, M.; Kotwicka, M.; Kotwicki, T. Methylation of Estrogen Receptor 1 Gene in the Paraspinal Muscles of Girls with Idiopathic Scoliosis and Its Association with Disease Severity. *Genes* **2021**, *12*, 790. [CrossRef] [PubMed]

76. Chmielewska, M.; Janusz, P.; Andrusiewicz, M.; Kotwicki, T.; Kotwicka, M. Methylation of estrogen receptor 2 (ESR2) in deep paravertebral muscles and its association with idiopathic scoliosis. *Sci. Rep.* **2020**, *10*, 22331. [CrossRef] [PubMed]

77. Liu, G.; Wang, L.; Wang, X.; Yan, Z.; Yang, X.; Lin, M.; Liu, S.; Zuo, Y.; Niu, Y.; Zhao, S.; et al. Whole-Genome Methylation Analysis of Phenotype Discordant Monozygotic Twins Reveals Novel Epigenetic Perturbation Contributing to the Pathogenesis of Adolescent Idiopathic Scoliosis. *Front. Bioeng. Biotechnol.* **2019**, *7*, 364. [CrossRef] [PubMed]

78. Meng, Y.; Lin, T.; Liang, S.; Gao, R.; Jiang, H.; Shao, W.; Yang, F.; Zhou, X. Value of DNA methylation in predicting curve progression in patients with adolescent idiopathic scoliosis. *EBioMedicine* **2018**, *36*, 489–496. [CrossRef]

79. Shands, A.R.J.; Eisberg, H.B. The incidence of scoliosis in the state of Delaware; a study of 50,000 minifilms of the chest made during a survey for tuberculosis. *J. Bone Jt. Surg. Am.* **1955**, *37*, 1243–1249. [CrossRef]

80. Sambrook, J.; Fritsch, E.F.; Maniatis, T. *Molecular Cloning: A Laboratory Manual*; Cold Spring Harbor Laboratory Press: Cold Spring Harbor, NY, USA, 1989.

81. Aryee, M.J.; Jaffe, A.E.; Corrada-Bravo, H.; Ladd-Acosta, C.; Feinberg, A.P.; Hansen, K.D.; Irizarry, R.A. Minfi: A flexible and comprehensive Bioconductor package for the analysis of Infinium DNA methylation microarrays. *Bioinformatics* **2014**, *30*, 1363–1369. [CrossRef]

82. Logue, M.W.; Smith, A.K.; Wolf, E.J.; Maniates, H.; Stone, A.; Schichman, S.A.; McGlinchey, R.E.; Milberg, W.; Miller, M.W. The correlation of methylation levels measured using Illumina 450K and EPIC BeadChips in blood samples. *Epigenomics* **2017**, *9*, 1363–1371. [CrossRef]

83. Leek, J.T.; Johnson, W.E.; Parker, H.S.; Jaffe, A.E.; Storey, J.D. The sva package for removing batch effects and other unwanted variation in high-throughput experiments. *Bioinformatics* **2012**, *28*, 882–883. [CrossRef]

84. Jaffe, A.E.; Irizarry, R.A. Accounting for cellular heterogeneity is critical in epigenome-wide association studies. *Genom. Biol.* **2014**, *15*, R31. [CrossRef]

85. Benjamini, Y.; Hochberg, Y. Controlling the false discovery rate: A practical and powerful approach to multiple testing. *J. R. Stat. Soc. Ser. B (Methodological)* **1995**, *57*, 289–300. [CrossRef]

86. Martorell-Marugan, J.; Gonzalez-Rumayor, V.; Carmona-Saez, P. mCSEA: Detecting subtle differentially methylated regions. *Bioinformatics* **2019**, *35*, 3257–3262. [CrossRef]

87. Mi, H.; Muruganujan, A.; Casagrande, J.T.; Thomas, P.D. Large-scale gene function analysis with the PANTHER classification system. *Nat. Protoc.* **2013**, *8*, 1551–1566. [CrossRef]

88. Mi, H.; Thomas, P. PANTHER pathway: An ontology-based pathway database coupled with data analysis tools. *Methods Mol. Biol.* **2009**, *563*, 123–140. [CrossRef]

89. Mi, H.; Huang, X.; Muruganujan, A.; Tang, H.; Mills, C.; Kang, D.; Thomas, P.D. PANTHER version 11: Expanded annotation data from Gene Ontology and Reactome pathways, and data analysis tool enhancements. *Nucleic Acids Res.* **2017**, *45*, D183–D189. [CrossRef]

90. Supek, F.; Bosnjak, M.; Skunca, N.; Smuc, T. REVIGO summarizes and visualizes long lists of gene ontology terms. *PLoS ONE* **2011**, *6*, e21800. [CrossRef]

91. Ebrahimi, P.; Luthman, H.; McGuigan, F.E.; Akesson, K.E. Epigenome-wide cross-tissue correlation of human bone and blood DNA methylation—can blood be used as a surrogate for bone? *Epigenetics* **2020**, *16*, 92–105. [CrossRef]

92. Paul, D.S.; Teschendorff, A.E.; Dang, M.A.; Lowe, R.; Hawa, M.I.; Ecker, S.; Beyan, H.; Cunningham, S.; Fouts, A.R.; Ramelius, A.; et al. Increased DNA methylation variability in type 1 diabetes across three immune effector cell types. *Nat. Commun.* **2016**, *7*, 13555. [CrossRef]

93. Webster, A.P.; Plant, D.; Ecker, S.; Zufferey, F.; Bell, J.T.; Feber, A.; Paul, D.S.; Beck, S.; Barton, A.; Williams, F.M.K.; et al. Increased DNA methylation variability in rheumatoid arthritis-discordant monozygotic twins. *Genom. Med.* **2018**, *10*, 64. [CrossRef]

94. Watanabe, K.; Takebayashi, H.; Bepari, A.K.; Esumi, S.; Yanagawa, Y.; Tamamaki, N. Dpy19l1, a multi-transmembrane protein, regulates the radial migration of glutamatergic neurons in the developing cerebral cortex. *Development* **2011**, *138*, 4979–4990. [CrossRef]

95. Watanabe, K.; Bizen, N.; Sato, N.; Takebayashi, H. Endoplasmic Reticulum-Localized Transmembrane Protein Dpy19L1 Is Required for Neurite Outgrowth. *PLoS ONE* **2016**, *11*, e0167985. [CrossRef]

96. Data, Z.H. Phenotype Annotation. 1994–2006. Available online: https://zfin.org/ZDB-FIG-070117-942 (accessed on 17 May 2021).

97. Perry, R.B.; Hezroni, H.; Goldrich, M.J.; Ulitsky, I. Regulation of Neuroregeneration by Long Noncoding RNAs. *Mol. Cell* **2018**, *72*, 553–567. [CrossRef]

98. Li, Z.; Li, X.; Shen, J.; Zhang, L.; Chan, M.T.V.; Wu, W.K.K. Emerging roles of non-coding RNAs in scoliosis. *Cell Prolif.* **2020**, *53*, e12736. [CrossRef]

99. Cunningham, T.J.; Duester, G. Mechanisms of retinoic acid signalling and its roles in organ and limb development. *Nat. Rev. Mol. Cell Biol.* **2015**, *16*, 110–123. [CrossRef]

100. Sirbu, I.O.; Duester, G. Retinoic-acid signalling in node ectoderm and posterior neural plate directs left-right patterning of somitic mesoderm. *Nat. Cell Biol.* **2006**, *8*, 271–277. [CrossRef]

101. Hitier, M.; Hamon, M.; Denise, P.; Lacoudre, J.; Thenint, M.A.; Mallet, J.F.; Moreau, S.; Quarck, G. Lateral Semicircular Canal Asymmetry in Idiopathic Scoliosis: An Early Link between Biomechanical, Hormonal and Neurosensory Theories? *PLoS ONE* **2015**, *10*, e0131120. [CrossRef]

102. Rousie, D.; Joly, O.; Berthoz, A. Posterior Basicranium asymmetry and idiopathic scoliosis. *arXiv* **2009**, arXiv:0911.5042.

103. Carry, P.M.; Duke, V.R.; Brazell, C.J.; Stence, N.; Scholes, M.; Rousie, D.L.; Hadley Miller, N. Lateral semi-circular canal asymmetry in females with idiopathic scoliosis. *PLoS ONE* **2020**, *15*, e0232417. [CrossRef]

104. Hino, H.; Araki, K.; Uyama, E.; Takeya, M.; Araki, M.; Yoshinobu, K.; Miike, K.; Kawazoe, Y.; Maeda, Y.; Uchino, M.; et al. Myopathy phenotype in transgenic mice expressing mutated PABPN1 as a model of oculopharyngeal muscular dystrophy. *Hum. Mol. Genet.* **2004**, *13*, 181–190. [CrossRef] [PubMed]

105. Joseph, R.M. Neuronatin gene: Imprinted and misfolded: Studies in Lafora disease, diabetes and cancer may implicate NNAT-aggregates as a common downstream participant in neuronal loss. *Genomics* **2014**, *103*, 183–188. [CrossRef] [PubMed]

106. Kikyo, N.; Williamson, C.M.; John, R.M.; Barton, S.C.; Beechey, C.V.; Ball, S.T.; Cattanach, B.M.; Surani, M.A.; Peters, J. Genetic and functional analysis of neuronatin in mice with maternal or paternal duplication of distal Chr 2. *Dev. Biol.* **1997**, *190*, 66–77. [CrossRef] [PubMed]

107. Evans, H.K.; Wylie, A.A.; Murphy, S.K.; Jirtle, R.L. The neuronatin gene resides in a "micro-imprinted" domain on human chromosome 20q11. 2. *Genomics* **2001**, *77*, 99–104. [CrossRef]

108. Lee, N.J.; Herzog, H. NPY regulation of bone remodelling. *Neuropeptides* **2009**, *43*, 457–463. [CrossRef]

109. Wu, J.Q.; Jiang, N.; Yu, B. Mechanisms of action of neuropeptide Y on stem cells and its potential applications in orthopaedic disorders. *World J. Stem. Cells* **2020**, *12*, 986–1000. [CrossRef]

110. Xu, M.; Horrell, J.; Snitow, M.; Cui, J.; Gochnauer, H.; Syrett, C.M.; Kallish, S.; Seykora, J.T.; Liu, F.; Gaillard, D.; et al. WNT10A mutation causes ectodermal dysplasia by impairing progenitor cell proliferation and KLF4-mediated differentiation. *Nat. Commun.* **2017**, *8*, 15397. [CrossRef]

111. Vink, C.P.; Ockeloen, C.W.; ten Kate, S.; Koolen, D.A.; Ploos van Amstel, J.K.; Kuijpers-Jagtman, A.M.; van Heumen, C.C.; Kleefstra, T.; Carels, C.E. Variability in dentofacial phenotypes in four families with WNT10A mutations. *Eur. J. Hum. Genet.* **2014**, *22*, 1063–1070. [CrossRef]

112. Cawthorn, W.P.; Bree, A.J.; Yao, Y.; Du, B.; Hemati, N.; Martinez-Santibanez, G.; MacDougald, O.A. Wnt6, Wnt10a and Wnt10b inhibit adipogenesis and stimulate osteoblastogenesis through a β-catenin-dependent mechanism. *Bone* **2012**, *50*, 477–489. [CrossRef]

113. Grimes, D.T.; Boswell, C.W.; Morante, N.F.; Henkelman, R.M.; Burdine, R.D.; Ciruna, B. Zebrafish models of idiopathic scoliosis link cerebrospinal fluid flow defects to spine curvature. *Science* **2016**, *352*, 1341–1344. [CrossRef]

114. Gurnett, C.A.; Alaee, F.; Kruse, L.M.; Desruisseau, D.M.; Hecht, J.T.; Wise, C.A.; Bowcock, A.M.; Dobbs, M.B. Asymmetric lower-limb malformations in individuals with homeobox PITX1 gene mutation. *Am. J. Hum. Genet.* **2008**, *83*, 616–622. [CrossRef]

115. Feinberg, A.P. Phenotypic plasticity and the epigenetics of human disease. *Nature* **2007**, *447*, 433–440. [CrossRef]

116. Hamilton, J.P. Epigenetics: Principles and practice. *Dig. Dis.* **2011**, *29*, 130–135. [CrossRef]

117. Baron, R.; Kneissel, M. WNT signaling in bone homeostasis and disease: From human mutations to treatments. *Nat. Med.* **2013**, *19*, 179–192. [CrossRef] [PubMed]

 *genes*

Article

Methylation of Estrogen Receptor 1 Gene in the Paraspinal Muscles of Girls with Idiopathic Scoliosis and Its Association with Disease Severity

Piotr Janusz [1,†], Małgorzata Chmielewska [2,*,†], Mirosław Andrusiewicz [2], Małgorzata Kotwicka [2] and Tomasz Kotwicki [1]

[1]   Department of Spine Disorders and Pediatric Orthopedics, Poznan University of Medical Sciences, 28 Czerwca 1956 r. Street 135/147, 61-545 Poznan, Poland; pjanusz@ump.edu.pl (P.J.); kotwicki@ump.edu.pl (T.K.)

[2]   Chair and Department of Cell Biology, Poznan University of Medical Sciences, Rokietnicka 5D, 60-806 Poznan, Poland; andrus@ump.edu.pl (M.A.); mkotwic@ump.edu.pl (M.K.)

*   Correspondence: mchmielewska@ump.edu.pl

†   These authors contributed equally to this work.

Abstract: Idiopathic scoliosis (IS) is a multifactorial disease with epigenetic modifications. Tissue dependent and differentially methylated regions (T-DMRs) may regulate tissue-specific expression of the estrogen receptor 1 gene (*ESR1*). This study aimed to analyze methylation levels within T-DMR1 and T-DMR2 and its concatenation with ESR1 expression of IS patients. The study involved 87 tissue samples (deep paravertebral muscles, both on the convex and the concave side of the curve, and from back superficial muscles) from 29 girls who underwent an operation due to IS. Patient subgroups were analyzed according to Cobb angle ≤70° vs. >70°. Methylation was significantly higher in the superficial muscles than in deep paravertebral muscles in half of the T-DMR1 CpGs and all T-DMR2 CpGs. The methylation level correlated with *ESR1* expression level on the concave, but not convex, side of the curvature in a majority of the T-DMR2 CpGs. The T-DMR2 methylation level in the deep paravertebral muscles on the curvature's concave side was significantly lower in patients with a Cobb angle ≤70° in four CpGs. DNA methylation of the T-DMRs is specific to muscle tissue location and may be related to *ESR1* expression regulation. Additionally, the difference in T-DMR2 methylation may be associated with IS severity.

Keywords: spinal curvatures; scoliosis; idiopathic; DNA methylation; pyrosequencing; estrogen receptor 1; ESR1; scoliosis progression; adolescent idiopathic scoliosis

Citation: Janusz, P.; Chmielewska, M.; Andrusiewicz, M.; Kotwicka, M.; Kotwicki, T. Methylation of Estrogen Receptor 1 Gene in the Paraspinal Muscles of Girls with Idiopathic Scoliosis and Its Association with Disease Severity. *Genes* **2021**, *12*, 790. https://doi.org/10.3390 /genes12060790

Academic Editor: Robert Winqvist

Received: 20 April 2021
Accepted: 20 May 2021
Published: 21 May 2021

Publisher's Note: MDPI stays neutral with regard to jurisdictional claims in published maps and institutional affiliations.

1. Introduction

The most common spine disorder in adolescents is idiopathic scoliosis (IS), affecting 1–3% of the population. It is a structural, three-dimensional spinal deformity characterized by lateral curvature of the spine, impaired kyphosis or lordosis, and vertebral rotation with a rib hump [1]. IS is a highly heterogeneous condition, with some patients having a rapidly progressive presentation, resulting in severe curves, and others progressing slowly to mild or moderate curves [2]. Progressive scoliosis may result in cosmetic deformity, back pain, and functional deficits as well as psychological problems and impaired social interactions. Severe curvatures are associated with cardiac dysfunction and pulmonary constraints [3–5]. Currently, clinical or radiological criteria cannot adequately predict which children who are diagnosed with mild disease may ultimately undergo subsequent curve progression that requires surgical intervention [6]. Identifying patients at risk of scoliosis, or those at risk of curve progression, is essential for early, appropriate treatment [1,7].

Despite the high prevalence of IS, its etiology remains poorly understood [8]. IS is considered a multifactorial disease with genetic susceptibilities [9]. Many candidate genes potentially associated with IS have been described in family linkage studies, single nucleotide polymorphisms association studies, and genome-wide association studies [8,10–12]. Results from these suggest IS is a complex genetic disorder [6]. It was postulated that genetic factors are more important in the occurrence of IS while environmental factors have a more significant impact on disease progression [13]. An epigenetic link between genetic and environmental factors may be involved in IS etiopathogenesis [14]. As a new area of research, only a few publications concerning the impact of DNA methylation on IS have been published [15–19]. However, none of these studies have evaluated this mechanism in paraspinal muscle tissues.

Due to the gender-related distribution of idiopathic scoliosis, the role of estrogen hormones in IS occurrence and progression has been suggested [20,21]. Previous studies have reported the effect of estrogens on skeletal muscles, in which the mRNA and protein expression of estrogen receptor 1 and 2 (*ESR1, ESR2*) has been demonstrated [22–24]. *ESR1* and *ESR2* expression was confirmed in the superficial and deep paravertebral muscles of patients with IS. Moreover, expression of *PELP1* (proline-, glutamic acid-, and leucine-rich protein) was significantly higher in the deep back muscles compared to superficial muscles. This protein participates in estrogen-induced signal transduction pathways. Additionally, *PELP1* expression level was correlated with both the Cobb angle value and *ESR1* expression [25].

DNA methylation is one of the most well-characterized epigenetic modifications. Methylation of CpG islands (CGI), located near the transcription start site of gene promoter and regulatory regions, is associated with altered gene expression [26,27]. Methylation at the gene promoter inhibits recognition and binding of transcription factors. This leads to the recruitment of proteins binding to methylated CpG dinucleotides which, in turn, interact with transcription repressors and activate chromatin condensation by recruiting histone deacetylases. As a result, DNA methylation in CpG site-rich regions, found in close proximity to the promoter region, is thought to play an essential role in gene silencing [28]. It has been indicated that a different methylation status is characteristic for particular types of tissues or the development phase [29]. Although the CpG islands in intragenic and regulatory regions of genes may display a tissue-dependent and differentially methylated region pattern [30], CGIs associated with transcription start sites rarely show tissue-specific patterns of methylation [31].

It was shown that in the case of ESR1, the level of methylation within the promoter is cell-specific [32]. Analysis of the C promoter (Figure 1) in in vitro study indicated that demethylation of this region is responsible for the increased expression of ESR1 [33]. The relationship between regulatory regions methylation of ESR1 and its expression was suggested. It has been reported that ESR1 has tissue-dependent and differentially methylated regions (T-DMRs; Figure 1), which are associated with tissue-specific gene expression [34]. A previous study showed that methylation at T-DMR1 and T-DMR2 is correlated with decreased ESR1 expression in the placenta and skin tissue but not in mammary glands and the endometrium [34]. Maekawa et al. suggested that ESR1 expression is tissue-specific and regulated by DNA methylation at T-DMR1 rather than by DNA methylation at the promoter region [34]. Thus, changes in ESR1 mRNA expression may not correspond with methylation of the ESR1 promoter. Moreover, it was indicated that, in the case of some breast cancer tissues, ESR1 expression might be modulated not only by DNA methylation at T-DMRs and promoter regions but also by different mechanisms that require clarification in future studies [34].

Taking into consideration that methylation level alterations among patients with different IS phenotypes may be associated with susceptibility to disease or disease progression, we therefore analyzed T-DMR1 and T-DMR2 methylation status. Subsequently, the expression level of *ESR1* in the superficial and paraspinal muscles on the convex and concave side of the IS curvature was analyzed and evaluated in relation to methylation status.

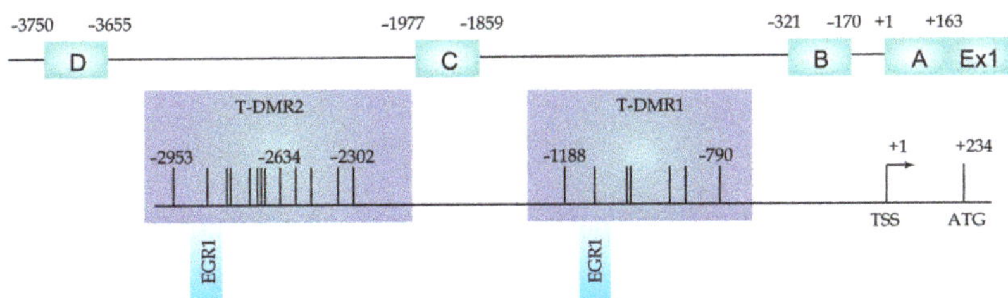

Figure 1. *ESR1* promoters (A-D) and T-DMR1, and T-DMR2 regions localization with respect to transcription start site (TSS) and translation start codon (ATG). EGR1 indicates transcription factor binding sites. Adapted from [34].

2. Results

2.1. Patient Characteristics

The study group consisted of 29 female IS patients (age at surgery: 12.1–17.9 years, mean age: 14.5 ± 1.5 years). The Cobb angle ranged from 52° to 115°, with a mean of 77.4 ± 16.1°. The mean age, number of curvatures, and Risser sign value did not differ significantly between the subgroups of patients with Cobb angles ≤70° and ≥70° (14.5 ± 1.3 vs. 14.7 ± 1.7, $p = 0.9$; 3 single: 7 double vs. 8 single: 11 double, $p = 0.7$; Me = 4 vs. Me = 4, $p = 0.7$, respectively). The mean Cobb angle value of patients with a Cobb angle ≤70° and those ≥70° was 61.1° ± 6° and 86° ± 12.7°, respectively.

2.2. DNA Methylation at the ESR1 T-DMR1 and T-DMR2

The methylation pattern within T-DMR1 and T-DMR2 of individual patients is shown in Figure 2.

The methylation level within the *ESR1* T-DMR1 region was significantly higher in the superficial muscle compared to the deep paravertebral muscles at the CpG1 ($p = 0.0001$; Figure 3; Supplementary Table S1) and CpG2 sites ($p < 0.0001$; Figure 3; Supplementary Table S1). The methylation level was significantly higher in the superficial muscle compared to both, the concave ($p < 0.05$; Figure 3; Supplementary Table S1) and convex side of the curvature ($p < 0.05$; Figure 3; Supplementary Table S1). Moreover, in the deep paravertebral muscles, methylation was decreased on the concave side in contrast to the convex side of the curvature. However, the difference was not statistically significant (Supplementary Table S1).

Significant differences in methylation levels of all CpG sites within the *ESR1* T-DMR2 region between superficial and deep paravertebral muscles were observed ($p < 0.05$; Figure 4; Supplementary Table S1). Methylation was found to be significantly higher in the superficial muscle versus the concave (at CpG sites 1–4 and 6–8; $p < 0.05$; Figure 4; Supplementary Table S1) and convex side of the curvature (at all CpG sites; $p < 0.05$; Figure 4; Supplementary Table S1). In contrast to the *ESR1* T-DMR1 region, the methylation level within the T-DMR2 region in the deep paravertebral muscles was lower on the convex side of the curvature in seven of eight CpGs compared to the concave side. However, the difference was not statistically significant (Figure 4; Supplementary Table S1).

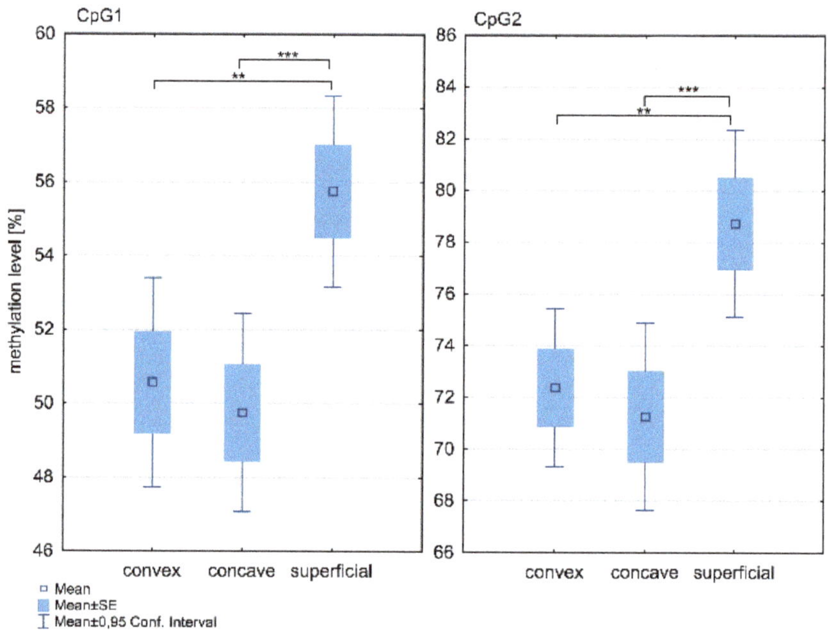

Figure 2. Dot plot of *ESR1* T-DMR1 and T-DMR2 regions methylation pattern of individual patients.

Figure 3. DNA methylation level within *ESR1* T-DMR1 region in deep paravertebral muscles and superficial muscles; ** $p < 0.01$, *** $p < 0.001$.

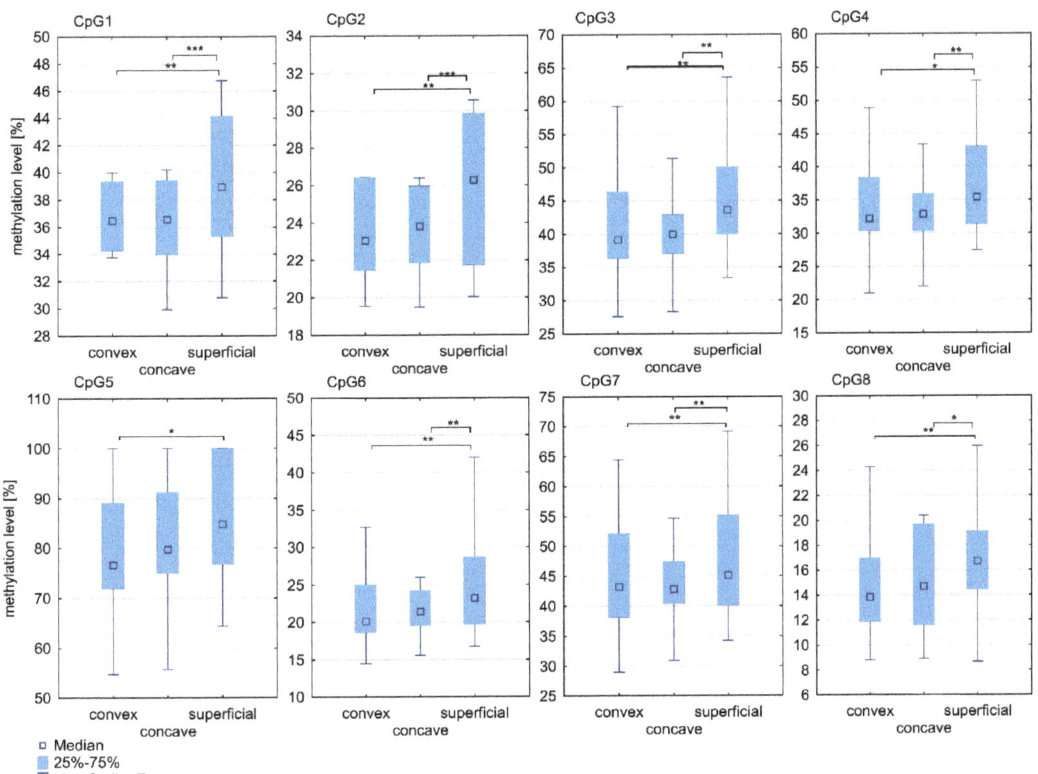

Figure 4. DNA methylation level within *ESR1* T-DMR2 region in deep paravertebral muscles and superficial muscles; * $p < 0.05$, ** $p < 0.01$, *** $p < 0.001$.

2.3. Correlation between ESR1 Methylation Levels and Relative Expression of the ESR1 Gene

The ESR1 relative expression did not differ significantly between the deep paravertebral muscles on both, convex and concave side, and superficial muscles ($p > 0.05$; Supplementary Figure S1).

On the concave side of the curvature, a significant, moderate, and positive correlation was observed between *ESR1* mRNA expression and methylation level at the CpG1 dinucleotide in the T-DMR1 region and at six CpG sites (CpG2-CpG8) in the T-DMR2 region (R ranged from 0.44 to 0.59; $p < 0.05$; Figure 5). No correlation between *ESR1* expression and methylation level within the T-DMR1 and T-DMR2 regions was found either in the superficial muscle or on the convex side of thoracic scoliosis ($p > 0.05$; Figure 5; Supplementary Figures S2 and S3).

	CpG	convex R	convex p-value	concave R	concave p-value	superficial R	superficial p-value
T-DMR1	1	-0.07	0.714	**0.50**	**0.006**	-0.09	0.634
	2	-0.06	0.752	0.28	0.138	-0.04	0.831
	3	-0.11	0.579	0.13	0.516	-0.02	0.902
	4	0.00	0.980	0.20	0.295	0.11	0.571
T-DMR2	1	0.21	0.285	0.33	0.080	-0.01	0.966
	2	0.23	0.220	0.32	0.089	-0.05	0.778
	3	0.25	0.183	**0.57**	**0.001**	-0.03	0.888
	4	0.24	0.202	**0.59**	**0.001**	0.04	0.840
	5	0.31	0.099	**0.57**	**0.001**	-0.04	0.834
	6	0.30	0.116	**0.44**	**0.017**	-0.02	0.904
	7	0.32	0.093	**0.54**	**0.003**	0.05	0.805
	8	0.36	0.056	**0.47**	**0.011**	-0.01	0.959

Figure 5. Correlation between *ESR1* expression and methylation level within T-DMR1 and T-DMR2 regions in deep paravertebral muscles and superficial muscles. R—Spearman rank correlation coefficient.

2.4. Association between Methylation Status of ESR1 and Cobb Angle

In the deep paravertebral muscle, the methylation level within the *ESR1* T-DMR2 region on the concave side of the curvature was significantly different between groups of patients with a Cobb angle >70° or ≤r0° at four CpG sites: CpG2 (p = 0.02; Figure 6; Supplementary Table S2), CpG3 (p = 0.04; Figure 6; Supplementary Table S2), CpG4 (p = 0.04; Figure 6; Supplementary Table S2) and CpG 6 (p = 0.005; Figure 6; Supplementary Table S2). There was no difference in the *ESR1* T-DMR1 region methylation level between groups of patients with a Cobb angle ≤70° or >70° (p > 0.05; Supplementary Table S2). No differences were observed in T-DMR1 methylation levels between groups of patients with Cobb angles ≤70° and >70° (p > 0.05; Supplementary Table S2).

No correlation was found between T-DMR1 region methylation level and Cobb angle value in either the superficial or deep paravertebral muscle tissues (r ranged from 0.02 to 0.23; p > 0.05). Examining the concave side of thoracic scoliosis, a significant, moderate and positive correlation between T-DMR2 methylation and Cobb angle was observed at CpG2 (R = 0.44; p = 0.02) and CpG6 (r = 0.5; p = 0.005). There was no significant correlation between T-DMR2 methylation and Cobb angle in the superficial muscles (CpG2 and CpG4-CpG8, R ranged from 0.03 to 0.27 (p > 0.05); CpG1 and CpG3, r ranged from 0.12 to 0.14 (p > 0.05)) or in the deep paravertebral muscles on the convex side of the curvature (CpG1, CpG2, CpG6 and CpG8, R ranged from 0.03 to 0.24 (p > 0.05); CpG3-CpG5 and CpG7, r ranged from 0.07 to 0.26 (p > 0.05)).

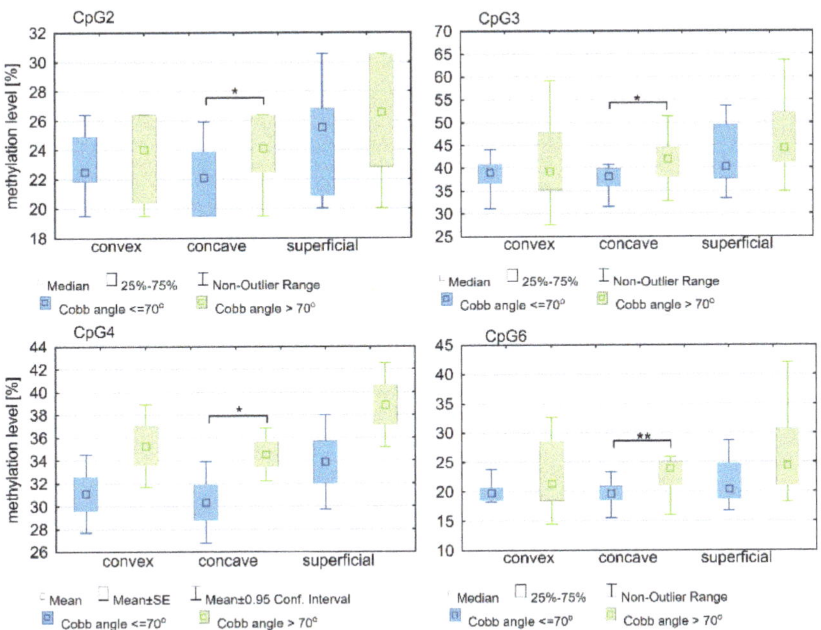

Figure 6. DNA methylation level within *ESR1* T-DMR2 region in deep paravertebral muscles and superficial muscles in patients with Cobb angles ≤70° and >70°; * $p < 0.05$, ** $p < 0.01$.

3. Discussion

Although the history of IS has been thoroughly described and treatment methods are established, the exact etiology and pathology have yet to be elucidated [1,8,35]. IS has been extensively analyzed with respect to susceptibility to scoliosis development and curvature progression, with various theories concerning IS etiology suggested. Recently, genetic studies revealed an important association between DNA polymorphisms and disease susceptibility and severity [36–38]. Despite promising results, these studies did not provide insight to any IS predisposition nor provide a molecular explanation of the disease. It has become increasingly apparent that many diseases are likely the result of interactions between genes and the environment [2]. According to Grauers et al., 38% of the variance in the liability of IS development is due to additive genetic effects and 62% to unique environmental effects [39]. Thus, one of the most interesting hypotheses regarding IS etiology is the linkage of genetic susceptibilities to environmental factors.

Several epigenetic studies concerning the etiopathogenesis of IS have been conducted. As of now, five of them described DNA methylation as an epigenetic mechanism associated with IS. Mao et al. evaluated methylation levels of the cartilage oligomeric matrix protein gene. They found that hypermethylation of the gene promoter correlated with adolescent idiopathic scoliosis (AIS) curve severity [16]. Shi et al. published two studies concerning DNA methylation in AIS and revealed an association of paired-like homeodomain 1 and protocadherin-10 gene methylation with IS susceptibility and curvature severity [17,18]. Meng et al. conducted analysis of the whole-genome methylation in two pairs of twins. They found an association between methylation levels at site cg01374129 and curve severity [19]. Liu et al. also performed whole-genome methylation analysis in a pair of twins. They discovered several signaling pathways potentially associated with AIS and a significantly higher methylated region in chromosome 15 of the AIS group [15]. All mentioned studies concerning DNA methylation were performed with peripheral blood samples. In the search for the molecular explanation of IS, we analyzed the local molecular predisposition to IS occurrence or progression at the apex of the curvature. We focused on paraspinal muscles as a possible target tissue for locally acting factors as these muscles play a key role in controlling spinal stability [2]. There is a hypothesis that dysfunctional paraspinal muscles may contribute the development of the scoliotic curve [2,40]. Additionally, reports have described functional and histological differences in the paraspinal muscles between the convex and the concave sides of the curve in IS patients [41,42].

An interesting feature of IS is the correlation of disease severity with gender, especially after puberty. The female/male ratio in mild scoliosis is reported to be 1.4/1, while in severe scoliosis, it is estimated to be 8.4/1 [13,20,43]. This shift suggests a relationship between sex hormones with a clinical manifestation of IS [43,44]. As *ESR1* and *ESR2* are known to mediate the effects of estrogens, they became the subject of genetic studies concerning DNA polymorphisms of *ESR1* and *ESR2* in IS. Although early studies were promising, they failed to be replicable in subsequent studies [45,46]. A recent cross-sectional study revealed that some *ESR1* and *ESR2* variants were associated with the occurrence risk of idiopathic scoliosis [12]. Meta-analysis performed by Sobhan et al. suggested that *ESR1* polymorphisms rs9340799 and rs2234693 are not related to the risk of IS occurrence. However, rs9340799 may be associated with the risk of developing AIS among the Asian population [47]. Due to the unknown role of estrogens and their receptors in IS etiology, we evaluated the gene methylation status of estrogen receptors in IS.

In our study, we found differences in methylation levels between the deep paravertebral muscles (*m. longissimus*) and the superficial muscles (*m. trapezius*) in two CpGs of T-DMR1 and in all CpGs of T-DMR2. We consider the superficial muscle as a control, due to its distance from deformation, anatomical borders (fascia layers) between it and the deep muscles, different embryogenesis and function from the deep muscles, and a disparate nerve supply. Thus, this observed difference in methylation supports the theory of distinctive methylation patterns depending on localization in the same tissue type. Slieker et al. identified, using genome-wide DNA methylation data, that there are T-DMRs in CpG-poor regions such as CGI shores or distal promoters, which are associated with aberrant transcription. Interestingly, the authors observed interindividual variation of DNA methylation for more than 8000 CpGs in the skeletal muscle tissue and within-individual methylation differences between muscle and blood tissues for over 2000 CpGs [48]. Therefore, the interpretation of methylation patterns for tissues representing cellular heterogeneity, such as skeletal muscle, is particularly complex. It is also a challenge in comparative tissue research [48–50]. According to Maekawa et al., *ESR1* has tissue-dependent and differentially methylated regions (T-DMRs), which are associated with tissue-specific gene expression [34].

Our results did not reveal a difference in DNA methylation between the concave and convex side of paravertebral muscles when all patients were considered together. However, we found a difference in DNA methylation between patients with a Cobb angle ≤70° and >70° in the T-DMR2 region at the concave side of the curvature. Moreover, in CpG2 and CpG6 in the T-DMR2 region, the level of methylation at the concave side of the curvature correlated with the Cobb angle value. According to the abovementioned studies concerning the differences between paraspinal muscles at the apex of the curvature, this side was significant in relation to the etiopathogenesis of IS [51–53]. The absence of this correlation in the superficial and paravertebral muscles on the convex side in any of the evaluated CpGs also supports the importance of the concave side of the curvature in predisposition to IS progression. Thus, we interpret these results as a lack of association of *ESR1* methylation with a predisposition to IS development and consider methylation status as an IS phenotype modifier rather than the direct molecular background. These results are in line with the opinion of Cheung et al. who state that some factors may contribute to curve progression while others contribute to curve initiation [54]. According to Leboeuf et al., estrogens may be considered contributing factors in the progression of scoliosis [55]. Our results support this hypothesis from an epigenetic point of view.

When debating on changes of the paravertebral muscles at the level of the concave side of the curvature, there exists a dilemma regarding the primary versus secondary nature due to impaired spine biomechanics (asymmetrical loading). It is possible that the difference in methylation may contribute to some asymmetry in muscle function and promote curvature progression. On the other hand, the presented results may be a consequence of exposure to different local mechanical conditions due to asymmetric loading or other unknown factors. In our opinion, the results of this study most likely reveal primary changes. The impact of asymmetrical loading or other factors on the methylation level should be detectable in all evaluated CpGs, not only in specific ones. To evaluate the direct impact of DNA methylation on IS progression, two patient subgroups were distinguished. The patients were divided according to disease severity and its possible impact on the patient health [35,56,57]. The skeletally mature patients with a Cobb angle between 50° and 70° require surgical scoliosis correction to avoid further curvature deterioration into adulthood [35,56,57]. Whereas severe curvatures can negatively impact patient health including outcomes such as decreased lung function, cardiac function, back pain, and degenerative spine disease [3,5,57]. There is no solid threshold for Cobb angle value when curvature significantly impacts the patients' health. Studies concerning surgical treatment of IS classify scoliosis as a severe when a curvature exceeds 70° in Cobb angle [58–60]. Thus, we used this value to categorize study subgroups.

Contrary to previous studies, we observed a positive correlation between *ESR1* expression and methylation level within regulatory regions. Maekawa et al. showed that *ESR1* expression was tissue specific and downregulated by DNA methylation at T-DMRs in normal tissues but not always in breast cancer. They have also evaluated the expression level of different *ESR1* variants and suggested that there is interplay between DNA methylation of T-DMRs and regions around upstream exons [34]. Our result, different from those of Maekawa [34], may be explained by the fact that we examined all transcription variants in one quantitative reaction. Additionally, we did not evaluate methylation of promoter regions but only both T-DMRs. Moreover, it is indicated that T-DMR methylation may modulate the availability of DNA sequences for methylation-dependent transcription factors [61]. Those findings are in line with results presented by Maekawa et al., who identified that EGR1 (early growth response protein 1) may be the potential transcription factor that binds to the T-DMRs and, as a result, upregulates *ESR1* expression [34]. It has also been suggested that there are T-DMRs negatively and positively correlated with gene expression depending on genomic localization [61].

Direct comparison of our results with other published studies concerning methylation of DNA in IS was challenging due to different tissue samples used for evaluation. Peripheral blood is a very good source of DNA when polymorphisms are considered. However, the methylation level obtained from the blood will only show the methylation level of whole DNA without specific local disturbances. Thus, a strong point of this study was the evaluation of tissues at the center of the pathology, thereby bringing forth new facts about the impact of *ESR1* DNA methylation on the IS phenotype.

Our study reveals new aspects concerning IS etiopathogenesis. It develops a further explanation of why some IS progress more often than others. A better understanding of the pathology improves the diagnosis and treatment methods. However, the direct clinical implications of this study are limited. Genetic studies aim to develop a test that may help distinguish the prognosis between patients with severe disease from a more benign condition. We hope that further studies beyond our results can be useful in the development of such a test.

The main limitation of our study is the lack of a healthy control group. It was impossible to obtain paraspinal muscles samples from healthy, age-matched females. We considered harvesting muscle samples from patients who have undergone a surgery due to degenerative spine disease. However, the vast difference in patient age and the muscle atrophy due to long-lasting degenerative spine disease may induce unknown methylation changes. This would be possible introduction of bias rather than a reliable evaluation of methylation impact on the etiology of IS. Another limitation in this study was sample size. However, it is comparable with other studies evaluating DNA methylation in IS even though the other studies were performed on blood samples [15–19].

4. Materials and Methods

4.1. Patient Population

The study group consisted of 29 girls who underwent an operation due to IS between January 2017 and December 2019 at the Department of Spine Disorders and Pediatric Orthopedics in Poznan University Hospital. All patients met the following inclusion criteria: (1) confirmed diagnosis of IS (other backgrounds of scoliosis were excluded); (2) no coexisting genetic, neurological, or orthopedic disorders; (3) thoracic location of the main curvature; (4) surgical treatment with posterior spinal instrumentation and fusion. All patients underwent clinical and radiological examinations, including long-cassette standing X-rays taken prior to surgery. Number, localization, and curvature size (Cobb angle) was measured [62]. Skeletal maturity was assessed by the Risser sign [63]. One experienced spine surgeon performed all measurements. Patients were divided into two subgroups according to final disease severity at skeletal maturity: scoliosis of equal to or less than $70°$ vs. greater than $70°$ as measured with the Cobb angle. The first subgroup (Cobb $\leq 70°$) consisted of 10 patients without a major risk of significant impact on cardio-pulmonary function in adulthood. The second subgroup (Cobb $> 70°$) consisted of 19 patients with severely progressive IS that possibly may impact cardio-pulmonary function.

4.2. Tissue Samples

During surgery, 1 cm^3 muscle tissue fragments were obtained from one deep paravertebral muscle (*m. longissimus thoracic*) on the convex and concave side of the curvature as well as from one superficial muscle (*m. trapezius*). Samples were stored at $-80\,°C$ in tubes containing nucleic acid preservation solution (Novazym, cat no. ST01; Poznan, Poland).

4.3. Genomic DNA Methylation Analysis

4.3.1. Genomic DNA Isolation and Bisulfite Conversion

Total genomic DNA was extracted using a silica matrix column kit (Zymo Research, cat no. D4069; Irvine, CA, USA) with a modified protocol. In short, 25 mg of tissue samples ground in liquid nitrogen were incubated overnight at 55 °C with proteinase K. The lysate was then centrifuged ($12,000 \times g$, 1 min, room temperature). Next, the procedure followed the isolation according to manufacturer's protocol. The gDNA quantity, purity, and integrity were assessed both spectrophotometrically and electrophoretically. One microgram of gDNA was bisulfite converted using an EZ DNA Methylation™ Kit (Zymo Research, cat no. D5002; Irvine, CA, USA) according to the manufacturer's protocol.

4.3.2. Polymerase Chain Reaction and Pyrosequencing Analysis

Bisulfite converted DNA served as the template for polymerase chain reaction (PCR) followed by pyrosequencing (PSQ). The primers for PCR and PSQ reactions were designed using PyroMark Assay Design software (version 2.0.1.15; Qiagen; Hilden, Germany). The input DNA sequences corresponded to the T-DMR1 and T-DMR2 regions of the *ESR1* gene (https://www.ncbi.nlm.nih.gov (accessed on 5 March 2019); GenBank N°: NG_008493.2). Sequencing, forward, and biotinylated reverse primers are presented in Table 1.

Polymerase chain reactions were performed using ZymoTaq™ PreMix (Zymo Research; cat no. E2004; Irvine, CA, USA) designed for the amplification of bisulfite-treated DNA. Reaction mixture components, concentrations, and thermal profile is presented in Table 2. Two microliters of the product were separated using a standard 2% agarose gel and compared to molecular mass marker (Novazym, cat no. MA1000-03; Poznan, Poland).

Table 1. Primer sequences and location.

	Primer	Sequence		Tm (°C)	GC (%)	PCR Product Size	Location with Respect to TSS	Location with Respect to ATG
ESR1 T-DMR1	→ PCR	GGGTGTATGTGAGTGTGTATGTTTAA	26	58.8	38.5		−1107	−1341
	← PCR [B]	ATAAAATATAACCTTTTCATACCAAACAT	29	56.8	20.7	256 bp	−851	−1085
	→ SEQ	GTATGTGAGTGTGTATGTTTAAT	23	44.7	30.4	-	−1105	−1337
ESR1 T-DMR2	→ PCR	GTTTTTATTGGGTGTTATGTGTTTTGG	27	56.8	24.1		−2886	−3120
	← PCR [B]	AAACCTTTCCATAAATAACTCAATTAACT	29	56.8	20.7	307 bp	−2579	−2813
	→ SEQ	GTTATGTGTTTTGGGAT	17	47.2	53.3	-	−2874	−3108

→ PCR—forward primer; ← PCR—reverse primer; [B]—biotinylated primer; Tm—melting temperature, GC—guanine-cytosine content; bp—base pairs; TSS—transcription start site; ATG—start codon; → SEQ—sequencing primer.

Table 2. PCR mixture content and thermal profile of the reactions.

Component	Initial Concentration	Volume Added	Final Concentration	Mixture Volume
ZymoTaqTMPremix	2×	5 µL	1×	
→PCR	10 µM	1 µL	1 µM	
←PCR	10 µM	1 µL	1 µM	10 µL
DNA	100 ng/µL	0.2 µL	2 ng/µL	
Nuclease-free water		2.8 µL		

	Thermal profile of the reactions	
Number of cycles	**Step**	**Duration, temperature**
1	Initial denaturation	10 min, 95 °C
37	Denaturation	30 s, 95 °C
	Annealing	30 s, 54 °C
	Extension	60 s, 72 °C
1	Final extension	7 min, 72 °C
1	Hold	∞, 4 °C

→PCR—forward primer; ←PCR—reverse primer; min—minutes, s—seconds.

PSQ analysis was performed using the PyroMark Q48 instrument (Qiagen; Hilden, Germany) according to CpG assays designed with Pyromark Q48 Autoprep 2.4.2 software (Qiagen; Hilden, Germany). Analysis of 4 and 8 CpG sites for T-DMR1 and T-DMR2, respectively, were performed (internal sodium bisulfite treatment quality control was included in each reaction). The methylation level was quantified using Pyromark Q48 Autoprep 2.4.2 software and expressed as a percentage ratio of methylated to non-methylated dinucleotides.

4.4. Analysis of ESR1 mRNA Expression

4.4.1. Total RNA Isolation and Reverse Transcription

Total cellular RNA was extracted using Renozol (GenoPlast Biochemicals, cat no. BNGPB1100-2; Rokocin, Poland) and Direct-zol RNA Miniprep Kit (Zymo Research, cat no. R2052; Irvine, CA, USA) following the manufacturer's protocol. RNA quantity and purity were assessed similarly to gDNA. RNA integrity was evaluated with 18S and 28S ribosomal RNA using 1% standard denaturing agarose gel electrophoresis.

Reverse transcription reactions were performed using M-MuLV-RT (Sigma-Aldrich, cat no.11785826001; Saint Louis, MO, USA) according to the manufacturer's protocol. The total reaction volume was 10 µL. In the first step, the mixture containing 500 ng of total RNA, water, 5 mmol/µL universal oligo(dT)$_{10}$ primer, and 300 nmol/µL random hexamer primer were denatured at 65 °C for 10 min then cooled on ice. Subsequently, 2 mmol/µL of each deoxynucleotide triphosphates, 1.5 U/reaction of *E. coli* Poly(A) Polymerase (Carolina Biosystems, cat no. PAPY-30; Prague, Czech Republic), 150 nm/µL deoxyadenosine triphosphates, 15U/reaction of ribonuclease inhibitor, 1X buffer M-MuLV-RT buffer, and 10 U/reaction of M-MuLV reverse transcriptase were added. Samples were incubated at 25 °C for 10 min, 55 °C for 60 min, then 5 min at 85 °C. Complementary DNA (cDNA) was either immediately used for quantitative polymerase chain reaction (qPCR) or stored at $-20\ ^\circ$C until further analysis (but no longer than seven days).

4.4.2. Quantitative Polymerase Chain Reaction

ESR1 mRNA was quantified using sequence-specific primers (sense: CCTTCTTCAA-GAGAAGTATTCAAGG and antisense: ATTCCCACTTCGTAGCATTTG) and the Roche Universal ProbeLibrary TaqMan® hydrolysis probe (#69, cat no. 04688686001) using the ProbeFinder Assay Design Center (https://lifescience.roche.com/en_pl/brands/universal-probe-library.html, accessed on 4 October 2016). The hypoxanthine-guanine phosphoribo-syltransferase (*HPRT*) gene was used as a reference gene (RealTime ready *HPRT*, Roche, cat no. 05532957001; Basel, Switzerland). The 20µL total volume reaction mixture contained 5 µL cDNA, 1X LightCycler® FastStart TaqMan® Probe Master (Roche, cat no. 04673417001; Basel, Switzerland), and 1X RealTime ready *HPRT* for reference gene or 200 nm of hy-drolysis probe #69 along with 400 nm of the primer mixture for the gene of interest, and nuclease-free water. qPCR reactions were performed using the LightCycler® 2.0 carousel glass capillary-based system (Roche). The thermal profile was performed as previously described [25]. Each sample was analyzed in duplicate with independently synthesized cDNA. The quantitative PCR results were assembled using the LightCycler Data Analysis Software version 5.0.0.38 (Roche; Basel, Switzerland), and the fluorescence measurement results were normalized to standard curves [25]. In each sample, *ESR1* expression was compared to reference gene expression in order to obtain a Cr value (concentration ratio) which corresponded to the relative *ESR1* expression level.

4.5. Statistical Analyses

Data analyses were performed using Statistica 13.3 software (TIBCO Software Inc.; Palo Alto, CA, USA) and PQStat 1.8.0.414 software (PQStat software; Poznan, Poland). The methylation level of specific CpG sites was analyzed in T-DMR1 and T-DMR2 separately for each CpG site in each region. The Shapiro–Wilk test was used for the normality of continuous variable distribution assessment. The differences in methylation levels between concave, convex, and superficial muscles was evaluated using repeated measures ANOVA or Friedman ANOVA with HSD Tukey and Dunn's Bonferroni post-hoc tests, respectively. Methylation between patient subgroups with a Cobb angle $\leq70^\circ$ or $>70^\circ$ was compared using an independent t-test or Mann–Whitney U test. The correlation coefficients were determined by Pearson's (r) or Spearman's rank test (R). Data are presented as mean $\pm$ SE (standard error) or median with interquartiles. Data was considered statistically significant when $p < 0.05$.

5. Conclusions

The DNA methylation level of *ESR1* regulatory regions is specific to the muscle tissue localization in patients with idiopathic scoliosis. The lack of significant asymmetry between the concave, compared to the convex, side of the spinal curvature suggests that *ESR1* methylation level does not signify predisposition to the occurrence of IS. The difference in *ESR1* T-DMR2 CpGs methylation of the deep paravertebral muscles on the concave side of the curvature may be associated with IS severity.

Supplementary Materials: The following are available online at https://www.mdpi.com/article/10.3390/genes12060790/s1, Supplementary Table S1. DNA methylation level (%) within *ESR1* T-DMR1 and T-DMR2 regions in deep paravertebral muscles and superficial muscles (n = 29). Supplementary Table S2. DNA methylation level (%) within *ESR1* T-DMR1 and T-DMR2 regions in deep paravertebral muscles and superficial muscles in the groups of patients with Cobb angles $\leq70^\circ$ (n = 10) and $>70^\circ$ (n = 19). Supplementary Figure S1. Relative *ESR1* expression level in deep paravertebral muscles and superficial muscles. Supplementary Figure S2. Scatter plots showing correlations between *ESR1* expression and methylation level within T-DMR1 region in deep paravertebral muscles and superficial muscles. Figure S3. Scatter plots showing correlations between *ESR1* expression and methylation level within T-DMR2 region in deep paravertebral muscles and superficial muscles.

Author Contributions: Conceptualization, P.J., M.C., M.A., M.K. and T.K.; methodology, P.J., M.C. and M.A.; validation, M.C. and M.A.; formal analysis, P.J., M.C., M.A. and M.K.; investigation, P.J., M.C. and M.A.; resources, P.J.; data curation, P.J., M.C. and M.A.; writing—original draft preparation, P.J., M.C., M.A. and M.K.; writing—review and editing, M.C., M.A. and T.K.; visualization, M.C. and M.A.; supervision, P.J., T.K.; project administration, P.J.; funding acquisition, P.J. All authors have read and agreed to the published version of the manuscript.

Funding: This research was funded by the National Science Centre, grant number 2016/23/D/NZ5/02606.

Institutional Review Board Statement: The study was conducted according to the guidelines of the Declaration of Helsinki and approved by the Institutional Review Board of Poznan University of Medical Sciences (protocol code No 546/17 and 741/19 and date of approval 5/11/2017 and 6/19/2019, respectively).

Informed Consent Statement: Informed consent was obtained from all the patients or their parents/legal guardians in the case of underage participants.

Data Availability Statement: The datasets used and analyzed during the current study are available from the corresponding author on reasonable request.

Acknowledgments: Not applicable.

Conflicts of Interest: The authors declare no conflict of interest.

Abbreviations

AIS: adolescent idiopathic scoliosis; cDNA: complementary DNA; ESR1: estrogen receptor 1; ESR2: estrogen receptor 2; *HPRT*: hypoxanthine-guanine phosphoribosyltransferase; IS: idiopathic scoliosis; PCR: polymerase chain reaction; PELP1: proline-, glutamic acid- and leucine-rich protein; qPCR: quantitative polymerase chain reaction; SE: standard error of mean; T-DMRs: differentially methylated regions.

References

1. Weinstein, S.L.; Dolan, L.A.; Cheng, J.C.Y.; Danielsson, A.; Morcuende, J.A. Adolescent idiopathic scoliosis. *Lancet* **2008**, *371*, 1527–1537. [CrossRef]
2. Newton Ede, M.M.P.; Jones, S.W. Adolescent idiopathic scoliosis: Evidence for intrinsic factors driving aetiology and progression. *Int. Orthop.* **2016**, *40*, 2075–2080. [CrossRef] [PubMed]
3. Weinstein, S.L.; Zavala, D.C.; Ponseti, I.V. Idiopathic scoliosis. Long-term follow-up and prognosis in untreated patients. *J. Bone Jt. Surg. Ser. A* **1981**, *63*, 702–712. [CrossRef]
4. Huh, S.; Eun, L.Y.; Kim, N.K.; Jung, J.W.; Choi, J.Y.; Kim, H.S. Cardiopulmonary function and scoliosis severity in idiopathic scoliosis children. *Korean J. Pediatr.* **2015**, *58*, 218–223. [CrossRef]
5. Nepple, J.J.; Lenke, L.G. Severe idiopathic scoliosis with respiratory insufficiency treated with preoperative traction and staged anteroposterior spinal fusion with a 2-level apical vertebrectomy. *Spine J.* **2009**, *9*, e9–e13. [CrossRef]
6. Noshchenko, A.; Hoffecker, L.; Lindley, E.M.; Burger, E.L.; Cain, C.M.J.; Patel, V.V.; Bradford, A.P. Predictors of spine deformity progression in adolescent idiopathic scoliosis: A systematic review with meta-analysis. *World J. Orthop.* **2015**, *6*, 537–558. [CrossRef]
7. Roye, B.D.; Wright, M.L.; Matsumoto, H.; Yorgova, P.; McCalla, D.; Hyman, J.E.; Roye, D.P.; Shah, S.A.; Vitale, M.G. An Independent Evaluation of the Validity of a DNA-Based Prognostic Test for Adolescent Idiopathic Scoliosis. *J. Bone Jt. Surg. Am.* **2015**, *97*, 1994–1998. [CrossRef]
8. Fadzan, M.; Bettany-Saltikov, J. Etiological Theories of Adolescent Idiopathic Scoliosis: Past and Present. *Open Orthop. J.* **2017**, *11*, 1466–1489. [CrossRef]
9. Lowe, T.G.; Edgar, M.; Margulies, J.Y.; Miller, N.H.; Raso, V.J.; Reinker, K.A.; Rivard, C.H. Etiology of idiopathic scoliosis: Current trends in research. *J. Bone Jt. Surg. Ser. A* **2000**, *82*, 1157–1168. [CrossRef]
10. Liu, G.; Liu, S.; Lin, M.; Li, X.; Chen, W.; Zuo, Y.; Liu, J.; Niu, Y.; Zhao, S.; Long, B.; et al. Genetic polymorphisms of GPR126 are functionally associated with PUMC classifications of adolescent idiopathic scoliosis in a Northern Han population. *J. Cell. Mol. Med.* **2018**, *22*, 1964–1971. [CrossRef]
11. Liu, G.; Liu, S.; Li, X.; Chen, J.; Chen, W.; Zuo, Y.; Liu, J.; Niu, Y.; Lin, M.; Zhao, S.; et al. Genetic polymorphisms of PAX1 are functionally associated with different PUMC types of adolescent idiopathic scoliosis in a northern Chinese Han population. *Gene* **2019**, *688*, 215–220. [CrossRef] [PubMed]
12. Wang, L.; Zhang, Y.; Zhao, S.; Dong, X.; Li, X.; You, Y.; Yan, Z.; Liu, G.; Tong, B.; Chen, Y.; et al. Estrogen receptors (ESRs) mutations in adolescent idiopathic scoliosis: A cross-sectional study. *Med. Sci. Monit.* **2020**, *26*, e921611-1–e921611-7. [PubMed]

Review

Models of Distal Arthrogryposis and Lethal Congenital Contracture Syndrome

Julia Whittle [1], Aaron Johnson [2], Matthew B. Dobbs [3] and Christina A. Gurnett [1,*]

[1]   Department of Neurology, Washington University in St Louis, St Louis, MO 63130, USA;
      juliawhittle@wustl.edu
[2]   Department of Developmental Biology, Washington University in St Louis, St Louis, MO 63130, USA;
      anjohnson@wustl.edu
[3]   Paley Orthopaedic and Spine Institute, West Palm Beach, FL 33407, USA; mdobbs@paleyinstitute.org
*   Correspondence: gurnettc@wustl.edu

Abstract: Distal arthrogryposis and lethal congenital contracture syndromes describe a broad group of disorders that share congenital limb contractures in common. While skeletal muscle sarcomeric genes comprise many of the first genes identified for Distal Arthrogyposis, other mechanisms of disease have been demonstrated, including key effects on peripheral nerve function. While Distal Arthrogryposis and Lethal Congenital Contracture Syndromes display superficial similarities in phenotype, the underlying mechanisms for these conditions are diverse but overlapping. In this review, we discuss the important insights gained into these human genetic diseases resulting from in vitro molecular studies and in vivo models in fruit fly, zebrafish, and mice.

Keywords: contracture; arthrogryposis; congenital

Citation: Whittle, J.; Johnson, A.; Dobbs, M.B.; Gurnett, C.A. Models of Distal Arthrogryposis and Lethal Congenital Contracture Syndrome. *Genes* **2021**, *12*, 943. https://doi.org/10.3390/genes12060943

Academic Editors: Philip Giampietro, Nancy Hadley-Miller and Cathy L. Raggio

Received: 17 May 2021
Accepted: 16 June 2021
Published: 20 June 2021

Publisher's Note: MDPI stays neutral with regard to jurisdictional claims in published maps and institutional affiliations.

1. Introduction

Arthrogryposis (arth = joint; grp = curved; osis = pathological state) describes a broad range of phenotypes consisting of multiple congenital joint contractures presenting at birth [1]. About 1 in 3000 live births presents with some form of arthrogryposis, many of which are nonprogressive and improve with physiotherapy. The core root of arthrogryposis is fetal akinesia, or lack of fetal movement, that results in contractures forming in the joints [1–3]. Movement is required for normal joint development; it influences the structure of the joints, as well as promoting cellular signaling that guides normal tissue development. Mechanical forces also influence bone morphology, affecting organization of chondrocytes, bone elongation, and differential growth, all affecting the shape of bones as they develop. Fetal akinesia impairs joint formation, which may lead to joint fusions. Furthermore, tension is required for normal tendon development, forming a connection between bone and muscle [4]. Arrested movement during development has significant impact on the formation of the skeleton, joints, muscle, and connective tissues.

The full range of joint movement in utero can be perturbed both intrinsically and extrinsically. Intrinsically, mutations affecting the muscle, bone, connective tissue, and neural system can affect the range of movement of joints. Currently, there are over 400 genes associated with arthrogryposis broadly, encapsulating a wide diversity of genes affecting different pathways, including genes associated with axon structure, circulatory development, or synaptic transmission [5]. Extrinsically, maternal disease or exposures, uterine space limitation, and decreased blood supply are also root causes for contraction defects [1,2]. Because joint motion is affected by many different systems, a wide range of issues during development can arrest joint motion.

A subset of arthrogryposis is described as Distal Arthrogryposis (DA), a group of genetically induced contractures that predominantly affect the joints of the distal limbs, including the hands, wrists, ankles, and feet. Clinically, the lower extremity manifestations commonly include clubfoot and vertical talus. There are currently 10 classifications of DA, including Sheldon-Hall syndrome (DA2B) and Freeman-Sheldon syndrome (DA2A) [6–9]. Freeman-Sheldon syndrome is considered the most severe form of DA, and also presents with facial contractures [9].

Currently, DA patients are offered supportive care to improve quality of life, including occupational therapy, physical therapy, and surgery [10]. While these treatments improve outcome for patients, they often fall short of complete restoration of range of motion in the joints and functionality. This strategy also fails to address underlying causes for DA, such as muscle weakness and impaired neurotransmission. Therefore, further investigation is necessary to understand the impact of disease variants which will allow us to determine the most effective treatment options for patients.

Lethal Congenital Contracture Syndromes (LCCS) are included in this review, as some of the same genes and disease mechanisms apply to this serious condition, which is typically fetal or neonatal lethal. LCCS presents with severe generalized contractures, along with many other typical features including incomplete lung development, and polyhydramnios [11]. In contrast to DA, which is most often inherited as an autosomal dominant condition, LCCS has only been described in the autosomal recessive state. Eleven subtypes have been described to date [10].

Various disease models have been developed to examine the mechanisms behind DA and to test therapeutic interventions (Table 1). Molecular and single-cell studies are useful for precisely examining the effects of DA-causing variants on the affected proteins and tissues. Access to human tissues, particularly from muscle biopsies, has facilitated molecular analysis for research, yet is clinically useful only in select cases [12,13]. In addition, protein modeling can help predict the impact of various amino acid substitutions on molecular interactions [14,15]. On the other hand, animal models are necessary to analyze the effect of single gene variants on organisms on scales larger than single cells. The effect of zygosity and gene dosage may also be better studied in animal models to assess interactions between normal and abnormal gene products. Animal models are also useful for studying experimental interventions that may improve patient quality of life and outcome, acting as stand-ins for potential human patients.

Table 1. List of genes and associated conditions and models of distal arthrogryposis (DA) and lethal congenital contracture syndrome (LCCS) used for study. Autosomal dominant (AD), Autosomal recessive (AR) [9,11,12,14,16–57].

Gene	Full Name	Disorder	Inheritance Pattern	Modeled in	Source Human	Models of Disease Source
ARTHROGRYPOSIS						
MYH3	Myosin, Heavy Polypeptide 3, Skeletal Muscle, Embryonic	DA1, DA2A, DA2B, DA8, Spondylocar-potarsal Syndrome	AD, AR	Zebrafish, Cell, Biochemical	Toydemir et al., 2006b [9]; Chong et al., 2015 [57]; Cameron-Christie et al., 2019 [21]	Racca et al., 2015 [12]; Walklate et al., 2016 [54]; Wang et al., 2019 [55]; Whittle et al., 2020 [16]; Guo et al., 2020 [14]; Das et al., 2019 [25]
TPM2	Tropomyosin 2	DA1, Cap Myopathy, Nemaline Myopathy	AD, AR	Drosophila, Biochemical	Sung et al., 2003 [50]	Williams et al., 2015 [56]; Borovikov et al., 2017 [20]; Matyushenko & Levitsky, 2020 [39];
MYLPF	Myosin Regulatory Light Chain 2, Skeletal Muscle Isoform	DA1, DA2B	AD, AR	Zebrafish	Chong et al., 2020 [23]	Chong et al., 2020 [23]
MYBPC1	Myosin-Binding Protein C, Slow-Type	DA1, DA2, LCCS4	AD	Zebrafish	Gurnett et al., 2010 [30]; Li et al., 2015 [58]; Ekhilevitch et al., 2016 [28]; Shashi et al., 2019 [49];	Ha et al., 2013 [31]
MYBPC2	Myosin-Binding Protein C, Fast-Type	DA (unspecified)	AD	Zebrafish	Bayram et al., 2016 [18]	Li et al., 2016 [37]
TNNT3	Troponin T3, Fast Skeletal Type	DA2B	AD, AR	Mouse	Sung et al., 2003 [50]; Sandaradura et al., 2018 [48]	Ju et al., 2013 [34]

Table 1. *Cont.*

Gene	Full Name	Disorder	Inheritance Pattern	Modeled in	Source Human	Models of Disease Source
TNNI2	Troponin I2, Fast Skeletal Type	DA2B	AD	Mouse, Drosophila	Sung et al., 2003 [50]	Zhu et al., 2014 [17]; Vigoreaux, 2001 [53]
PIEZO2	Piezo Type Mechanosensitive Ion Channel Component 2	DA3, DA5	AR	Cell	McMillin et al., 2014 [40]	Coste et al., 2013 [24]; McMillin et al., 2014 [40]
ECEL1	Endothelin Converting Enzyme Like 1	DA5 (or DA5D)	AR	-	McMillin et al., 2013 [41]	-
MYH8	Myosin, Heavy Polypeptide 8, Skeletal Muscle, Fetal	DA7	AD	-	Toydemir et al., 2006a; [51] Veugelers et al., 2004 [52]	-
LETHAL CONGENITAL CONTRATURE SYNDROME						
GLE1	GLE1 RNA Export Mediator	LCCS1	AR	Zebrafish, Cell, Biochemical	Jao et al., 2012 [33]	Folkmann et al., 2013 [29]; Jao et al., 2012 [33]
ERBB3	ERB-B2 Receptor Tyrosine Kinase 3	LCCS2	AR	Mouse	Narkis et al., 2007 [44]	Riethmacher et al., 1997 [47]
PIP5K1C	Phosphatidylinositol 4-Phosphate 5-Kinase, type 1, gamma	LCCS3	AR	Mouse	Narkis et al., 2007 [43]	DiPaolo et al., 2004 [26]
MYBPC1	Myosin-Binding Protein C, Slow-Type	LCCS4, DA1, DA2	AD, AR	Zebrafish	Markus et al., 2012 [11]	Ha et al., 2013 [31]
DNM2	Dynamin, 2	LCCS5, Centronuclear Myopathy, CMT2M, CMT Intermed	AD, AR	Mouse	Koutsopoulos et al., 2013 [35]	Durieux et al., 2010 [27]; Koutsopoulos et al., 2013 [35]
ZBTB42	Zinc finger-and BTB Domain-containing Protein 42	LCCS6	AR	Zebrafish	Patel et al., 2014 [45]	Patel et al., 2014 [45]
CNTNAP1	Contactin-associated protein 1	LCCS7, Congenital Hypomyelinating Neuropathy	AR	Mouse	Laquerriere et al., 2014 [36]	Bhat et al., 2001 [19]
ADCY6	Adenylyl cyclase 6	LCCS8	AR	Zebrafish	Laquerriere et al., 2014 [36]	Laquerriere et al., 2014 [36]
ADGRG6	Adhesion G-protein coupled receptor G6 or GPR126	LCCS9	AR	Zebrafish	Ravenscroft et al., 2015 [46]	Monk et al., 2009 [42]
NEK9	Nima-related kinase 1	LCCS10	AR	-	Casey et al., 2016 [22]	-
GLDN	Gliomedin	LCCS11	AR	-	Maluenda et al., 2016 [38]	-

Models of human disease are rapidly becoming more sophisticated, with the ability to knock-in single nucleotide variants and create conditional (tissue specific or time-dependent) knockouts [16,17]. Loss-of-function alleles, which are often easier to generate, provide critical information about gene function, but may not fully explain autosomal dominant phenotypes in which gain-of-function or dominant negative effects can cause markedly different phenotypes. Conditional knockouts, while very helpful in defining gene function, rarely replicate the human phenotype in its entirety, but may be required when early lethality limits further study. These methods allow researchers to design models that more accurately represent these human conditions, and replicate pathogenic effects broadly or in specific tissues.

This review will examine genetic models of DA and LCCS, and the impact they have had in understanding the underlying pathophysiology. While we will describe both in vitro and in vivo approaches, we will focus primarily on vertebrate models, as these have the potential to provide insight into the multifaceted effects of disease variants on the multiple tissue types that contribute to these complex human phenotypes. We will also examine the current trajectory of DA research, and how these research strategies can help those afflicted by DA.

2. Muscle-Related Distal Arthrogryposis

2.1. MYH3

Missense mutations in *MYH3*, the earliest expressed embryonic myosin heavy chain gene that is predominantly expressed in myotubes destined to become fast-twitch myofibers [9], are strongly associated with DA clinically, and contribute to multiple subtypes including DA1, DA2A, and DA2B with varying degrees of severity [7–9,59,60]. *MYH3*-associated DA is almost always caused by single missense variants, as frameshift knockout or premature stop mutations are frequently observed in healthy population controls [9]. However, nonsense *MYH3* variants may contribute to autosomal recessive spondylocarpotarsal syndrome in the compound heterozygous state when presenting along with a missense allele [21], and have also been described with autosomal dominant spondylocarpotarsal syndrome [61]. DA-associated pathogenic variants cluster in the motor domain, but have also been found in the tail region of the protein [9]. Many of these missense variants are de novo, but some segregate in families with complete, or nearly complete penetrance [9]. *MYH3* appears to be one of the most common genes associated with DA, therefore various in vitro and in vivo studies, including protein modeling, cell models, and vertebrate studies, have been performed to elucidate the effects of *MYH3* variants on muscle function and the subsequent effects on the joints and skeleton.

2.1.1. Biochemical and Cell Models for *MYH3*-Associated Distal Arthrogryposis

Single molecule and single cell studies are useful to examine the precise impact of a variant on protein function. The effects of amino acid substitutions are difficult to predict without mechanistic examination. Single-molecule studies facilitate understanding the effect of a missense variant on protein function and can later be translated into an understanding of how small mechanical differences affect tissues and whole body systems. Missense variants can be studied in human skeletal muscle biopsies. However, these are not routinely performed for DA diagnosis, which makes these studies challenging. To study this mechanistic link between DA phenotype and gene variant, Racca et al. performed contractility studies on isolated muscle cells and myofibrils derived from biopsied muscle tissue from DA2A patients [12]. They found that a DA2A-associated *MYH3* variant inhibited cross-bridge detachment, thereby slowing muscle relaxation and lowering force production. A later study replicated these results while examining multiple *MYH3* variants associated with DA2 [54]. In addition to slower actin-myosin detachment, ATP binding and ATPase activity were lower in variant MYH3 molecules.

Development of single cell models, in which *MYH3* variants are exogenously expressed or overexpressed from a plasmid or virus, have been limited by the difficulties of expressing large genes, like *MYH3*, in vitro. Other key challenges of in vitro modeling include the paucity of skeletal muscle cell lines other than C2C12 cells, and the propensity of muscle cells to form a syncytium. In addition, difficulties in differentiating human induced pluripotent stem cells (iPSCs) into skeletal muscle have also limited their use for arthrogryposis disease modeling. Likewise, because many features of DA are due to complex relationships between different cell types, co-cultures of muscle cells with tenocytes and bone may be required to recapitulate the human condition. Thus, many investigators have preferred to study DA genes in whole organisms.

2.1.2. Invertebrate Models for *MYH3*-Associated Distal Arthrogryposis

Drosophila melanogaster (fruit flies) are useful tools for studying muscle function and myofibril assembly, particularly as introduction of single variants are traditionally simpler in this system compared to other models. *MYH3* and the *Drosophila* myosin heavy chain gene, *Mhc*, are highly conserved. *Drosophila* have the advantage of having only a single myosin heavy chain, which eliminates the possible obscuration of effects due to compensation by other myosin heavy chain analogs. Therefore, the effect of variants on protein function can be examined in a setting without other myosin heavy chain isoforms.

Drosophila transgenic models have been generated by overexpressing *Mhc* constructs containing DA variants [14,25]. Guo et al. predicted that a DA1 variant would perturb a hydrophobic interaction, while a DA2B mutation would introduce a hydrogen bond that was not present in the wild type. The effect of these predicted interactions was tested mechanistically in *Drosophila* models. Muscle fibers containing the DA alleles were extracted and found to have lower actin affinity, reduced power output, and increased stiffness, which may explain the motor deficits [14].

Morphological studies of *Drosophila* skeletal muscle expressing three DA2A *Mhc* transgenes (R672H, R672C, and T178I) showed branching and splitting defects, which were most severe in the R672C variant, which caused Z-discs to be split and malformed. In addition, Z-disc distance was shorter in the transgenic flies indicating an overall shortening of the sarcomeres, perhaps due to an enhanced contractile state of the myofibers. Presumably, the shortened sarcomeres observed in *Drosophila* contribute to the formation of contractures in human patients. Indeed, ATPase activity was reduced in these transgenic flies, leading to functional defects in muscle activity [25].

In studying the intact adult *Drosophila*, Guo et al. showed that the Mhc^{F437I} mutants had a much longer lifespan than Mhc^{A234T} mutants, consistent with the less severe phenotype of DA1 patients compared to DA2B patients [14]. In addition, the researchers found that both mutants displayed aberrant myofibril assembly, as well as misaligned sarcomere structure including distorted M and Z lines [14]. Again, this phenotype was more severe in A234T mutants than in the F437I mutants. Mhc^{F437I} mutants displayed essentially normal myofibers and sarcomeres, while Mhc^{A234T} mutants had small myofibers with disrupted morphology, as well as abnormal sarcomeres [14]. Das et al. also found that decreased climbing capability of adult flies also correlated with the phenotypic severity in humans [25].

Like *Drosophila*, *Caenorhabditis elegans* (*C. elegans*) has also been used to study myosin heavy chain genes [62]. There are many advantages of *C. elegans* and *Drosophila* for disease modeling, including large numbers of progeny, knowledge of ontogeny of individual cells, and ease of functional studies for drug screening. However, as described in Gil-Galvez et al., the evolutionary distance, and differences in number of myosin heavy chain genes between invertebrates and humans makes it difficult to determine whether the gene being studied has the same function, particularly in terms of spatial and temporal expression, as its human counterpart. Furthermore, Gil-Galvarez et al. caution against overexpression studies in *C. elegans* broadly, citing interference with muscle cell function overall [62]. The major drawback of invertebrate models is, quite obviously, the lack of skeletal structures which limits their use in understanding the complex relationship between muscle, nerve, and bone.

2.1.3. Vertebrate Models for *MYH3*-Associated Distal Arthrogryposis

Germline loss of *Myh3* in mice results in altered muscle fiber size, fiber number, fiber type, and misregulation of genes, and adult *Myh3* null mice develop scoliosis [32]. However, the molecular defect may make these mice a better model for recessive *MYH3* spondylocarpotarsal synostosis syndromes than for autosomal dominant DA [21,61]. Notably, many patients with spondylocarpotarsal synostosis syndrome also have congenital contractures, which highlights the phenotypic overlap between DA and spondylocarpotarsal synostosis syndrome. Interestingly, *MYH3* was also shown to be expressed in bone, which the authors state may explain the effects of *MYH3* variants on both skeletal muscle and bone, particularly for patients with spondylocarpotarsal synostosis syndrome and bony fusions [61].

To more accurately model DA2A, Whittle et al. recently introduced one of the most common Freeman-Sheldon syndrome *MYH3* variants, R672H, into an analogous gene in zebrafish (*Danio rerio*) (*smyhc1*^{R673H}) [16]. Zebrafish breed profusely and are cost-effective compared to mice. They also mature quickly, develop in vitro, and are transparent in the first few days of life, which facilitates imaging. Zebrafish are also vertebrates, making them more closely related to humans than *Drosophila* or *C. elegans*. Because two zebrafish lines were created, including *smyhc1* null and *smyhc1*^{R673H} lines, gene dosage effects were studied by examining the variant in the context of different zygosities. Indeed, *smyhc1*^{R673H} homozygotes displayed severe, early lethal phenotype compared to *smyhc1*^{R673H} heterozygotes, indicating that the *smyhc1*^{R673H} mutation acts as a hypermorph [16]. This result suggests human fetal lethality if a DA missense variant occurs in the homozygous state, which has not yet been described.

Zebrafish larvae harboring the *smyhc1*^{R673H} variant demonstrated severe notochord kinks [1], and adults had vertebral fusions that were similar to those seen in patients autosomal dominant spondylocarpotarsal synostosis due to *MYH3* variants. On histological examination, skeletal muscle showed severely shortened and misshapen muscle fibers. Similar to studies in *Drosophila*, the somite length was reduced in *smyhc1*^{R673H} mutants, consistent with shortening of the sarcomere.

A major advantage of zebrafish is the ease with which drugs can be administered for therapeutic investigations and drug screening. Based on knowledge that myosin ATPase inhibitors are now being evaluated to treat human cardiomyopathy due to similar variants in cardiac myosin genes [63], Whittle et al. preemptively treated embryos with para-aminoblebbistatin to prevent contractures from forming in larvae. Para-aminoblebbistatin inhibits myosin heavy chain ATPase activity, which chemically relaxed the skeletal muscle and prevented the curved phenotype of the treated *smyhc1* mutant fish (Figure 1) [16]. Previous molecular and single fiber studies predicted this mechanistic effect. Based on this experimental work, myosin ATPase inhibitors may be a viable avenue for *MYH3*-associated DA treatment, but will most likely require development of skeletal-muscle-specific inhibitors and treatment at an appropriately early developmental window.

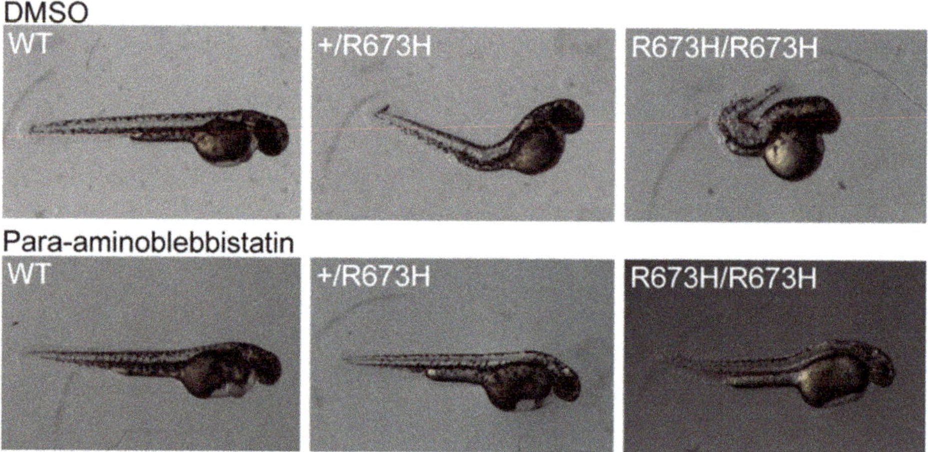

Figure 1. The curved spinal phenotype associated with both *smyhc1*$^{+/R673H}$ and *smyhc1*$^{R673H/R673H}$ genotypes is normalized with the myosin inhibitor para-aminoblebbistatin. Embryos were treated from 24–48 hpf and photographed at 48 hpf. Treated embryos are shown below DMSO treated controls. Unlike the newer myosin inhibitors that are being developed, para-aminoblebbistatin has many toxic effects, including lethal cardiac edema, which limits its use as a human therapeutic. These images are similar to those published in [16].

2.2. MYBPC1 and MYBPC2

Strong evidence now exists linking variants in the slow skeletal muscle myosin binding protein C1 (*MYBPC1*) to dominantly inherited DA1 [30], DA2 [58], arthrogryposis multiplex congenita [28], myopathy with tremor [49], and, in the recessive state, to lethal congenital contracture syndrome LCCS4 [11].

Morpholino knockdown of *mybpc1* in zebrafish resulted in embryos with severe body curvature, as well as impaired motor excitation with defective myofibril organization and reduced sarcomere numbers [31]. Furthermore, overexpression of human *MYBPC1* DA1-associated variants in zebrafish resulted in hypermorphic effects with body curvature, decreased motor activity, and impaired survival. No effect was seen with overexpression of wild-type transcripts, suggesting that overexpression studies in zebrafish could be an efficient model for future functional testing of the human variants of uncertain clinical significance.

In contrast to *MYBPC1,* which is strongly implicated in human disease, the role of fast skeletal muscle myosin binding protein C2 (*MYBPC2*) in DA is less clear, as there is only a single report of *MYBPC2* variants in DA patients in whom other known arthrogryposis gene variants were also observed, suggesting a possible role as a modifier [18]. Knockdown of *MYBPC2* with morpholino oligonucleotides produced a myopathic phenotype [37], but single variants have not yet been studied.

2.3. TPM2

TPM2 variants cause a spectrum of phenotypes, including DA1, DA2, as well as nemaline myopathy and cap myopathy (reviewed in (Tajsharghi et al., 2012)) [64]. All are autosomal dominant with the exception of a pathogenic null variant identified in a consanguineous family with Escobar variant of multiple pterygium syndrome that was observed in the recessive state [65]. Biochemical studies were undertaken to study *TPM2* gain-of-function phenotypes, including in vitro motility assays, which showed variable effects on calcium sensitivity and tropomyosin flexibility [20,39]. In addition, *TPM2* was recently shown to have noncanonical roles other than its sarcomeric function, where it binds thin filament actin to regulate muscle contraction. In this work, *TPM2* directly regulated muscle morphogenesis by directing myotubes toward tendon attachment sites [56]. Muscle morphology was disrupted in both flies and zebrafish expressing DA1-associated *TPM2* variants, likely by causing myofiber hypercontraction (Figure 2).

2.4. TNNI2

While the function of the fast skeletal muscle Troponin I (*TNNI2*) has been described in flightless *Drosophila* models [53], only a single DA disease-associated missense variant has been modeled in mice, which accurately recapitulated the human disease [17]. However, the small body size of mice carrying the *TNNI2* DA variant could not be explained by the direct effect of the variant on skeletal muscle morphology or function. Rather, *TNNI2* was shown to be expressed in osteoblasts and chondrocytes of long bone growth plates, through which its effects on growth was predicted to occur. Therefore, like the studies described above for *MYH3*, this model provides evidence that some DA phenotypes may be directly attributable to expression in non-muscle tissue, such as bone.

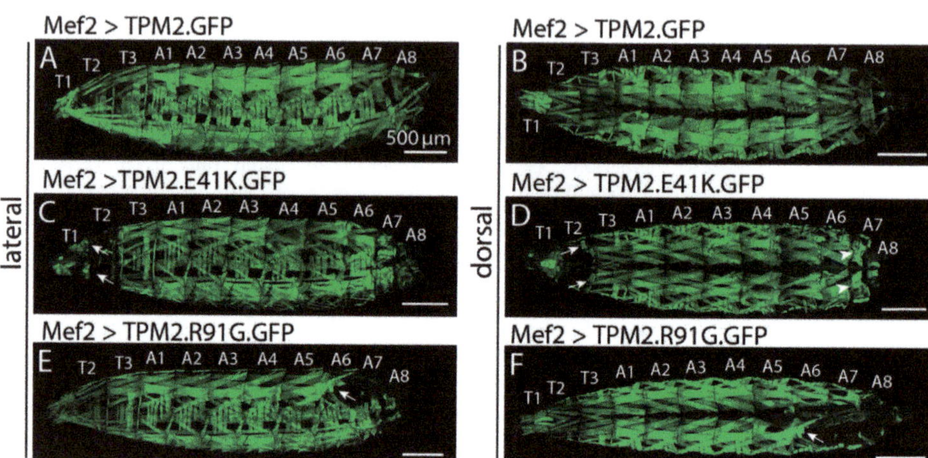

Figure 2. DA associated *TPM2* variants cause muscle phenotypes in *Drosophila*. Confocal micrographs of live L3 larva that express GFP-tagged TPM2 variants in skeletal muscles (body wall muscles). Mef2.Gal4 was used to activate UAS.TPM2 transgenes. Lateral and dorsal views are shown for each genotype. (**A,B**) Larva that express TPM2.GFP showed normal muscle histology. Larva that expresses TPM2.E41K.GFP (**C,D**) or TPM2.R91G.GFP (**E,F**). GFP have rounded myofibers that appear to result from internal tears (arrows; note affected muscles remain associated with tendons at segment boundaries) and shortened segments that could be due to hypercontractile muscles (arrowheads). Thoracic segments (T1–T3) and abdominal segments (A1–A8) are labeled. Scale bars, 500 mM. Previously unpublished data.

2.5. TNNT3

DA-associated variants in fast skeletal muscle Troponin T (*TNNT3*) are also dominantly inherited missense variants, and therefore, like those in many genes described previously, cause disease through a gain-of-function manner and therefore cannot be adequately modeled using a simple knockout approach. Therefore, while knockout approaches have shown a critical function of *TNNT3* during vertebral development [34], no models with disease-specific missense variants have been generated, and future studies are needed.

2.6. MYLPF

Exome sequencing recently identified *MYLPF*, a phosphorylatable fast skeletal muscle regulatory light chain, as a cause of DA [23]. Some affected individuals were homozygous for rare variants in the gene, while other individuals have autosomal dominant disease, a finding similar to what was described for *MYH3*-related disorders. However, unlike *MYH3*-related disorders, the phenotypes for *MYLPF* autosomal dominant and recessive conditions are apparently indistinguishable. Protein modeling of *MYLPF* alleles suggested that the autosomal dominant pathogenic variants cause disease through their direct interaction with myosin, while the recessive alleles only indirectly affect the interaction with myosin [23].

Previously described *mylpf* knockout mice did not develop either fast or slow type skeletal muscle mass, which resulted in early death before or after delivery, presumably due to respiratory failure [66]. Similarly, an individual with recessive *MYLPF*-associated DA1 was found to have absent skeletal muscle in an amputated foot, which could best be described as amyoplasia. The recessive *MYLPF* pathogenic variant in this individual was hypothesized to be hypomorphic. Therefore, to further model hypomorphic alleles of *MYLPF*, a zebrafish *mylpfa* mutant was characterized. Because zebrafish have 2 *MYLPF* genes, *mylpfa* and *mylpfb*, the more prominently expressed gene (*mlfpfa*) was chosen to model hypomorphic *MYLPF* alleles [23]. Of note, zebrafish had an evolutionary genome duplication event that has resulted in many human genes being represented twice in the zebrafish genome. Some of these genes have subsequently evolved to occupy additional temporal or spatial expression patterns and/or adopted new developmental roles. While this duplication event may complicate identification of the relevant human gene, gene duplications can also be an advantage, allowing genes to be studied whose knockout is early embryonic lethal in other species.

Mylpfa zebrafish null mutants were found to have paralyzed pectoral fins, an impaired escape response, and consistently lower trunk muscle force compared to wildtype [23]. In in vitro studies, myosin extracted from *mylpfa* mutant larvae propelled significantly more slowly than their wild type protein. Skeletal muscle fibers were also found to degenerate in mutant larvae, collapsing and losing structure, developing membrane abnormalities, indicating that *mylpf* is necessary to maintain cellular integrity in muscle cells [23]. Thus, *MYLPF* may be unique among the DA genes in also causing amyoplasia, which may have important implications for personalized therapeutic strategies.

3. Neural-Related Distal Arthrogryposis

3.1. PIEZO2

DA is not only associated with muscle proteins. Whole genome sequencing was performed on individuals with DA5, in which arthrogryposis occurs in combination with ptosis, ophthalmoplegia, and facial dysmorphism, and gain of function variants (I802F and E2727del) were discovered in *PIEZO2* [24], a mechanosensitive cation channel responsible for mediating cation currents in primary sensory neurons [24,40,55]. Because of their mechanosensitivity, these channels are also termed "stretch-activated ion channels". *PIEZO2* was found to affect the skeleton non-autonomously in mice. Loss of *PIEZO2* specifically in skeletal tissue did not affect bone development; however, loss of gene function in proprioceptive neurons caused spine malalignment [67]. This suggests both that the neural system is necessary in maintaining normal skeletal development, and that *PIEZO2* is a critical gene in this process.

To test the mechanistic effects of these pathogenic variants, one group recently transfected human embryonic kidney cells with wild type *PIEZO2*, or *PIEZO2* with missense variants encoding I802F, or E2727del [24]. Both of these disease-associated variants caused the channel to recover more quickly from inactivation and resulted in increased channel activity following a mechanical stimulus. This supports the hypothesis that DA5 is caused by gain of function mutations that alter mechanosensory nerves. Although additional studies are needed, overstimulation may directly affect the neuromuscular pathway that controls muscle tone in developing fetuses, perhaps causing near-constitutive contractions that constrain the developing joint.

3.2. ECEL1

Variants in *ECEL1*, which encodes the endothelin-converting enzyme like-1, were identified in the recessive state in several families with DA5D, a rare form of arthrogryposis in which affected individuals have contractures as well as distinctive facial features and ptosis [41]. *ECEL1*, which is expressed in brain and nerve, is required for post-natal development in mice. Loss of *Ecel1* in mice results in abnormal terminal branching of motor neurons at the skeletal muscle endplate [68]. The mechanism by which *ECEL1* directs motor neuron branching is currently unknown; however, the resulting contractures in patients with *ECEL1* variants were proposed to be caused by a similar mechanism to those caused by genes such as *CHRNG* that causes multiple pterygium syndrome that impairs neurotransmission at the neuromuscular junction [69].

4. Lethal Congenital Contracture Syndrome

In contrast with DA, which are more common and often autosomal dominant, Lethal Congenital Contracture Syndromes (LCCSs) are a group of rare autosomal recessive forms of arthrogryposis. LCCS are characterized by lack of fetal movement (akinesia), micrognathia, incomplete lung development, polyhydramnios, characteristic contractures of the limbs (clubfoot, hyperextended knees, elbow and wrist flexion contractures) and motoneuron degeneration. Eleven subtypes of LCCS have been characterized. However, there are likely to be many more genes that result in these conditions as more genetic studies are performed on products of conception due to spontaneous abortion or stillbirth. LCCS is more common in communities with high rates of consanguinity consistent with the recessive inheritance pattern. Variable expression of LCCS phenotypes may be due to residual gene function in patients with missense variants or modifier genes. The recessive phenotypes of LCCS have made them more amenable to study by complete knockdown of gene expression.

4.1. Nuclear mRNA Export (GLE1, ERBB3, and PIP5K1C)

The first three LCCS subtypes may all act through a similar pathway by supporting nuclear mRNA export. LCCS type 1 (LCCS1) is caused by mutations in *GLE1* RNA Export Mediator (*GLE1*), a regulator of post-transcriptional gene expression [70]. *GLE1* acts as an mRNA export factor, as well as by mediating translation initiation and termination [15]. In mice, in situ hybridization showed marked expression in the neural tube of 11 dpf embryos, specifically in the ventral portion from which motoneurons generate [70]. In zebrafish, the gene is expressed prominently in the central nervous system during development [33].

A mutation in *GLE1*, Fin$_{Major}$, has been linked to LCCS1 by causing a splice-site mutation that results in a 3 amino acid insertion in the coiled-coil domain [70]. The coiled-coil domain is required for the protein to self-associates to form oligomers, and one group examined the effect of the Fin$_{Major}$ mutation on polypeptide self-association in vitro and in vivo [29]. Both in vitro and in living cells, the GLE1 protein self-aggregated, and Fin$_{Major}$ mutant oligomers were malformed. In human cell culture and in the yeast model, these malformed oligomers were found to perturb mRNA export from the nucleus [29].

Because the Fin$_{Major}$ mutation reduced function of the GLE1 protein in mRNA transport, *gle1* knockdown and knockouts were studied in zebrafish to understand its effects on development. Knockouts developed with small eyes and underdeveloped jaws and pectoral fins [33]. Cell death was also observed in the head and spinal cord, and there were fewer motoneurons than in wild type fish. Motoneurons also exhibit aberrant branching that worsened with age. Maternal *gle1* mRNA is loaded into the yolk sac of oocytes, where it contributes to zebrafish embryogenesis; therefore, morpholino oligonucleotides were also used to knock down expression of the mRNA in embryos. This exacerbated the phenotype, with CNS cell death becoming apparent earlier in development, at 1 dpf, which suggests an important role for *gle1* for early development. Notably, this phenotype is rescued in morphants injected with human wild type *GLE1*, but not when injected with the Fin$_{Major}$ allele [33]. Thus, this zebrafish model may be a viable tool for screening and determining the pathogenicity of human alleles.

LCCS2 is due to loss-of-function mutations in Erb-B2 Receptor Tyrosine Kinase 3 (*ERBB3*), which encodes HER3, a known modulator of the phosphatidylinositol pathway [44]. Interestingly, variants in LCCS3 were found to be due to variants in Phosphatidylinositol-4-Phosphate 5-Kinase Type 1 (*PIP5K1C*), which encodes the enzyme PIPK-gamma of the phosphatidylinositol pathway [43]. Nouslainen et al. realized that both *ERBB3* and *PIP5K1C* are involved in the synthesis of inositol hexakisphosphate, which binds directly to yeast *Gle1*, activating *Dbp5* for mRNA transport [70]. Because Gle1 is expressed in the neural tube during development, pathogenic variants in this gene can be devastating to development of the nervous system, as *Gle1* is integral to mRNA transport [70].

4.2. Peripheral Nerve (CNTNAP1, ADGRG6, GLDN)

The genes responsible for LCCS7, LCCS9, and LCCS11 are all highly expressed in peripheral nerves and required for proper peripheral nerve function. *Contactin Associated Protein 1 (CNTNAP1)*, which causes LCCS7, is a contactin-associated protein that is required for localization of the paranodal junction proteins contactin and neurofascin. *CNTNAP1* is also required for the normal spatial expression patterns of neuronal sodium and potassium channels [19]. Likewise, the causative gene for LCCS11, *gliomedin (GLDN)*, is a ligand for neurofascin and Nrcam, which are axonal immunoglobulin cell adhesion molecules critical for association with sodium channels at the nodes of Ranvier [71]. *Adhesion G Protein Coupled Receptor G6 (ADGRG6)*, which is also known as GPR126, is required for normal Schwann cell development. Thus, defects in all three genes likely result in similar peripheral nerve dysfunction at very early stages in development that leads to the LCCS phenotype.

5. Conclusions

Many techniques and organisms have been used for modeling arthrogryposis, each of which provides complementary information that is essential for understanding basic mechanisms and will yield translational benefits to human patients. There is an expanding list of genes that are associated with limb contractures, as one of many clinical features, beyond those discussed in this review article. Other genes are yet to be discovered, and disease models are often needed to provide evidence of causality. Furthermore, as exome sequencing becomes standard care, disease models may be helpful to facilitate variant interpretation. However, it will be essential to develop more efficient methods for introducing and studying large numbers of individual variants.

Although most genes responsible for distal arthrogryposis and LCCS are skeletal muscle sarcomeric genes or genes critical for neuronal function and neuromuscular transmission, crucial aspects remain to be established using disease models. It is important to determine whether common pathways and mechanisms supported by the genetic data will predict a unifying approach to therapy. Furthermore, now that gene therapies are becoming viable treatment mechanisms, where and when the defect needs to be corrected to prevent development of the DA or LCCS phenotype needs to be elucidated. Disease models will be essential to improve treatment for these challenging disorders.

Author Contributions: J.W. and C.A.G. wrote the first draft of the manuscript. J.W., A.J., M.B.D. and C.A.G. wrote or critically reviewed the manuscript, and reviewed and approved the final version. All authors have read and agreed to the published version of the manuscript.

Funding: Research reported in this publication was supported by National Institute of Arthritis and Musculoskeletal and Skin Diseases under Award Numbers R01AR067715 and R01AR070299, Eunice Kennedy Shriver National Institutes of Child Health and Human Development of the National Institutes of Health under the Award Numbers R03 HD104065 an P01 HD084387, Washington University Institute of Clinical and Translational Sciences grant UL1 TR002345 from the National Center for Advancing Translational Sciences of the National Institutes of Health, Washington University Musculoskeletal Research Center (NIH/NIAMS P30 AR057235) (NIH/NIAMS P30 AR074992), and the Eunice Kennedy Shriver National Institute of Child Health & Human Development of the National Institutes of Health under Award Number P50 HD103525 to the Intellectual and Developmental Disabilities Research Center at Washington University. This study was funded with support from Shriners Hospital for Children, and the Children's Discovery Institute of Washington University and St Louis Children's Hospital.

Conflicts of Interest: The authors declare no conflict of interest.

References

1. Hall, J.G. Arthrogryposis (multiple congenital contractures): Diagnostic approach to etiology, classification, genetics, and general principles. *Eur. J. Med. Genet.* **2014**, *57*, 464–472. [CrossRef]
2. Hall, J.G. Arthrogryposis multiplex congenita: Etiology, genetics, classification, diagnostic approach, and general aspects. *J. Pediatric Orthop.* **1997**, *6*, 159–166. [CrossRef]
3. Ravenscroft, G.; Clayton, J.S.; Faiz, F.; Sivadorai, P.; Milnes, D.; Cincotta, R.; Moon, P.; Kamien, B.; Edwards, M.; Delatycki, M. Neurogenetic fetal akinesia and arthrogryposis: Genetics, expanding genotype-phenotypes and functional genomics. *J. Med. Genet.* **2020**. [CrossRef] [PubMed]
4. Felsenthal, N.; Zelzer, E. Mechanical regulation of musculoskeletal system development. *Development* **2017**, *144*, 4271–4283. [CrossRef] [PubMed]

5. Kiefer, J.; Hall, J.G. Gene ontology analysis of arthrogryposis (multiple congenital contractures). In *American Journal of Medical Genetics Part C: Seminars in Medical Genetics*; Wiley Online Library: Hoboken, NJ, USA, 2019; pp. 310–326.

6. Bamshad, M.; Van Heest, A.E.; Pleasure, D. Arthrogryposis: A review and update. *J. Bone. Jt. Surgery. Am. Vol.* **2009**, *91*, 40. [CrossRef] [PubMed]

7. Beck, A.E.; McMillin, M.J.; Gildersleeve, H.I.; Shively, K.; Tang, A.; Bamshad, M.J. Genotype-phenotype relationships in Freeman–Sheldon syndrome. *Am. J. Med. Genet. Part A* **2014**, *164*, 2808–2813. [CrossRef]

8. Scala, M.; Accogli, A.; De Grandis, E.; Allegri, A.; Bagowski, C.P.; Shoukier, M.; Maghnie, M.; Capra, V. A novel pathogenic MYH3 mutation in a child with Sheldon–Hall syndrome and vertebral fusions. *Am. J. Med. Genet. Part A* **2018**, *176*, 663–667. [CrossRef] [PubMed]

9. Toydemir, R.M.; Rutherford, A.; Whitby, F.G.; Jorde, L.B.; Carey, J.C.; Bamshad, M.J. Mutations in embryonic myosin heavy chain (MYH3) cause Freeman-Sheldon syndrome and Sheldon-Hall syndrome. *Nat. Genet.* **2006**, *38*, 561–566. [CrossRef]

10. Desai, D.; Stiene, D.; Song, T.; Sadayappan, S. Distal Arthrogryposis and Lethal Congenital Contracture Syndrome—An Overview. *Front. Physiol.* **2020**, *11*, 689. [CrossRef]

11. Markus, B.; Narkis, G.; Landau, D.; Birk, R.Z.; Cohen, I.; Birk, O.S. Autosomal recessive lethal congenital contractural syndrome type 4 (LCCS4) caused by a mutation in MYBPC1. *Hum. Mutat.* **2012**, *33*, 1435–1438. [CrossRef]

12. Racca, A.W.; Beck, A.E.; McMillin, M.J.; Korte, F.S.; Bamshad, M.J.; Regnier, M. The embryonic myosin R672C mutation that underlies Freeman-Sheldon syndrome impairs cross-bridge detachment and cycling in adult skeletal muscle. *Hum. Mol. Genet.* **2015**, *24*, 3348–3358. [CrossRef]

13. Dieterich, K.; Le Tanno, P.; Kimber, E.; Jouk, P.S.; Hall, J.; Giampietro, P. The diagnostic workup in a patient with AMC: Overview of the clinical evaluation and paraclinical analyses with review of the literature. In *American Journal of Medical Genetics Part C: Seminars in Medical Genetics*; Wiley Online Library: Hoboken, NJ, USA, 2019; pp. 337–344.

14. Guo, Y.; Kronert, W.A.; Hsu, K.H.; Huang, A.; Sarsoza, F.; Bell, K.M.; Suggs, J.A.; Swank, D.M.; Bernstein, S.I. Drosophila myosin mutants model the disparate severity of type 1 and type 2B distal arthrogryposis and indicate an enhanced actin affinity mechanism. *Skelet. Muscle* **2020**, *10*, 1–18. [CrossRef]

15. Folkmann, A.W.; Dawson, T.R.; Wente, S.R. Insights into mRNA export-linked molecular mechanisms of human disease through a Gle1 structure–function analysis. *Adv. Biol. Regul.* **2014**, *54*, 74–91. [CrossRef] [PubMed]

16. Whittle, J.; Antunes, L.; Harris, M.; Upshaw, Z.; Sepich, D.S.; Johnson, A.N.; Mokalled, M.; Solnica-Krezel, L.; Dobbs, M.B.; Gurnett, C.A. MYH 3-associated distal arthrogryposis zebrafish model is normalized with para-aminoblebbistatin. *EMBO Mol. Med.* **2020**, *12*, e12356. [CrossRef] [PubMed]

17. Zhu, X.; Wang, F.; Zhao, Y.; Yang, P.; Chen, J.; Sun, H.; Liu, L.; Li, W.; Pan, L.; Guo, Y. A gain-of-function mutation in Tnni2 impeded bone development through increasing Hif3a expression in DA2B mice. *PLoS Genet.* **2014**, *10*, e1004589. [CrossRef] [PubMed]

18. Bayram, Y.; Karaca, E.; Akdemir, Z.C.; Yilmaz, E.O.; Tayfun, G.A.; Aydin, H.; Torun, D.; Bozdogan, S.T.; Gezdirici, A.; Isikay, S. Molecular etiology of arthrogryposis in multiple families of mostly Turkish origin. *J. Clin. Investig.* **2016**, *126*, 762–778. [CrossRef] [PubMed]

19. Bhat, M.A.; Rios, J.C.; Lu, Y.; Garcia-Fresco, G.P.; Ching, W.; Martin, M.S.; Li, J.; Einheber, S.; Chesler, M.; Rosenbluth, J. Axon-glia interactions and the domain organization of myelinated axons requires neurexin IV/Caspr/Paranodin. *Neuron* **2001**, *30*, 369–383. [CrossRef]

20. Borovikov, Y.S.; Simonyan, A.O.; Karpicheva, O.E.; Avrova, S.V.; Rysev, N.A.; Sirenko, V.V.; Piers, A.; Redwood, C.S.J.B. The reason for a high Ca2+-sensitivity associated with Arg91Gly substitution in TPM2 gene is the abnormal behavior and high flexibility of tropomyosin during the ATPase cycle. *Biochem. Biophys. Res. Commun.* **2017**, *494*, 681–686. [CrossRef]

21. Cameron-Christie, S.R.; Wells, C.F.; Simon, M.; Wessels, M.; Tang, C.Z.; Wei, W.; Takei, R.; Aarts-Tesselaar, C.; Sandaradura, S.; Sillence, D.O. Recessive Spondylocarpotarsal synostosis syndrome due to compound heterozygosity for variants in MYH3. *Am. J. Hum. Genet.* **2018**, *102*, 1115–1125. [CrossRef]

22. Casey, J.P.; Brennan, K.; Scheidel, N.; McGettigan, P.; Lavin, P.T.; Carter, S.; Ennis, S.; Dorkins, H.; Ghali, N.; Blacque, O.E.; et al. Recessive NEK9 mutation causes a lethal skeletal dysplasia with evidence of cell cycle and ciliary defects. *Hum. Mol. Genet.* **2016**, *25*, 1824–1835. [CrossRef]

23. Chong, J.X.; Talbot, J.C.; Teets, E.M.; Previs, S.; Martin, B.L.; Shively, K.M.; Marvin, C.T.; Aylsworth, A.S.; Saadeh-Haddad, R.; Schatz, U.A. Mutations in MYLPF cause a novel segmental amyoplasia that manifests as distal arthrogryposis. *Am. J. Hum. Genet.* **2020**, *107*, 293–310. [CrossRef]

24. Coste, B.; Houge, G.; Murray, M.F.; Stitziel, N.; Bandell, M.; Giovanni, M.A.; Philippakis, A.; Hoischen, A.; Riemer, G.; Steen, U. Gain-of-function mutations in the mechanically activated ion channel PIEZO2 cause a subtype of Distal Arthrogryposis. *Proc. Natl. Acad. Sci. USA* **2013**, *110*, 4667–4672. [CrossRef]

25. Das, S.; Kumar, P.; Verma, A.; Maiti, T.K.; Mathew, S.J. Myosin heavy chain mutations that cause Freeman-Sheldon syndrome lead to muscle structural and functional defects in Drosophila. *Dev. Biol.* **2019**, *449*, 90–98. [CrossRef]

26. Di Paolo, G.; Moskowitz, H.S.; Gipson, K.; Wenk, M.R.; Voronov, S.; Obayashi, M.; Flavell, R.; Fitzsimonds, R.M.; Ryan, T.A.; De Camilli, P.J.N. Impaired PtdIns (4, 5) P 2 synthesis in nerve terminals produces defects in synaptic vesicle trafficking. *Nature* **2004**, *431*, 415–422. [CrossRef]

27. Durieux, A.-C.; Vignaud, A.; Prudhon, B.; Viou, M.T.; Beuvin, M.; Vassilopoulos, S.; Fraysse, B.; Ferry, A.; Lainé, J.; Romero, N.B.J.H.m.g. A centronuclear myopathy-dynamin 2 mutation impairs skeletal muscle structure and function in mice. *Hum. Mol. Genet.* **2010**, *19*, 4820–4836. [CrossRef]

28. Ekhilevitch, N.; Kurolap, A.; Oz-Levi, D.; Mory, A.; Hershkovitz, T.; Ast, G.; Mandel, H.; Baris, H. Expanding the MYBPC1 phenotypic spectrum: A novel homozygous mutation causes arthrogryposis multiplex congenita. *Clin. Genet.* **2016**, *90*, 84–89. [CrossRef]

29. Folkmann, A.W.; Collier, S.E.; Zhan, X.; Ohi, M.D.; Wente, S.R. Gle1 functions during mRNA export in an oligomeric complex that is altered in human disease. *Cell* **2013**, *155*, 582–593. [CrossRef] [PubMed]

30. Gurnett, C.A.; Desruisseau, D.M.; McCall, K.; Choi, R.; Meyer, Z.I.; Talerico, M.; Miller, S.E.; Ju, J.-S.; Pestronk, A.; Connolly, A.M. Myosin binding protein C1: A novel gene for autosomal dominant distal arthrogryposis type 1. *Hum. Mol. Genet.* **2010**, *19*, 1165–1173. [CrossRef] [PubMed]

31. Ha, K.; Buchan, J.G.; Alvarado, D.M.; Mccall, K.; Vydyanath, A.; Luther, P.K.; Goldsmith, M.I.; Dobbs, M.B.; Gurnett, C.A. MYBPC1 mutations impair skeletal muscle function in zebrafish models of arthrogryposis. *Hum. Mol. Genet.* **2013**, *22*, 4967–4977. [CrossRef]

32. Agarwal, M.; Sharma, A.; Kumar, P.; Kumar, A.; Bharadwaj, A.; Saini, M.; Kardon, G.; Mathew, S.J. Myosin heavy chain-embryonic regulates skeletal muscle differentiation during mammalian development. *Development* **2020**, *147*, dev184507. [CrossRef] [PubMed]

33. Jao, L.-E.; Appel, B.; Wente, S.R. A zebrafish model of lethal congenital contracture syndrome 1 reveals Gle1 function in spinal neural precursor survival and motor axon arborization. *Development* **2012**, *139*, 1316–1326. [CrossRef] [PubMed]

34. Ju, Y.; Li, J.; Xie, C.; Ritchlin, C.T.; Xing, L.; Hilton, M.J.; Schwarz, E.M. Troponin T3 expression in skeletal and smooth muscle is required for growth and postnatal survival: Characterization of Tnnt3tm2a (KOMP) Wtsi mice. *Genesis* **2013**, *51*, 667–675. [CrossRef] [PubMed]

35. Koutsopoulos, O.S.; Kretz, C.; Weller, C.M.; Roux, A.; Mojzisova, H.; Böhm, J.; Koch, C.; Toussaint, A.; Heckel, E.; Stemkens, D.; et al. Dynamin 2 homozygous mutation in humans with a lethal congenital syndrome. *Eur. J. Hum. Genet.* **2013**, *21*, 637–642. [CrossRef] [PubMed]

36. Laquérriere, A.; Maluenda, J.; Camus, A.; Fontenas, L.; Dieterich, K.; Nolent, F.; Zhou, J.; Monnier, N.; Latour, P.; Gentil, D.; et al. Mutations in CNTNAP1 and ADCY6 are responsible for severe arthrogryposis multiplex congenita with axoglial defects. *Hum. Mol. Genet.* **2014**, *23*, 2279–2289. [CrossRef] [PubMed]

37. Li, M.; Andersson-Lendahl, M.; Sejersen, T.; Arner, A. Knockdown of fast skeletal myosin-binding protein C in zebrafish results in a severe skeletal myopathy. *J. Gen. Physiol.* **2016**, *147*, 309–322. [CrossRef] [PubMed]

38. Maluenda, J.; Manso, C.; Quevarec, L.; Vivanti, A.; Marguet, F.; Gonzales, M.; Guimiot, F.; Petit, F.; Toutain, A.; Whalen, S.; et al. Mutations in GLDN, encoding gliomedin, a critical component of the nodes of ranvier, are responsible for lethal arthrogryposis. *Am. J. Hum. Genet.* **2016**, *99*, 928–933. [CrossRef] [PubMed]

39. Matyushenko, A.; Levitsky, D. Molecular mechanisms of pathologies of skeletal and cardiac muscles caused by point mutations in the tropomyosin genes. *Biochemistry* **2020**, *85*, 20–33. [CrossRef]

40. McMillin, M.J.; Beck, A.E.; Chong, J.X.; Shively, K.M.; Buckingham, K.J.; Gildersleeve, H.I.; Aracena, M.I.; Aylsworth, A.S.; Bitoun, P.; Carey, J.C. Mutations in PIEZO2 cause Gordon syndrome, Marden-Walker syndrome, and distal arthrogryposis type 5. *Am. J. Hum. Genet.* **2014**, *94*, 734–744. [CrossRef]

41. McMillin, M.J.; Below, J.E.; Shively, K.M.; Beck, A.E.; Gildersleeve, H.I.; Pinner, J.; Gogola, G.R.; Hecht, J.T.; Grange, D.K.; Harris, D.J. Mutations in ECEL1 cause distal arthrogryposis type 5D. *Am. J. Hum. Genet.* **2013**, *92*, 150–156. [CrossRef]

42. Monk, K.R.; Naylor, S.G.; Glenn, T.D.; Mercurio, S.; Perlin, J.R.; Dominguez, C.; Moens, C.B.; Talbot, W.S. AG protein–coupled receptor is essential for Schwann cells to initiate myelination. *Science* **2009**, *325*, 1402–1405. [CrossRef]

43. Narkis, G.; Ofir, R.; Landau, D.; Manor, E.; Volokita, M.; Hershkowitz, R.; Elbedour, K.; Birk, O.S. Lethal contractural syndrome type 3 (LCCS3) is caused by a mutation in PIP5K1C, which encodes PIPKIγ of the phophatidylinsitol pathway. *Am. J. Hum. Genet.* **2007**, *81*, 530–539. [CrossRef]

44. Narkis, G.; Ofir, R.; Manor, E.; Landau, D.; Elbedour, K.; Birk, O.S. Lethal congenital contractural syndrome type 2 (LCCS2) is caused by a mutation in ERBB3 (Her3), a modulator of the phosphatidylinositol-3-kinase/Akt pathway. *Am. J. Hum. Genet.* **2007**, *81*, 589–595. [CrossRef]

45. Patel, N.; Smith, L.L.; Faqeih, E.; Mohamed, J.; Gupta, V.A.; Alkuraya, F.S. ZBTB42 mutation defines a novel lethal congenital contracture syndrome (LCCS6). *Hum. Mol. Genet.* **2014**, *23*, 6584–6593. [CrossRef]

46. Ravenscroft, G.; Nolent, F.; Rajagopalan, S.; Meireles, A.M.; Paavola, K.J.; Gaillard, D.; Alanio, E.; Buckland, M.; Arbuckle, S.; Krivanek, M.; et al. Mutations of GPR126 are responsible for severe arthrogryposis multiplex congenita. *Am. J. Hum. Genet.* **2015**, *96*, 955–961. [CrossRef] [PubMed]

47. Riethmacher, D.; Sonnenberg-Riethmacher, E.; Brinkmann, V.; Yamaai, T.; Lewin, G.R.; Birchmeier, C. Severe neuropathies in mice with targeted mutations in the ErbB3 receptor. *Nature* **1997**, *389*, 725–730. [CrossRef] [PubMed]

48. Sandaradura, S.A.; Bournazos, A.; Mallawaarachchi, A.; Cummings, B.B.; Waddell, L.B.; Jones, K.J.; Troedson, C.; Sudarsanam, A.; Nash, B.M.; Peters, G.B.; et al. Nemaline myopathy and distal arthrogryposis associated with an autosomal recessive TNNT3 splice variant. *Hum. Mutat.* **2018**, *39*, 383–388. [CrossRef] [PubMed]

49. Shashi, V.; Geist, J.; Lee, Y.; Yoo, Y.; Shin, U.; Schoch, K.; Sullivan, J.; Stong, N.; Smith, E.; Jasien, J. Heterozygous variants in MYBPC1 are associated with an expanded neuromuscular phenotype beyond arthrogryposis. *Hum. Mutat.* **2019**, *40*, 1115–1126. [CrossRef] [PubMed]

50. Sung, S.S.; Brassington, A.-M.E.; Grannatt, K.; Rutherford, A.; Whitby, F.G.; Krakowiak, P.A.; Jorde, L.B.; Carey, J.C.; Bamshad, M. Mutations in genes encoding fast-twitch contractile proteins cause distal arthrogryposis syndromes. *Am. J. Hum. Genetics* **2003**, *72*, 681–690. [CrossRef] [PubMed]

51. Toydemir, R.M.; Chen, H.; Proud, V.K.; Martin, R.; van Bokhoven, H.; Hamel, B.C.; Tuerlings, J.H.; Stratakis, C.A.; Jorde, L.B.; Bamshad, M.J. Trismus-pseudocamptodactyly syndrome is caused by recurrent mutation of MYH8. *Am. J. Med Genet. Part A* **2006**, *140*, 2387–2393. [CrossRef] [PubMed]

52. Veugelers, M.; Bressan, M.; McDermott, D.A.; Weremowicz, S.; Morton, C.C.; Mabry, C.C.; Lefaivre, J.-F.; Zunamon, A.; Destree, A.; Chaudron, J.-M.; et al. Mutation of perinatal myosin heavy chain associated with a Carney complex variant. *N. Engl. J. Med.* **2004**, *351*, 460–469. [CrossRef]

53. Vigoreaux, J.O. Genetics of the Drosophila flight muscle myofibril: A window into the biology of complex systems. *Bioessays* **2001**, *23*, 1047–1063. [CrossRef] [PubMed]

54. Walklate, J.; Vera, C.; Bloemink, M.J.; Geeves, M.A.; Leinwand, L. The most prevalent Freeman-Sheldon Syndrome mutations in the embryonic myosin motor share functional defects. *J. Biol. Chem.* **2016**, *291*, 10318–10331. [CrossRef] [PubMed]

55. Wang, L.; Zhou, H.; Zhang, M.; Liu, W.; Deng, T.; Zhao, Q.; Li, Y.; Lei, J.; Li, X.; Xiao, B. Structure and mechanogating of the mammalian tactile channel PIEZO2. *Nature* **2019**, *573*, 225–229. [CrossRef]

56. Williams, J.; Boin, N.G.; Valera, J.M.; Johnson, A.N. Noncanonical roles for Tropomyosin during myogenesis. *Development* **2015**, *142*, 3440–3452. [PubMed]

57. Chong, J.X.; Burrage, L.C.; Beck, A.E.; Marvin, C.T.; McMillin, M.J.; Shively, K.M.; Harrell, T.M.; Buckingham, K.J.; Bacino, C.A.; Jain, M.; et al. Autosomal-dominant multiple pterygium syndrome is caused by mutations in MYH3. *Am. J. Hum. Genet.* **2015**, *96*, 841–849. [CrossRef]

58. Li, X.; Zhong, B.; Han, W.; Zhao, N.; Liu, W.; Sui, Y.; Wang, Y.; Lu, Y.; Wang, H.; Li, J. Two novel mutations in myosin binding protein C slow causing distal arthrogryposis type 2 in two large Han Chinese families may suggest important functional role of immunoglobulin domain C2. *PLoS ONE* **2015**, *10*, e0117158. [CrossRef]

59. Alvarado, D.M.; Buchan, J.G.; Gurnett, C.A.; Dobbs, M.B. Exome sequencing identifies an MYH3 mutation in a family with distal arthrogryposis type 1. *J. Bone Jt. Surg. Am. Vol.* **2011**, *93*, 1045. [CrossRef] [PubMed]

60. Beck, A.E.; McMillin, M.J.; Gildersleeve, H.I.; Kezele, P.R.; Shively, K.M.; Carey, J.C.; Regnier, M.; Bamshad, M.J. Spectrum of mutations that cause distal arthrogryposis types 1 and 2B. *Am. J. Med. Genet. Part A* **2013**, *161*, 550–555. [CrossRef] [PubMed]

61. Zieba, J.; Zhang, W.; Chong, J.X.; Forlenza, K.N.; Martin, J.H.; Heard, K.; Grange, D.K.; Butler, M.G.; Kleefstra, T.; Lachman, R.S. A postnatal role for embryonic myosin revealed by MYH3 mutations that alter TGFβ signaling and cause autosomal dominant spondylocarpotarsal synostosis. *Sci. Rep.* **2017**, *7*, 1–9. [CrossRef]

62. Gil-Gálvez, A.; Carbonell-Corvillo, P.; Paradas, C.; Miranda-Vizuete, A. Cautionary note on the use of Caenorhabditis elegans to study muscle phenotypes caused by mutations in the human MYH7 gene. *BioTechniques* **2020**, *68*, 296–299. [CrossRef]

63. Repetti, G.G.; Toepfer, C.N.; Seidman, J.G.; Seidman, C.E. Novel therapies for prevention and early treatment of cardiomyopathies: Now and in the future. *Circ. Res.* **2019**, *124*, 1536–1550. [CrossRef]

64. Tajsharghi, H.; Ohlsson, M.; Palm, L.; Oldfors, A. Myopathies associated with β-tropomyosin mutations. *Neuromuscul. Disord.* **2012**, *22*, 923–933. [CrossRef]

65. Monnier, N.; Lunardi, J.; Marty, I.; Mezin, P.; Labarre-Vila, A.; Dieterich, K.; Jouk, P.S. Absence of β-tropomyosin is a new cause of Escobar syndrome associated with nemaline myopathy. *Neuromuscul. Disord.* **2009**, *19*, 118–123. [CrossRef] [PubMed]

66. Wang, Y.; Szczesna-Cordary, D.; Craig, R.; Diaz-Perez, Z.; Guzman, G.; Miller, T.; Potter, J.D. Fast skeletal muscle regulatory light chain is required for fast and slow skeletal muscle development. *FASEB J.* **2007**, *21*, 2205–2214. [CrossRef]

67. Assaraf, E.; Blecher, R.; Heinemann-Yerushalmi, L.; Krief, S.; Vinestock, R.C.; Biton, I.E.; Brumfeld, V.; Rotkopf, R.; Avisar, E.; Agar, G.; et al. Piezo2 expressed in proprioceptive neurons is essential for skeletal integrity. *Nat. Commun.* **2020**, *11*, 1–15. [CrossRef] [PubMed]

68. Schweizer, A.; Valdenaire, O.; Köster, A.; Lang, Y.; Schmitt, G.; Lenz, B.; Bluethmann, H.; Rohrer, J. Neonatal lethality in mice deficient in XCE, a novel member of the endothelin-converting enzyme and neutral endopeptidase family. *J. Biol. Chem.* **1999**, *274*, 20450–20456. [CrossRef] [PubMed]

69. Morgan, N.V.; Brueton, L.A.; Cox, P.; Greally, M.T.; Tolmie, J.; Pasha, S.; Aligianis, I.A.; van Bokhoven, H.; Marton, T.; Al-Gazali, L. Mutations in the embryonal subunit of the acetylcholine receptor (CHRNG) cause lethal and Escobar variants of multiple pterygium syndrome. *Am. J. Hum. Genet.* **2006**, *79*, 390–395. [CrossRef]

70. Nousiainen, H.O.; Kestilä, M.; Pakkasjärvi, N.; Honkala, H.; Kuure, S.; Tallila, J.; Vuopala, K.; Ignatius, J.; Herva, R.; Peltonen, L. Mutations in mRNA export mediator GLE1 result in a fetal motoneuron disease. *Nat. Genet.* **2008**, *40*, 155–157. [CrossRef]

71. Eshed, Y.; Feinberg, K.; Poliak, S.; Sabanay, H.; Sarig-Nadir, O.; Spiegel, I.; Bermingham, J.R., Jr.; Peles, E. Gliomedin mediates Schwann cell-axon interaction and the molecular assembly of the nodes of Ranvier. *Neuron* **2005**, *47*, 215–229. [CrossRef]

Article

Prevalence of *POC5* Coding Variants in French-Canadian and British AIS Cohort

Hélène Mathieu [1], Aurélia Spataru [1], José Antonio Aragon-Martin [2], Anne Child [3], Soraya Barchi [1], Carole Fortin [4,5], Stefan Parent [1,6] and Florina Moldovan [1,7,*]

1   CHU Sainte-Justine Research Center, Montreal, QC H3T1C5, Canada; helene.mathieu.1@umontreal.ca (H.M.); spataruaurelia@gmail.com (A.S.); soraya.barchi.hsj@ssss.gouv.qc.ca (S.B.); stefan.parent@umontreal.ca (S.P.)
2   National Heart and Lung Institute (NHLI), Imperial College, Guy Scadding Building, London SW3 6LY, UK; j.aragon-martin@imperial.ac.uk
3   Marfan Trust, National Heart and Lung Institute (NHLI), Imperial College, Guy Scadding Building, London SW3 6LY, UK; annechildgenetics@outlook.com
4   École de Réadaptation, Faculté de Médecine, Université de Montréal, Montreal, QC H3C 3J7, Canada; carole.fortin@umontreal.ca
5   Centre de Recherche, CHU Sainte-Justine, Montreal, QC H1T1C5, Canada
6   Faculty of Surgery, Université de Montréal, Montreal, QC H3T 1J4, Canada
7   Faculty of Dentistry, Université de Montréal, Montreal, QC H3T 1J4, Canada
*   Correspondence: florina.moldovan@umontreal.ca; Tel.: +1-5143454931 (ext. 5746)

Abstract: Adolescent idiopathic scoliosis (AIS) is a complex common disorder of multifactorial etiology defined by a deviation of the spine in three dimensions that affects approximately 2% to 4% of adolescents. Risk factors include other affected family members, suggesting a genetic component to the disease. The *POC5* gene was identified as one of the first ciliary candidate genes for AIS, as three variants were identified in large families with multiple members affected with idiopathic scoliosis. To assess the prevalence of p.(A429V), p.(A446T), and p.(A455P) *POC5* variants in patients with AIS, we used next-generation sequencing in our cohort of French-Canadian and British families and sporadic cases. Our study highlighted a prevalence of 13% for *POC5* variants, 7.5% for p.(A429V), and 6.4% for p.(A446T). These results suggest a higher prevalence of the aforementioned *POC5* coding variants in patients with AIS compared to the general population.

Keywords: *POC5*; adolescent idiopathic scoliosis; cilia; genetics; spine deformity

Citation: Mathieu, H.; Spataru, A.; Aragon-Martin, J.A.; Child, A.; Barchi, S.; Fortin, C.; Parent, S.; Moldovan, F. Prevalence of *POC5* Coding Variants in French-Canadian and British AIS Cohort. *Genes* **2021**, *12*, 1032. https://doi.org/10.3390/genes12071032

Academic Editors: Philip Giampietro, Nancy Hadley-Miller, Cathy L. Raggio and Zhousheng Xiao

Received: 17 May 2021
Accepted: 28 June 2021
Published: 1 July 2021

Publisher's Note: MDPI stays neutral with regard to jurisdictional claims in published maps and institutional affiliations.

1. Introduction

Adolescent idiopathic scoliosis (AIS) is a common disorder characterized by a combination of deviations of the spine in the sagittal and the coronal plane, with vertebral rotation. It affects approximatively 3% of the adolescent population [1,2]; affects females more than males, with a ratio ranging from 1.5:1 to 3:1 [2,3]; and is more prevalent in northern latitudes [4]. The etiology of AIS remains not fully understood, but it is now widely accepted that this disorder has a genetic component, as supported by family history, and higher concordance rates for monozygotic twins compared to dizygotic twins [5–7]. Approximatively 40% of AIS patients have a family history [8,9]. The genetic model for AIS remains unclear; indeed, several studies have suggested that it is a polygenic and multifactorial disease [9]. However, other analyses suggest mendelian inheritance, such as autosomal dominant or sex-related, could show with incomplete penetrance [10–12]. Since the advent of next-generation sequencing, candidate-gene analysis using pedigrees and population-based genome-wide association studies (GWAS) have been widely used to assess the genetic etiology of AIS. Despite all these efforts, only a few of the candidate genes have been functionally linked to the development of AIS. In 2015, Patten et al. [13] performed a linkage analysis followed by exome sequencing, and identified coding variants in the centrosomal protein gene *POC5* (NM_001099271) in a multiplex four-generation

AIS French family [13]. Indeed, a rare SNP, p.(A446T), was found to perfectly segregate with AIS in four families from a pool of 41 AIS French families. Two additional rare variants, p.(A429V) and p.(A455P), were found in the *POC5* gene in AIS sporadic cases. Moreover, all three *POC5* coding variants were functionally related to AIS using a zebrafish model [13]. More recently, a fourth SNP (rs6892146) was identified to be associated with AIS development in a Chinese population [14], but the three other SNPs were not found in this cohort. POC5 is a centriolar protein that is essential for cell cycle progression, cilia elongation [15], centriole elongation, and maturation. Since the identification of the *POC5* gene, a ciliary gene that is strongly associated with AIS, the ciliary pathway has been thoroughly investigated and has revealed promising results [12,15–18].

To investigate the prevalence of *POC5* genetic variants in AIS, French-Canadian and British AIS patients were screened by targeted or whole-exome sequencing followed by Sanger analyses of DNA.

2. Materials and Methods

2.1. Patients

One hundred and seventy-seven AIS patients with a Cobb angle of at least 10°, 73 patients from a British cohort (63 unrelated consecutives IS individuals and 10 families), and 104 patients from a French-Canadian cohort (30 unrelated consecutive AIS individuals and 74 patients from 43 families), were recruited. Genomic DNA was extracted from saliva (Cat. RU49000, Norgen, Thorold, ON, Canada) or blood following the protocol provided by the company.

Samples were collected in accordance with the policies regarding the ethical use of human tissues for research. The protocol used in this study was approved by the Centre hospitalier universitaire Sainte-Justine Ethics Committee (#3704).

The control population consisted of an in-house cohort of 1268 individuals with similar ancestry (French, French-Canadian, or European) and was not screened for the presence of AIS [13].

2.2. Targeted Exome Sequencing

A library was generated from 10 ng of genomic DNA to perform targeted sequencing of the *POC5* gene using the Ion AmpliSeq (Life Technologies). Sequencing of the 12 exons of *POC5* of 63 British unrelated AIS individuals and 10 families, and 18 French-Canadian families affected with autosomal dominant AIS, was performed by the Centre de Génomique Clinique Pédiatrique intégré CHU Sainte-Justine. The library was prepared using the Ion AmpliSeq DNA and RNA Library Preparation (MAN0006735, Rev. B.0, Ion Torrent, Life Technologies) prior to the exome sequencing following the Ion PGM IC 200 Kit (MAN0007661, Rev. B.0) protocol. Sequencing reads were aligned to the reference human genome sequence (hg19) [19], and the SNPs were identified by Ion Reporter (Ion Torrent). Identified variants were annotated using ANNOVAR [20], which is implemented in the VarAFT software [21] that we used to select exonic and splicing variants with a MAF (minor allele frequency) $\leq$ 1%. The identified SNPs were then compared to the EVS (Exome Variant Server) database, the Genome Aggregation Database (gnomAD: https://gnomad.broadinstitute.org (accessed on 21 March 2020)), and our in-house control cohort (n = 1268).

2.3. Whole-Exome Sequencing

A library was generated from 10 ng of genomic DNA to perform whole-exome sequencing using the HiSeq 4000 sequencing machine from the Centre de Génomique Clinique Pédiatrique intégré CHU Sainte-Justine. The gDNA extracted from saliva or blood was sheared to a mean fragment size of 200 pb (Covaris E220 Montreal, PQ, Canada), and gDNA fragments were used for DNA library preparation following the protocol for the SeqCap EZ HyperCap (Roche-NimbleGen, Pleasanton, CA, USA). Enriched DNA fragments were sequenced with 100 pb paired-end reads (HiSeq 4000, Illumina, Vancouver,

BC, Canada). Sequencing reads were converted to FASTQ using bcl2fastq software (Illumina) and trimmed by Trimmomatic. The reads were then aligned to the reference human genome sequence (hg19) using the Burrows–Wheeler transform (BWT), followed by a local alignment for indels using Genome Analysis ToolKit software (Broad Institute). Duplicate sequencing reads were excluded by Picard software, and SNPs were identified using the GATK Unified Genotyper and annotated by ANNOVAR software [20], implemented in the VarAFT software [21] that we used to select exonic and splicing variants with a MAF (minor allele frequency) $\leq$ 1% in the *POC5* gene. The identified SNPs were then compared to the EVS (Exome Variant Server) database, the Genome Aggregation Database (gnomAD: https://gnomad.broadinstitute.org (accessed on 21 March 2020)), and our in-house control cohort (n = 1268).

2.4. Validation with Sanger Sequencing

All *POC5* coding variants identified by WES or targeted exome sequencing were validated by Sanger sequencing. The segregation of the coding variants was also completed using Sanger sequencing for the take-out-concerned families studied. PCR amplification was performed using the TransStart FastPfu FLY DNA Polymerase (AP231, Civic Bioscience) following the instructions of the manufacturer with primers FWD-5′-GGACCAAACTTTAGCCAGTATG-3′ and RV-5′-TCTCGATCTCCTGACCTCGT-3′. Sanger sequencing of amplicons was performed on an ABI 3730xl DNA Analyzer (Applied Biosystems, Louisville, KY, USA) at Eurofins Genomics (Louisville, KY, USA).

2.5. Statistics

The allelic frequency of the *POC5* coding variants was compared to the control population using a one-tailed Fischer's exact test. A p-value < 0.05 was considered statistically significant.

3. Results

3.1. Patient Enrolment

Since the identification of *POC5* as a candidate gene for adolescent idiopathic scoliosis and its functional validation, we have analyzed the prevalence of *POC5* coding variants within the AIS population, and also have sought to identify new candidate genes. Ninety AIS French-Canadian families were recruited demonstrating different types of transmission: autosomal dominant or recessive, and 30 AIS sporadic cases. For this study, 43 families were selected. Our French-Canadian cohort was supplemented with the 73 UK AIS patients of our collaborators from London, United Kingdom.

3.2. POC5 Variants Prevalence Using Next-Generation Sequencing

Among the 53 French-Canadian and British AIS families and 94 unrelated AIS patients, 177 AIS patients were screened for *POC5* coding variants. The combination of whole-exome sequencing and targeted sequencing by AmpliSeq using a targeted Amplicon chip followed by a confirmation with Sanger sequencing revealed that 13% (p < 0.0001, Fischer's exact test) of AIS patients with or without family history were carrying one of the three variants of *POC5*, previously identified as the first causative gene [13]. Indeed, 11 of them carried the A429V variant (6 families and 5 sporadic cases); i.e., 7.5% (p < 0.0001), and 11 were found to carry the A446T variant (2 familial, 6 sporadic cases); i.e., 6.4% (p = 0.0052). No patient with the A455P variant was reported (Table 1).

Table 1. *POC5* coding variant distribution among the French-Canadian and British AIS cohort compared to 1268 controls. The number of patients from families or sporadic cases that were carrying *POC5* coding variants and the frequency for each of the 2 variants (p.(A446T) and p.(A429V)) are reported.

Data	Families (n = 53; 83 AIS Patients)	AIS Cases with Unknown Pedigree data (n = 94)	Controls Matched for Ethnicity with Families and Cases (n = 1268)	Comparison of Allelic Frequency of the Rare Variants in AIS Cases vs. Controls (Fisher's Exact Test, One Tailed)
Sequencing Methods	WES + Targeted Exome	WES + Targeted Exome	WES + Sanger	
p.(A446T)	2/53 3.8%	6/94 6.4%	19/1268 1.5%	$p = 0.0052$
p.(A429V)	6/53 11.3%	5/94 5.3%	9/1268 0.7%	$p < 0.0001$
***POC5* coding variants**	8/53 15.1%	11/94 11.7%	28/1268 2.2%	$p < 0.0001$

The EVS (Exome Variant Server) database [22] reported a minor allele frequency (MAF) in the European-American population of 1.2% for the p.A429V variant and 1.5% for p.A446T. The Genome Aggregation Database (gnomAD: https://gnomad.broadinstitute.org (accessed on 21 March 2020)) [23] reported a MAF of 1.1% for the p.A429V (rs146984380) variant and 1.6% for p.A446T (rs34678567).

3.3. Segragation Analysis of POC5 Coding Variants with AIS

The segregation analysis of the *POC5* coding variants with the disease for the AIS families was then performed using Sanger sequencing (Figure 1). The variant p.(A446T) was found in two families (F62 and F80), and showed a perfect segregation with the disease in family 80. Sadly, DNA was not able for the rest of family 62. For the families that were found to carry the variant p.(A429V)—F02, F18, F37, F57, F58, and F66—the segregation analysis showed incomplete penetrance (Figure 1).

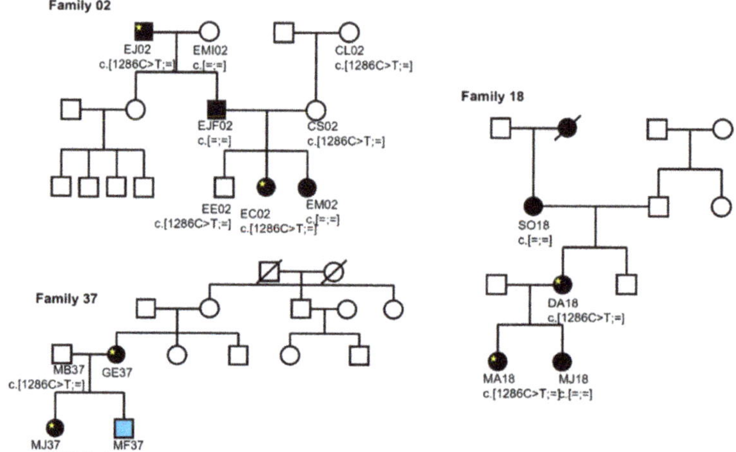

Figure 1. *Cont.*

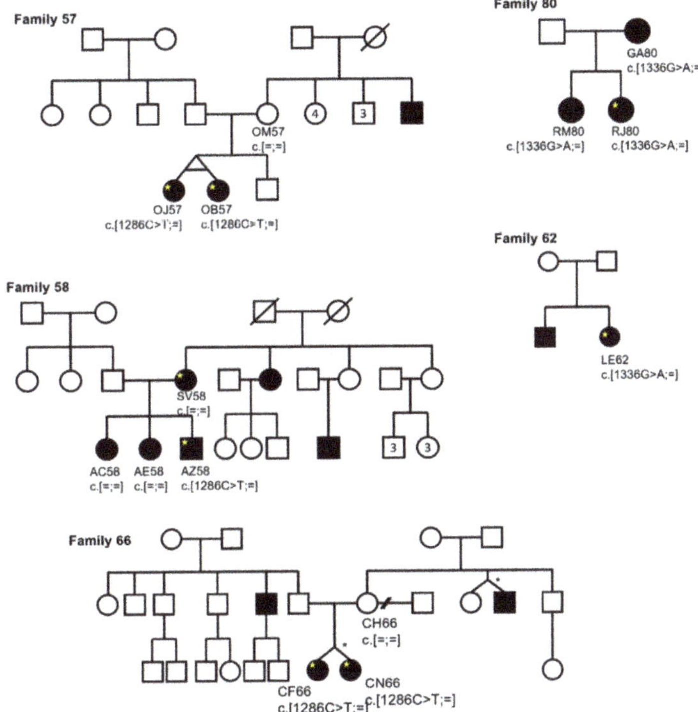

Figure 1. Pedigrees of French-Canadian families showing the co-segregation of POC5 variants (c. 1286C > T (p.(A429V) and c.1336G > A (p.(A446T)) with the disease. Open circles and squares indicate unaffected individuals, Blackened circles and squares indicate affected females and males, respectively. Blue circles and squares indicate juvenile females and males. Yellows stars indicate exomed AIS patients. * Incomplete penetrance.

4. Discussion

We reported the prevalence of *POC5* gene variants in 13% of AIS patients with or without a family history of this condition; that is, six times more frequent than in our in-house control cohort that matched for ethnicity. Two methods were used for the screening of *POC5* gene in both French-Canadian and British families with AIS. This study confirmed the previously reported data by Patten et al. [13], which evidenced three rare SNPs (p.A446T, p.A429V, and p.A455P) in the *POC5* gene in four families of a pool of 41 AIS French families and 150 IS cases from France. We also confirmed the autosomal dominant transmission pattern of *POC5* coding variants in families with AIS.

Adolescent idiopathic scoliosis is a complex disease with a multifactorial etiology including genetic, environmental, and hormonal factors, but the pathogenesis of this disease remains poorly understood. To decipher the genetic implication in AIS, different approaches were used: association studies and linkage analyses to identify causative genes or genes that may impact AIS susceptibility and/or disease progression. Many association studies were performed using GWAS technology and genome-wide linkage analysis, followed by exome sequencing and highlighted multiple-locus candidate, especially on chromosomes 6, 9, 16, 17 [24], and 19 [25]. *POC5*, locus 5q13.3, was the first unambiguously causative gene that was identified and functionally related to the diseases using a zebrafish (*Danio rerio*) model. The three causative variants of the *POC5* gene (p.A429V, p.A446T, and p.A455P) were found in exon 10, and all three corresponded to the substitution of an alanine to another amino acid, suggesting that the alanine in exon 10 of *POC5* could play an important role in the pathogenesis of AIS. Our study supported the importance of *POC5* variants in the AIS population. In AIS patients from French-Canadian and British families, we found *POC5* variants in 13% of AIS cases, but not all the variants showed a perfect segregation with the disease, highlighting the fact that AIS is a complex disease, and very likely a polygenic disorder. In our cohort of French-Canadian subjects, the previously identified variant in IS cases, namely p.A429V, was the most predominant.

To date, these three variants of *POC5* have not been found in the Chinese population [14]; however, in the cohort of this Chinese study, AIS patients were not screened for their family history. It is important to underline that *POC5* gene variants were first identified in a huge multiplex family in which variants and the disease were transmitted in an autosomal dominant manner through four generations. This was consistent with the hypothesis of the initiating role of POC5 in the development of the familial form of AIS.

POC5, a conserved protein, is essential for centriole assembly, elongation, and cell cycle [15,26]. Recent studies identified POC5 as interacting with POC1B, FAM161A, and centrin-2 to build an inner scaffold with an helicoidal assembly, which provides the structural flexibility and strength to maintain microtubule and ciliary cohesion [27]. It can be hypothesized that those alanines in exon 10 of *POC5* are important for the proper interaction of POC5 with its protein partners, and therefore for the proper ciliary integrity. Several osteogenic pathways are hosted on primary cilia, including Sonic Hedgehog, Wnt, or the calcium-signaling pathway, and could play a part in AIS pathogenesis. More recently, primary cilia have been found to be related to bone-mass reduction through the microtubule (MT) network disorganization caused by the reduction of MT anchorage to the basal body [28]. Altogether, these findings could explain the low bone mineral density observed in AIS patients [29].

5. Conclusions

The *POC5* gene was one of the first pieces of the puzzle of the genetic etiology of AIS, and since its identification, other genes coding for components of the primary cilia have been found to be linked to this disease [12,16]. Our study confirmed a higher prevalence of *POC5* variants in patients with AIS compared to the general population, as Patten et al. [13] already reported, and this reinforces that *POC5* plays a role in the pathogenesis of AIS. The functionality of those variants also was previously reported [13], and was related to ciliary functionality [15]. Further investigations are necessary in order to identify additional genes and finally draw complete pathways that relate the primary cilia to AIS. Finding causative genes for AIS and understanding the molecular consequences of these gene variants is necessary to improve the knowledge about this disease, especially by deciphering the genetic involvement.

Author Contributions: H.M.: methodology, investigation, validation, writing—original draft preparation, writing—review and editing; A.S.: investigation, formal analysis; J.A.A.-M.: recruitment of the patients, funding acquisition; A.C.: recruitment of the patients, funding acquisition, editing; S.B.: recruitment of the patients, writing—original draft preparation; C.F.: recruitment of the patients; S.P.: recruitment of the patients; F.M.: project administration, funding acquisition, visualization, writing—original draft preparation, writing—review and editing. All authors have read and agreed to the published version of the manuscript.

Funding: This study was supported by the Yves Cotrel Foundation (to F.M., S.P., and A.C.), the Scoliosis Research Society (to F.M.), the Marfan Trust UK (to A.C. and J.A.A.-M.), the Peter and Sonia Field Charitable Trust (to A.C.), and St. George's University of London (to A.C. and J.A.A.-M.). This study also was supported by RSBO (Réseau de recherché en Santé Buccodentaire et Osseuse), CHU Sainte-Justine Foundation (to H.M.), and FRQS (to H.M.).

Institutional Review Board Statement: The study was conducted according to the guidelines of the Declaration of Helsinki, and approved by the Centre hospitalier universitaire Sainte-Justine Ethics Committee (protocol code #3704, July 2016).

Informed Consent Statement: Written informed consent was obtained from all subjects involved in this study and from legally authorized representative of minor participants prior to enrollment.

Data Availability Statement: The data presented in this study are available in doi: 10.3390/genes1 2071032.

Acknowledgments: Special thanks to the Clinical Research Orthopaedic Unit (URCO) team, who contributed to the recruitment of the patients. We also thank Philippe Campeau for his help with the exome analysis.

Conflicts of Interest: The authors declare no conflict of interest. The funders had no role in the design of the study; in the collection, analyses, or interpretation of data; in the writing of the manuscript; or in the decision to publish the results.

References

1. Edery, P.; Margaritte-Jeannin, P.; Biot, B.; Labalme, A.; Bernard, J.C.; Chastang, J.; Kassai, B.; Plais, M.H.; Moldovan, F.; Clerget-Darpoux, F. New disease gene location and high genetic heterogeneity in idiopathic scoliosis. *Eur. J. Hum. Genet.* **2011**, *19*, 865–869. [CrossRef] [PubMed]
2. Fadzan, M.; Bettany-Saltikov, J. Etiological theories of adolescent idiopathic scoliosis: past and present. *Open Orthop. J.* **2017**, *11*, 1466–1489. [CrossRef]
3. Konieczny, M.R.; Senyurt, H.; Krauspe, R. Epidemiology of adolescent idiopathic scoliosis. *J. Child. Orthop.* **2013**, *7*, 3–9. [CrossRef]
4. Grivas, T.B.; Vasiliadis, E.; Mouzakis, V.; Mihas, C.; Koufopoulos, G. Association between adolescent idiopathic scoliosis prevalence and age at menarche in different geographic latitudes. *Scoliosis* **2006**, *1*, 9. [CrossRef]
5. Mongird-Nakonieczna, J.; Kozlowski, B. Functional and structural scoliosis in monozygotic twins. *Chir. Narzadow Ruchu Ortop. Pol.* **1976**, *41*, 34–36.
6. Grauers, A.; Rahman, I.; Gerdhem, P. Heritability of scoliosis. *Eur. Spine J.* **2012**, *21*, 1069–1074. [CrossRef] [PubMed]
7. Zaydman, A.M.; Strokova, E.L.; Pahomova, N.Y.; Gusev, A.F.; Mikhaylovskiy, M.V.; Shevchenko, A.I.; Zaidman, M.N.; Shilo, A.R.; Subbotin, V.M. Etiopathogenesis of adolescent idiopathic scoliosis: Review of the literature and new epigenetic hypothesis on altered neural crest cells migration in early embryogenesis as the key event. *Med. Hypotheses* **2021**, *151*, 110585. [CrossRef]
8. Mongird-Nakonieczna, J.; Kozlowski, B. Familial occurrence of idiopathic scoliosis. *Chir. Narzadow Ruchu Ortop. Pol.* **1976**, *41*, 161–165.
9. Miller, N.H. Genetics of familial idiopathic scoliosis. *Clin. Orthop. Relat. Res.* **2007**, *462*, 6–10. [CrossRef] [PubMed]
10. Axenovich, T.I.; Zaidman, A.M.; Zorkoltseva, I.V.; Tregubova, I.L.; Borodin, P.M. Segregation analysis of idiopathic scoliosis: demonstration of a major gene effect. *Am. J. Med. Genet.* **1999**, *86*, 389–394. [CrossRef]
11. Justice, C.M.; Miller, N.H.; Marosy, B.; Zhang, J.; Wilson, A.F. Familial idiopathic scoliosis: evidence of an X-linked susceptibility locus. *Spine* **2003**, *28*, 589–594. [CrossRef]
12. Mathieu, H.; Patten, S.A.; Aragon-Martin, J.A.; Ocaka, L.; Simpson, M.; Child, A.; Moldovan, F. Genetic variant of TTLL11 gene and subsequent ciliary defects are associated with idiopathic scoliosis in a 5-generation UK family. *Sci. Rep.* **2021**, *11*, 11026. [CrossRef]
13. Patten, S.A.; Margaritte-Jeannin, P.; Bernard, J.C.; Alix, E.; Labalme, A.; Besson, A.; Girard, S.L.; Fendri, K.; Fraisse, N.; Biot, B.; et al. Functional variants of POC5 identified in patients with idiopathic scoliosis. *J. Clin. Investig.* **2015**, *125*, 1124–1128. [CrossRef] [PubMed]
14. Xu, L.; Sheng, F.; Xia, C.; Li, Y.; Feng, Z.; Qiu, Y.; Zhu, Z. Common variant of POC5 is associated with the susceptibility of adolescent idiopathic scoliosis. *Spine* **2018**, *43*, E683–E688. [CrossRef]
15. Hassan, A.; Parent, S.; Mathieu, H.; Zaouter, C.; Molidperee, S.; Bagu, E.T.; Barchi, S.; Villemure, I.; Patten, S.A.; Moldovan, F. Adolescent idiopathic scoliosis associated POC5 mutation impairs cell cycle, cilia length and centrosome protein interactions. *PLoS ONE* **2019**, *14*, e0213269. [CrossRef] [PubMed]
16. Oliazadeh, N.; Gorman, K.F.; Eveleigh, R.; Bourque, G.; Moreau, A. Identification of elongated primary cilia with impaired mechanotransduction in idiopathic scoliosis patients. *Sci. Rep.* **2017**, *7*, 44260. [CrossRef] [PubMed]
17. Grimes, D.T.; Boswell, C.W.; Morante, N.F.; Henkelman, R.M.; Burdine, R.D.; Ciruna, B. Zebrafish models of idiopathic scoliosis link cerebrospinal fluid flow defects to spine curvature. *Science* **2016**, *352*, 1341–1344. [CrossRef]
18. Laura, M.-H.; Yasmine, C.-B.; Raphaël, P.; Lotfi, S.; Hugues, P.-M. The orthopedic characterization of cfap298tm304 mutants validate zebrafish to faithfully model human AIS. *Sci. Rep.* **2021**, *11*, 7392. [CrossRef]
19. Bearce, E.A.; Grimes, D.T. On being the right shape: Roles for motile cilia and cerebrospinal fluid flow in body and spine morphology. *Semin. Cell Dev. Biol.* **2021**, *110*, 104–112. [CrossRef] [PubMed]
20. Wang, K.; Li, M.; Hakonarson, H. ANNOVAR: Functional annotation of genetic variants from high-throughput sequencing data. *Nucleic Acids Res.* **2010**, *38*, e164. [CrossRef]
21. Desvignes, J.P.; Bartoli, M.; Delague, V.; Krahn, M.; Miltgen, M.; Beroud, C.; Salgado, D. VarAFT: A variant annotation and filtration system for human next generation sequencing data. *Nucleic Acids Res.* **2018**, *46*, W545–W553. [CrossRef]

22. Exome Variant Server, NHLBI GO Exome Sequencing Project (ESP), Seattle, WA, USA. Available online: http://evs.gs.washington.edu/EVS/) (accessed on 21 March 2020).

23. Karczewski, K.J.; Francioli, L.C.; Tiao, G.; Cummings, B.B.; Alföldi, J.; Wang, Q.; Collins, R.L.; Laricchia, K.M.; Ganna, A.; Birnbaum, D.P.; et al. The mutational constraint spectrum quantified from variation in 141,456 humans. *Nature* **2020**, *581*, 434–443. [CrossRef] [PubMed]

24. Miller, N.H.; Justice, C.M.; Marosy, B.; Doheny, K.F.; Pugh, E.; Zhang, J.; Dietz, H.C., 3rd; Wilson, A.F. Identification of candidate regions for familial idiopathic scoliosis. *Spine* **2005**, *30*, 1181–1187. [CrossRef] [PubMed]

25. Chan, V.; Fong, G.C.; Luk, K.D.; Yip, B.; Lee, M.K.; Wong, M.S.; Lu, D.D.; Chan, T.K. A genetic locus for adolescent idiopathic scoliosis linked to chromosome 19p13.3. *Am. J. Hum. Genet.* **2002**, *71*, 401–406. [CrossRef] [PubMed]

26. Azimzadeh, J.; Hergert, P.; Delouvee, A.; Euteneuer, U.; Formstecher, E.; Khodjakov, A.; Bornens, M. hPOC5 is a centrin-binding protein required for assembly of full-length centrioles. *J. Cell Biol.* **2009**, *185*, 101–114. [CrossRef]

27. Le Guennec, M.; Klena, N.; Gambarotto, D.; Laporte, M.H.; Tassin, A.M.; van den Hoek, H.; Erdmann, P.S.; Schaffer, M.; Kovacik, L.; Borgers, S.; et al. A helical inner scaffold provides a structural basis for centriole cohesion. *Sci. Adv.* **2020**, *6*, eaaz4137. [CrossRef] [PubMed]

28. Shi, W.; Zhang, Y.; Chen, K.; He, J.; Feng, X.; Wei, W.; Hua, J.; Wang, J. Primary cilia act as microgravity sensors by depolymerizing microtubules to inhibit osteoblastic differentiation and mineralization. *Bone* **2020**, *136*, 115346. [CrossRef] [PubMed]

29. Beling, A.; Hresko, M.T.; DeWitt, L.; Miller, P.E.; Pitts, S.A.; Emans, J.B.; Hedequist, D.J.; Glotzbecker, M.P. Vitamin D levels and pain outcomes in adolescent idiopathic scoliosis patients undergoing spine fusion. *Spine Deform.* **2021**. [CrossRef]

genes

MDPI

The Mutational Landscape of *PTK7* in Congenital Scoliosis and Adolescent Idiopathic Scoliosis

Zhe Su [1,2,†], Yang Yang [1,3,4,†], Shengru Wang [1,3,4], Sen Zhao [1,4], Hengqiang Zhao [1,2,3], Xiaoxin Li [3,4,5], Yuchen Niu [3,4,5], Deciphering Disorders Involving Scoliosis and COmorbidities (DISCO) Study Group [‡], Guixing Qiu [1,3,4], Zhihong Wu [3,4,5], Nan Wu [1,3,4] and Terry Jianguo Zhang [1,3,4,*]

[1] State Key Laboratory of Complex Severe and Rare Diseases, Department of Orthopedic Surgery, Peking Union Medical College Hospital, Peking Union Medical College and Chinese Academy of Medical Sciences, Beijing 100730, China; suesu0092@126.com (Z.S.); kaido137@163.com (Y.Y.); wangshengru@foxmail.com (S.W.); zhaosen830@163.com (S.Z.); zhaohq921@gmail.com (H.Z.); qiuguixingpumch@126.com (G.Q.); dr.wunan@pumch.cn (N.W.)
[2] Graduate School, Peking Union Medical College, Beijing 100005, China
[3] Beijing Key Laboratory for Genetic Research of Skeletal Deformity, Beijing 100730, China; ldx217@yeah.net (X.L.); nhtniuyuchen@126.com (Y.N.); wuzh3000@126.com (Z.W.)
[4] Key Laboratory of Big Data for Spinal Deformities, Chinese Academy of Medical Sciences, Beijing 100730, China
[5] Medical Research Center, State Key Laboratory of Complex Severe and Rare Diseases, Peking Union Medical College Hospital, Peking Union Medical College and Chinese Academy of Medical Sciences, Beijing 100730, China
* Correspondence: jgzhang_pumch@yahoo.com
† These authors contributed equally to this work.
‡ Membership of the DISCO Study Group is provided in the Acknowledgments.

Citation: Su, Z.; Yang, Y.; Wang, S.; Zhao, S.; Zhao, H.; Li, X.; Niu, Y.; Deciphering Disorders Involving Scoliosis and COmorbidities (DISCO) Study Group; Qiu, G.; Wu, Z.; et al. The Mutational Landscape of *PTK7* in Congenital Scoliosis and Adolescent Idiopathic Scoliosis. *Genes* **2021**, *12*, 1791. https://doi.org/10.3390/genes12111791

Academic Editors: Philip Giampietro, Nancy Hadley-Miller, Cathy L. Raggio and Donato Gemmati

Received: 11 July 2021
Accepted: 8 November 2021
Published: 12 November 2021

Publisher's Note: MDPI stays neutral with regard to jurisdictional claims in published maps and institutional affiliations.

Abstract: Depletion of *ptk7* is associated with both congenital scoliosis (CS) and adolescent idiopathic scoliosis (AIS) in zebrafish models. However, only one human variant of *PTK7* has been reported previously in a patient with AIS. In this study, we systemically investigated the variant landscape of *PTK7* in 583 patients with CS and 302 patients with AIS from the Deciphering Disorders Involving Scoliosis and COmorbidities (DISCO) study. We identified a total of four rare variants in CS and four variants in AIS, including one protein truncating variant (c.464_465delAC) in a patient with CS. We then explored the effects of these variants on protein expression and sub-cellular location. We confirmed that the c.464_465delAC variant causes loss-of-function (LoF) of *PTK7*. In addition, the c.353C>T and c.2290G>A variants identified in two patients with AIS led to reduced protein expression of PTK7 as compared to that of the wild type. In conclusion, LoF and hypomorphic variants are associated with CS and AIS, respectively.

Keywords: protein tyrosine kinase 7 (*PTK7*); congenital scoliosis; adolescent idiopathic scoliosis; whole exome sequencing

1. Introduction

Protein tyrosine kinase (*PTK7*), also known as colon carcinomakinase-4 (*CCK-4*), is an evolutionarily conserved atypical receptor tyrosine kinase. The PTK7 protein consists of seven extracellular immunoglobulin (Ig)-like domains, one transmembrane domain and one catalytically inactive kinase domain [1]. *PTK7* plays an important role in vertebrate canonical and non-canonical Wnt (planar cell polarity, PCP), Semaphorin/Plexin and vascular endothelial growth factor (VEGF) signaling pathways [2–5]. These signaling pathways are important for embryonic developmental processes, including tissue specification, axial morphogenesis, formation of the cardiovascular, endocrine and immune systems, and regulation of neural crest migration and tumorigenesis [6–10].

Zebrafish models depleted of *ptk7* presented various spinal curve phenotypes. Maternal zygotic *ptk7* (MZ*ptk7*) and zygotic *ptk7* (Z*ptk7*) mutant zebrafish develop spinal curvatures that model congenital scoliosis (CS) and adolescent idiopathic scoliosis (AIS), respectively, due to differential timing of ptk7 loss-of-function [2,11]. Meanwhile, a novel sequence variant $PTK7^{P545A}$ has been reported in a patient with AIS, but without further in vitro investigation [11]. The association between human *PTK7* variants and scoliotic phenotypes continue to be understudied.

In this study, we analyzed variants in *PTK7* identified in a mixed cohort of patients with congenital scoliosis and adolescent idiopathic scoliosis, then performed in vitro experiments to determine the effects of these variants on protein expression and sub-cellular location.

2. Materials and Methods

2.1. Human Subjects

A total of 885 Han Chinese individuals who received a diagnosis of congenital scoliosis (CS, $n = 583$) and adolescent idiopathic scoliosis (AIS, $n = 302$) were recruited between 2009 and 2018 at Peking Union Medical College Hospital (PUMCH) for the Deciphering disorders Involving Scoliosis and COmorbidities (DISCO, http://discostudy.org/, accessed on 1 November 2021) project. Clinical manifestations, physical examination results, and detailed medical histories were obtained with the patients' informed consent. Clinical diagnoses were confirmed by radiology imaging, including X-ray and computed tomography (CT). The criteria for the diagnosis of congenital scoliosis and adolescent idiopathic scoliosis were as follow: congenital scoliosis was caused by vertebral defects, and may be associated with rib anomalies, while idiopathic scoliosis was diagnosed by spinal curvature exceeding $10°$ on a plain antero-posterior X-ray image, with no other identifiable underlying disease. For a diagnosis of adolescent idiopathic scoliosis, patients were required to have an onset age of 10–18 years old. All radiographic evaluations were conducted by trained spine surgeons, while clinical reviews were performed by alternate observers blinded to the radiographic assessment. Patients with a prior molecular diagnosis such as a disease-causing genetic variant were excluded. Blood was obtained from all subjects and whole-exome sequencing (WES) was performed.

Written informed consent for both clinical data and the genetic exome sequencing was obtained from each participant prior to study participation. This study was approved by the Department of Scientific Research and Ethics Committee of PUMCH in China.

2.2. Bioinformatic Analysis and Variant Interpretation

Two DNA extraction and purification kits, Red Blood Cell (RBC) Lysis Buffer (R1010, Solarbio) and Circulating Nucleic Acid Kit (55114, Qiagen), were used in accordance with the manufacturers' protocols. Approximately 4 mL of peripheral blood was transferred to an Eppendorf safe lock tube after sufficient centrifugation. 10 mL RBC lysis solution was added to each centrifuge tube for efficient lysis. 4.5 mL cell lysis solution and 250 µL proteinase K solution were added to each tube and placed at 56.5 °C constant temperature shaker digestion overnight. 1.5 mL protein precipitation solution was added to each tube and allowed to incubate for 10 min at -20 °C. After centrifugation, the supernatant was taken, and 7 mL precooled isopropanol was added into the supernatant until floccule was precipitated. Finally, 1 mL of 75% ethanol was used to wash the DNA pellet after inverting the tube several times, followed by centrifugation at $17,000 \times g$ for 10 min. The quality and quantity of the DNA was evaluated using a spectrophotometer (NanoPhotometer Pearl, Denville Scientific, Inc., Holliston, MA, USA) and fluorometer (Qubit® dsDNA High Sensitivity and dsDNA Broad Range assay, Life Technologies Corporation, Waltham, MA, USA). DNA samples were prepared in Illumina libraries and then underwent whole-exome capture with the SureSelect Human All Exon V6 + UTR r2 core design (91 Mb, Agilent, Santa Clara, CA, USA), followed by sequencing on the Illumina HiSeq 4000 platform in 150-bp paired-end reads mode (Illumina, San Diego, CA, USA). WES data processing was performed with the Peking Union Medical College Hospital Pipeline (PUMP) based on the reference genome GRCh37-v1.6 [12,13]. Combined Annotation Dependent Depletion (CADD PHRED-score) [14] and Polyphen-2 [15] were used to predict the pathogenicity of candidate variants. Genotype was filtered for read-depth (DP > 10×), genotype quality (GQ > 20), quality by depth (QD < 2), strand odds ratio (SOR > 9), and allele balance (AB > 0.25). The populational frequency of each QC-passed variant was obtained from the public population databases, including the 1000 Genomes Project, the Exome Sequencing Project [16], the Genome Aggregation Database (gnomAD) [17], and the in-house database of DISCO (Deciphering disorders Involving Scoliosis and COmorbidities, http://discostudy.org/, ≈8000 exomes/genomes, accessed on 1 November 2021) study. Rare variants (minor allele frequency < 0.001) were retained for further filtering. From these rare variants, we included the protein-altering or splice-region variants for subsequent analysis. Potential spicing variants were predicted using SpliceAI [18].

Candidate variants of *PTK7* were selected based on the following criteria:

(1)   Predicted to alter the protein sequence;
(2)   Either absent or with a low frequency (<0.001) from the public database mentioned above.
(3)   The missense variants have a CADD score ≥15.

2.3. Site-Directed Mutagenesis

pcDNA3.1+ with C-terminal flag-tagged wild type (WT) and variant *PTK7* cDNA (NM_152881.3) plasmids were acquired from Beijing Hitrobio Biotechnology (Beijing, China). The variant constructs were sequenced on both strands to verify nucleotide changes.

2.4. Cell Culture and Transfection Assay

HEK293T cells were cultured in Dulbecco's Modified Eagle's medium (DMEM, Invitrogen, Carlsbad, CA, USA) supplemented with 10% fetal bovine serum (Gibco, Waltham, MA, USA), penicillin (50 U/mL), and streptomycin (50 µg/mL) in six-well plates. HEK293T cells were transfected with full-length WT or variant *PTK7* constructs using Lipofectamine 3000 (Invitrogen).

2.5. Western Blot Assay

After a 48 h transfection, HEK293T cells were harvested in RIPA lysis buffer (Solarbio, Beijing, China) and whole-cell lysate was resolved on gels under reducing conditions, transferred to a Nitrocellulose Transfer (NC) membrane, blocked with non-fat milk for 30 min at room temperature, and probed with primary antibodies: mouse anti-DDK(FLAG) antibody (1:1000, ZSGB-BIO, TA-05) and mouse anti-β-actin antibody (1:5000, Proteintech, 66009-1-Ig) over-night at 4 °C, and then with a goat anti-mouse horseradish peroxidase-conjugated secondary antibody (1:5000, ZSGB-BIO, ZB-2305). Immunoreactivity was visualized using Western Lighting chemiluminescence reagent (Beyotime, Shanghai, China). All Western blotting experiments were repeated three times.

2.6. Immuno-Fluorescence

Cells were grown on 35 mm glass bottom cell culture dishes for 48 h after transfection, washed three times, fixed in 4% paraformaldehyde, permeabilized with 0.1% Triton X-100, and incubated with primary anti-DDDDK-tag mouse antibody (1:1000, MBL, M185-3L) at 4 °C overnight. Cells were incubated and stained with secondary antibody Alexa FlourTM 488 goat anti-mouse IgG (1:2000, Invitrogen, 1911843), then covered with DAPI (Solarbio, Beijing, China).

2.7. Statistical Analysis

The overall protein expression levels were normalized to WT (set as 1.0) and mean values of variant versus WT from all three experiments were compared using unpaired *t*-test. All charts are drawn and analyzed using GraphPad Prism 8 and $p < 0.05$ was considered significant for all analyses.

3. Results

A total of 885 genomes from patients with scoliosis were sequenced and eight *PTK7* variants in nine patients were found. The mean age of the included nine patients with variants was 11.11 ± 5.51 years. In five CS patients, the mean Cobb angle of the coronal plane was 59.94° ± 25.85°. Among them, three patients displayed kyphosis with a mean angle of 50.53° ± 8.05°. The mean Cobb angle of structural curve in four AIS patients was 48.95° ± 4.77°. In the CS group (*n* = 583), four possibly deleterious variants were revealed in five patients, including one frameshift variant and three missense variants (c.464_465delAC, c.1394A>G, c.1879G>A, c. 1955G>T) (Table 1). One of the missense variants (c.1955G>T) was identified in two patients. In the AIS group (*n* = 302), four deleterious missense variants (c.49C>T, c.353C>T, c.2290G>A, c.2384G>A) were identified (Figure 1A, Table 2). No peripheral blood samples from the patients' families were obtained, and no similar family history of spinal deformity was found after follow-up.

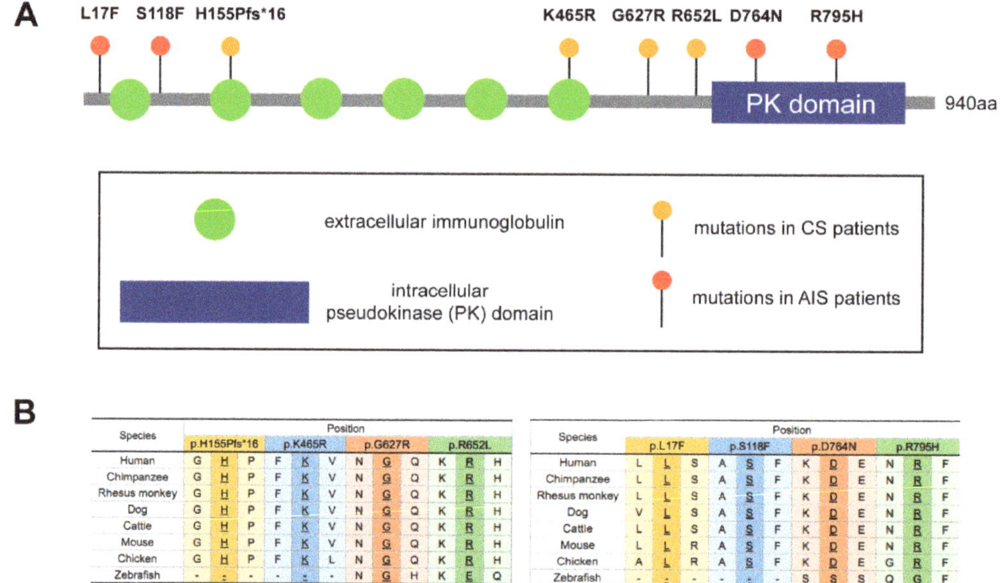

Species	Position											
	p.H155Pfs*16			p.K465R			p.G627R			p.R652L		
Human	G	H	P	F	K	V	N	G	Q	K	R	H
Chimpanzee	G	H	P	F	K	V	N	G	Q	K	R	H
Rhesus monkey	G	H	P	F	K	V	N	G	Q	K	R	H
Dog	G	H	P	F	K	V	N	G	Q	K	R	H
Cattle	G	H	P	F	K	V	N	G	Q	K	R	H
Mouse	G	H	P	F	K	V	N	G	Q	K	R	H
Chicken	G	H	P	F	K	L	N	G	Q	K	R	H
Zebrafish	-	-	-	-	-	-	N	G	H	K	E	Q

Species	Position											
	p.L17F			p.S118F			p.D764N			p.R795H		
Human	L	L	S	A	S	F	K	D	E	N	R	F
Chimpanzee	L	L	S	A	S	F	K	D	E	N	R	F
Rhesus monkey	L	L	S	A	S	F	K	D	E	N	R	F
Dog	V	L	S	A	S	F	K	D	E	N	R	F
Cattle	L	L	S	A	S	F	K	D	E	N	R	F
Mouse	L	L	R	A	S	F	K	D	E	N	R	F
Chicken	A	L	R	A	S	F	K	D	E	G	R	F
Zebrafish	-	-	-	-	-	-	S	S	S	Q	G	F

Figure 1. Mapping and conservation analysis of the identified variants. (**A**) Mapping of eight *PTK7* variants revealed that p.L17F, p.S118F, p.H155Pfs*16 and p.K465R are located in the extracellular region of PTK7 protein. p.D764N and p.R795H are located in the intracellular pseudokinase domain. (**B**) The conservation of variants in human and other vertebrate species.

3.1. Variant and Phenotypic Characteristics

Patient SCO1905P0038 is a 13-year-old male with T12 butterfly vertebra and T9-T10 segmentation defect. The spinal plain radiograph shows not only a coronal curve to the left, but also has a severe thoracolumbar kyphosis in the sagittal plane with a 65° Cobb angle, both results of continuous deformity in the vertebral body (Figure 2A). The patient has a heterozygous deletion between nucleotide 464 and 465 (c.464_465delAC, p.H155Pfs*16). This frameshift variant was mapped to the extracellular immunoglobulin region of PTK7 protein (Figure 1A) and is predicted to cause the early termination of mRNA translation. It is a novel variant, previously undescribed in mutational databases and is highly conserved across different vertebral species except zebrafish (Figure 1B).

Table 1. Summary of clinical features of the *PTK7* variant carriers.

Patient Number	Variant	Gender	Age of Onset	Diagnosis	Vertebral Malformations	Other skeletal Malformations or Deformities	Cardiac Abnormalities	Spinal Canal Deformities	Visceral Abnormalities
SCO1905P0038	c.464_465delAC	M	13	CS	T12 butterfly vertebra, T9-T10 failure of segmentation	None	None	None	None
SCO2003P2127	c.1394A>G	F	21	CS	None	Fusion of 4th and 5th ribs, 12th ribs absent	Postoperative of patent ductus arteriosus	None	None
SCO2003P0372	c.1879G>A	F	8	CS	L2 Hemivertebra, L2-L3 fusion	None	None	None	None
SCO1908P0053	c.1955G>T	F	1	CS	T11, T12 segmented wedge vertebrae	None	None	None	None
SCO2003P0541	c.1955G>T	F	7	CS	T7-T11 failure of segmentation	Chest deformity	None	Diastematomyelia	None
SCO1907P0150	c.49C>T	F	11	AIS	None	None	Ventricular septal defect	None	None
SCO2003P0632	c.353C>T	F	12	AIS	None	None	None	None	None
SCO2003P2288	c.2290G>A	F	14	AIS	None	None	None	None	None
SCO2003P2237	c.2384G>A	F	13	AIS	None	None	None	None	None

M, male; F, female; CS, congenital scoliosis; AIS, adolescent idiopathic scoliosis.

Table 2. Sequence variants identified in *PTK7*.

Patient Number	cDNA Change	Protein Change	Variant Type	Position	Gnomad_Exome_ALL	Gnomad_Exome_EAS	Gnomad_Genome_ALL	Gnomad_Genome_EAS	CADD	Ployphen-2 HDV Score	Evidence of Pathogenicity by ACMG	ACMG Classification
SCO1905P0038	c.464_465delAC	p.H155Pfs*16	Frameshift	43097560	0	0	0	0	NA	NA	PVS1+PM2+PP3	Pathogenic
SCO2003P2127	c.1394A>G	p.K465R	Missense	43109684	0	0	0.00003247	0.0006	26.2	0.997	PP3	Uncertain significance
SCO2003P0372	c.1879G>A	p.G627R	Missense	43112206	0.00004873	0.0006	0.00003232	0.0006	26.9	0.986	PP3	Uncertain significance
SCO1908P0053 SCO2003P0541	c.1955G>T	p.R652L	Missense	43112282	0.00007718	0.0005	0	0	23.6	0.947	PP3	Uncertain significance
SCO1907P0150	c.49C>T	p.L17F	Missense	43044275	0	0	0	0	19.95	0.997	PM2	Uncertain significance
SCO2003P0632	c.353C>T	p.S118F	Missense	43096988	0	0	0	0	25.6	0.982	PM2	Uncertain significance
SCO2003P2288	c.2290G>A	p.D764N	Missense	43114395	0.00000814	0.000058	0	0	23.8	0.114	PM1+PP3	Uncertain significance
SCO2003P2237	c.2384G>A	p.R795H	Missense	43126607	0.00000406	0	0.00003231	0	30	0.986	PM1+ +PP3	Uncertain significance

The RefSeq transcript sequence used for PTK7 is NM_152881.3. Abbreviations: NA, not available. ACMG, American College of Medical Genetics and Genomics. PVS, pathogenic very strong. PM, pathogenic moderate. PP, supporting pathogenicity.

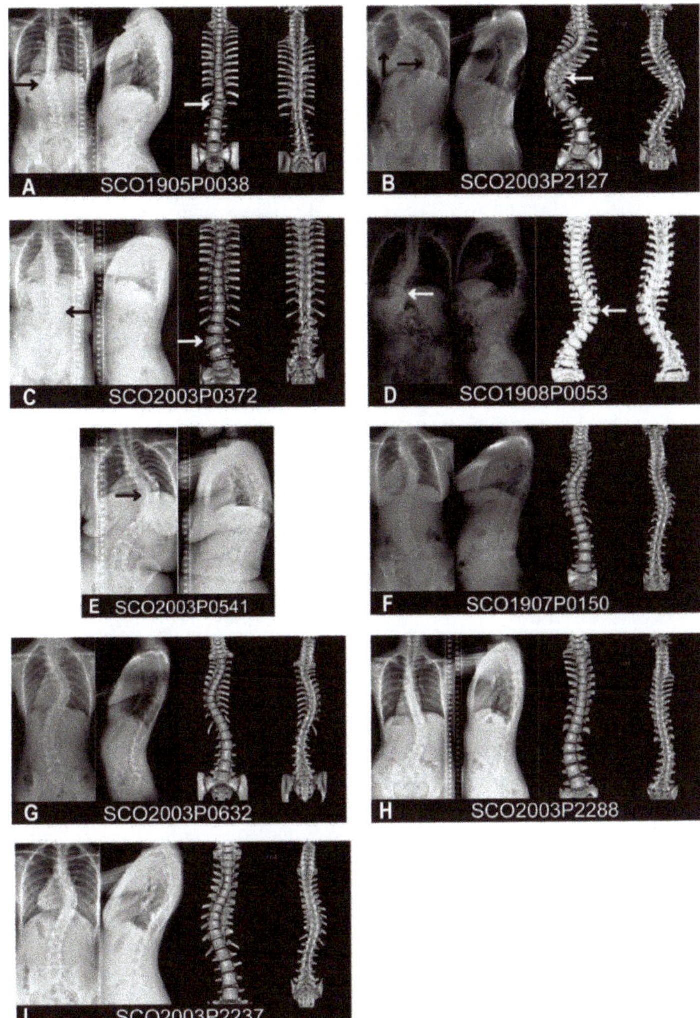

Figure 2. Antero-posterior, lateral spinal X-ray and the spinal three-dimensional CT reconstruction of patient SCO1905P0038 (**A**), SCO2003P2127 (**B**), SCO2003P0372 (**C**), SCO1908P0053 (**D**), SCO1907P0150 (**F**), SCO2003P0632 (**G**), SCO2003P2288 (**H**), SCO2003P2237 (**I**). Antero-posterior, lateral spinal X-ray of patient SCO2003P0541 (**E**). Arrowheads point to the vertebral or rib deformities.

Patient SCO2003P2127 is a 21-year-old female with a diagnosis of congenital scoliosis. This patient presents with the failure of segmentation in the concave side of T9-T11, fusion of the 4th and 5th ribs, and absence of 12th ribs. As a result, the patient has a severe imbalance of the spine in the coronal plane, with a 92° Cobb angle of the main curve at T5-L1 and a compensatory curve of 45° at the lumbar level (Figure 2B). She also has a history of patent ductus arteriosus. The heterozygous missense variant c.1394A>G (p.K465R) of the *PTK7* gene is mapped to the extracellular immunoglobulin region of the PTK7 protein (Figure 1A). This variant is previously unreported in mutational databases and is highly conserved (Figure 1B). It is predicted by CADD and PolyPhen-2 to be deleterious.

Patient SCO2003P0372 is an 8-year-old female with L2 hemivertebrae and L2-L3 segmentation defect. Her spine has a 38° right curve at the lumbar region with the hemivertebrae as its apex in the coronal plane, and a slight compensatory curve at the thoracic region (Figure 2C). She has a heterozygous missense variant c.1879G>A (p.G627R) of *PTK7*. This variant is located in the intracellular domain and adjacent to the transmembrane region of the PTK7 protein (Figure 1A). This variant is highly conserved (Figure 1B) and has been reported in the gnomAD database with low frequency. It is predicted by CADD and PolyPhen-2 to be deleterious.

Patient SCO1908P0053 is a 1-year-old female detected segmented wedge vertebrae in T11 and T12. This child displays severe imbalance in both the coronal and sagittal planes, in which the left curve reached 78° with the deformed vertebral body as the apex in the coronal plane, and severe kyphosis of 48° in the thoracolumbar region in the sagittal plane (Figure 2D). Due to her young age, long segment involvement, and the combination of thoracolumbar scoliosis and kyphosis, she received growth rod implantation and repeated growth rod extension processes over the past few years. Patient SCO2003P0541 is a 7-year-old female with segmentation failure from T7 to T11. Although the spinal deformity was discovered at the age of 7, the patient did not receive proper treatment until adulthood, leading to the continued progression of kyphoscoliosis. Similar to the patient SCO1908P0053, this patient has severe imbalance on both the coronal and sagittal planes in the thoracic spine, with a right curve of 87° and a thoracic kyphosis of 62°, which compared with the normal physiological thoracic kyphosis ranged from 10° to 40° (Figure 2E). She also suffered from chest deformity and diastematomyelia from MRI scans, which is consistant with the reported association between *PTK7* variants and neural tube defects (NTDs) [19]. The patient showed no dyspnea, sensory or motor disorders in the lower limbs, nor did she show urinal or excretory dysfunction. These two individuals carry the same heterozygous missense variant c. 1955G>T (p.R652L). This variant is located in the intracellular domain of PTK7, close to the pseudokinase (PK) domain (Figure 1A). This variant is conserved across vertebral species besides zebrafish (Figure 1B), and has been reported in gnomAD database with low frequency.

Patients SCO1907P0150, SCO2003P0632, SCO2003P2288 and SCO2003P2237 all suffer from adolescent idiopathic scoliosis (AIS). Patient SCO1907P0150 is an 11-year-old female. The patient has two curves on the coronal plane, a 47° left curve in the upper thoracic segment (T2-T6) and a 51° right curve in the thoracic segment (T6-T12) (Figure 2F). She had a congenital ventricular septal defect that was surgically treated. WES analysis reveals a missense variant c. 49C>T (p.L17F) in *PTK7*, which has not been reported in the mutational databases. This variant is located at the beginning of the extracellular portion of the PTK7 protein (Figure 1A). It is predicted by CADD and PolyPhen-2 score to be deleterious.

Patient SCO2003P0632 displays a 50° right curve and a 42° left curve at the thoracic and thoracolumbar levels, respectively, without significant trunk deviation (Figure 2G). A missense variant c. 353C>T (p.S118F) is identified in patient SCO2003P0632, a 12-year-old girl, which is also absent from mutational databases. It is predicted to be located at the junction between the first and second Ig domains of the extracellular part of PTK7 protein (Figure 1A). This variant is predicted by CADD and Polyphen-2 scores to be deleterious.

Patient SCO2003P2288 is a 14-year-old female with a c. 2290G>A (p.D764N) variant in *PTK7*. X-rays showed that the main curve was a 46° right-sided lumbar scoliosis, with a long compensatory curve of 25° at T1-T11 to maintain basic balance of the trunk (Figure 2H). This missense variant has been reported in gnomAD database with low frequency and the altered amino acid is highly conserved in vertebrates except zebrafish (Figure 1B). The variant is mapped to the PK domain of the intracellular portion of the PTK7 protein (Figure 1A). The pathogenicity assessment has contradictory results using CADD and Polyphen-2 scores.

Patient SCO2003P2237 is a 13-year-old girl. On the coronal plane, the patient presents a single thoracic curve from T7 to T12 with a Cobb Angle of 52°. Due to the severe vertebral rotation, a relatively obvious razor-back deformity is seen on the sagittal plane (Figure 2I). WES analysis reveals that she has a missense variant c.2384G>A (p.R795H) in *PTK7*. The variant amino acid is highly conserved in vertebrates (Figure 1B) and mapped to the PK domain of the intracellular section of the PTK7 protein (Figure 1A). Reported in gnomAD database with low frequency, it is predicted by both CADD and Polyphen-2 scores to be deleterious.

3.2. Western Blot and Immunocytochemistry Analyses

To identify the influence of the identified variants on PTK7 protein function, we evaluated overall protein expression and sub-cellular location of the variants of PTK7 compared to the WT. Western blotting analysis identified the immunoreactive-specific band for flag-tagged WT PTK7 at 150kDa and β-actin at 42kDa. The overall expression level was quantified by estimating bands from PNGase-treated samples and normalizing to WT. As anticipated, the overall expression of the frameshift variant (c.464_465delAC, p.H155Pfs*16) was significantly decreased compared to that of WT ($p < 0.0001$) (Figure 3A), indicating the loss-of-function effect of this variant. The expression level of two missense variants (p.S118F and p.D764N) were partially reduced ($p = 0.0061$ and $p = 0.0293$, respectively) (Figure 3A). Interestingly, these two variants were both identified in patients with AIS but not CS. There were no significant differences in the overall protein expression of the other missense variants (p.K465R, p.G627R, p.R562L, p.L17F, p.R795H) compared with the WT.

We also performed immunocytochemistry (ICC) assays and found there was a faintest signal from the frameshift variant of the PTK7 protein (Figure 3B), which supported the results of Western blotting assays. However, the sub-cellular location of all missense variant PTK7 protein did not change compared to WT (Figure 3B).

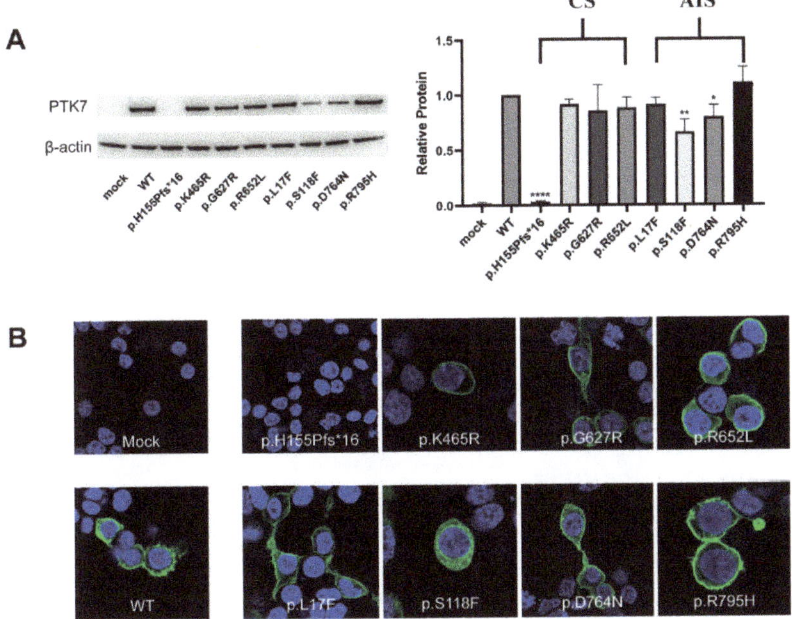

Figure 3. Western blot and ICC analyses. (**A**) HEK293T cells transiently transfected with WT or variant *PTK7* constructs revealed diminished PTK7 protein expression levels of p.H155Pfs*16, p.S118F and p.D764N. * p value < 0.05, ** p value < 0.01, **** p value < 0.0001. (**B**) ICC assays showed a faint signal of the frameshift variant PTK7 protein. No difference in PTK7 protein sub-cellular location was detected.

4. Discussion

Here, we performed WES on the genomes of 885 scoliosis patients, including 583 CS patients and 302 AIS patients, and identified seven missense variants and one frameshift variant in *PTK7*. Loss-of-function of PTK7 resulted in skeletal phenotypes in both zebrafish and mice. In zebrafish models, embryos with the *PTK7^{hsc9}* variant (a mutant allele harboring a 10bp deletion that results in a frameshift and the incorporation of multiple premature termination codons) had defects in the convergence of both neuroectoderm and axial mesoderm tissues, and abnormal three-dimensional spinal curvature with growth and development [2]. In 2016, Grimes et al. reported that mutated *ptk7* zebrafish exhibited spinal curvature as well as hydrocephalus, ependymal cell (EC) ciliary dysfunction and abnormal rate and pattern of the cerebrospinal fluid (CSF) flow [20]. The *chuzhoi* mutant mice embryo showed several congenital abnormalities including neural tube, heart and lung defects caused by the disruption of PTK7 protein expression [21]. In human, LoF mutations in *PTK7* and the double-heterozygous mutation of *PTK7* and *VANGL2* were associated with spina bifida and increased the genetic risk of NTDs [19]. However, only one case of scoliosis with a *PTK7* mutation has been previously reported [11], and our identification of these eight variants expands the variant and phenotypic spectrum of *PTK7*.

The PTK7 protein consists of an extracellular continuous immunoglobulin domain, a transmembrane domain and an intracellular PK domain [1]. PTK7 interacts with several molecules in both canonical and non-canonical Wnt signaling as a co-receptor. These molecules include Wnt ligands, Wnt receptors as well as intracellular components such as Dvls and β-catenin [22]. According previous studies, the extracellular domains and intracellular domains of PTK7 play distinct roles. By constructing and studying *PTK7* with deletion of different domains, Hayes et al. found that the plasma membrane-tethered *Ptk7* extracellular domain (*Ptk7*ΔICD) was sufficient to promote normal PCP as well as the inhibition of canonical Wnt signaling [2], while the intracellular domains may play a specific role in oriented cell division, radial intercalation and cilia orientation [23–26]. The intracellular PK domain of PTK7 may act as a scaffold promoting the binding of intracellular proteins and other receptors. In *Xenopus*, the ptk7 intracellular domain was stimulated by Wnt5A and induced its translocation into the nucleus, which promoted cell and tissue movements [27]. Additionally, the C-terminal PTK7 species up-regulated cadherin-11 expressed in the mesoderm-derived tissues, and which regulates osteogenesis. It is possible that PTK7 and Cadherin-11 might interact in embryogenesis and regulate similar developmental processes [28,29]. In our study, the frameshift variant and three missense variants (p.L17F, p.S118F, and p.K465R) were located in the extracellular domains of the PTK7 protein, while the other four missense variants (p.G627R, p.R652L, p.D764N, and p.R795H) were mapped to the intracellular portion, with p.D764N and p.R795H variants being located in the pseudokinase domain. In our expression assay, one variant in the extracellular domain and one variant in the intracellular domain were shown to alter the expression of *PTK7*, suggesting that both domains are critical for the integrity of PTK7 protein.

Interestingly, the frameshift variant that caused almost no expression of PTK7 was found in the CS patient, while the two missense variants that resulted in the significant decreases in protein levels were both found in patients with AIS. The patient carrying the truncating variant was diagnosed to have a mixed subtype of CS, with failure of formation (T12 butterfly vertebra) and segmental disorder (T9-T10) and no other developmental malformations. The two patients with the missense variants (p.S118F and p.D764N) did not have congenital developmental deformities of the vertebral body or other systems, presenting simple thoracolumbar scoliosis. However, due to the complex etiology of AIS, the role of PTK7 in AIS still warrants further validation and investigation.

Taken together, we hypothesize that the pathogenesis of vertebral deformities and the onset time of spinal curvature may be due to the effect of *PTK7* variants on the protein expression. In other words, congenital and adolescent idiopathic scoliosis may have a common genetic basis. Although some missense variants in our study did not show abnormalities in protein expression or sub-cellular location, it is possible that the altered amino acids can affect the structure of the protein and subsequently affect downstream signaling pathways. Therefore, it is necessary to further explore the possible downstream pathways of PTK7, such as the canonical and non-canonical Wnt pathways.

5. Conclusions

In conclusion, we identified eight *PTK7* variants in our mixed scoliosis cohort, including one frameshift variant (c.464_465delAC) and seven missense variants (c.49C>T, c.353C>T, c.1394A>G, c.1879G>A, c. 1955G>T, c.2290G>A, c.2384G>A). The frameshift variant resulted in a depleted expression of PTK7 protein, and two of missense variants caused reduced expression of PTK7. Our study extended the variant and phenotype spectrums of *PTK7* and suggested a common genetic basis of CS and AIS.

Author Contributions: Conceptualization, Z.S., N.W. and S.Z.; methodology, Z.S., S.Z. and H.Z.; software, S.Z. and Y.N.; validation, Z.S., S.W., S.Z. and X.L.; formal analysis, Z.S. and Y.Y.; investigation, Z.S.; resources, S.W., Y.Y. and Deciphering Disorders Involving Scoliosis and COmorbidities (DISCO) Study Group; data curation, Z.S., S.Z. and H.Z.; writing—original draft preparation, Z.S., Y.Y.; writing—review and editing, S.W., S.Z., X.L. and N.W.; visualization, Z.S.; supervision, N.W., Z.W., G.Q. and T.J.Z.; project administration, N.W., Z.W., G.Q. and T.J.Z.; funding acquisition, N.W., Z.W., G.Q. and T.J.Z. All authors have read and agreed to the published version of the manuscript.

Funding: This research was funded in part by the National Natural Science Foundation of China (81902178 to S.W., 81972037 and 82172382 to J.Z., 81822030 and 82072391 to N.W., 81930068 and 81772299 to Z.W., 81772301 and 81972132 to G.Q.), Beijing Natural Science Foundation (No. L192015 to J.Z., JQ20032 to N.W., 7191007 to Z.W.), Capital's Funds for Health Improvement and Research (2020-4-40114 to N.W.), Tsinghua University-Peking Union Medical College Hospital Initiative Scientific Research Program, Non-profit Central Research Institute Fund of Chinese Academy of Medical Sciences (No. 2019PT320025), and the PUMC Youth Fund & the Fundamental Research Funds for the Central Universities (No. 3332019021 to Shengru W.).

Institutional Review Board Statement: Approval for the study was obtained from the ethics committee at Peking Union Medical College Hospital (JS-098).

Informed Consent Statement: Written informed consent was provided by each participant.

Data Availability Statement: Data are available upon reasonable request. The datasets analyzed during the current study are available from the corresponding author on reasonable request.

Acknowledgments: We thank the patients and their families who participated in this research. We also thank the Nanjing Geneseq Technology Inc. for the technical help in sequencing, the Ekitech Technology Inc. for the technical support in database and data management. The members of the DISCO study group are provided in http://discostudy.org/.

Conflicts of Interest: The authors declare no conflict of interest.

References

1. Mossie, K.; Jallal, B.; Alves, F.; Sures, I.; Plowman, G.D.; Ullrich, A. Colon carcinoma kinase-4 defines a new subclass of the receptor tyrosine kinase family. *Oncogene* **1995**, *11*, 2179–2184.
2. Hayes, M.; Naito, M.; Daulat, A.; Angers, S.; Ciruna, B. Ptk7 promotes non-canonical Wnt/PCP-mediated morphogenesis and inhibits Wnt/beta-catenin-dependent cell fate decisions during vertebrate development. *Development* **2013**, *140*, 1807–1818. [CrossRef]
3. Lee, H.K.; Chauhan, S.K.; Kay, E.; Dana, R. Flt-1 regulates vascular endothelial cell migration via a protein tyrosine kinase-7-dependent pathway. *Blood* **2011**, *117*, 5762–5771. [CrossRef] [PubMed]
4. Lu, X.; Borchers, A.G.; Jolicoeur, C.; Rayburn, H.; Baker, J.C.; Tessier-Lavigne, M. PTK7/CCK-4 is a novel regulator of planar cell polarity in vertebrates. *Nature* **2004**, *430*, 93–98. [CrossRef] [PubMed]
5. Wagner, G.; Peradziryi, H.; Wehner, P.; Borchers, A. PlexinA1 interacts with PTK7 and is required for neural crest migration. *Biochem. Biophys. Res. Commun.* **2010**, *402*, 402–407. [CrossRef] [PubMed]
6. Capparuccia, L.; Tamagnone, L. Semaphorin signaling in cancer cells and in cells of the tumor microenvironment–two sides of a coin. *J. Cell Sci.* **2009**, *122*, 1723–1736. [CrossRef]

7. Gay, C.M.; Zygmunt, T.; Torres-Vazquez, J. Diverse functions for the semaphorin receptor PlexinD1 in development and disease. *Dev. Biol.* **2011**, *349*, 1–19. [CrossRef]

8. Neufeld, G.; Kessler, O. The semaphorins: Versatile regulators of tumour progression and tumour angiogenesis. *Nat. Rev. Cancer* **2008**, *8*, 632–645. [CrossRef]

9. Moon, R.T.; Brown, J.D.; Torres, M. WNTs modulate cell fate and behavior during vertebrate development. *Trends Genet.* **1997**, *13*, 157–162. [CrossRef]

10. Wodarz, A.; Nusse, R. Mechanisms of Wnt signaling in development. *Annu. Rev. Cell Dev. Biol.* **1998**, *14*, 59–88. [CrossRef]

11. Hayes, M.; Gao, X.; Yu, L.X.; Paria, N.; Henkelman, R.M.; Wise, C.A.; Ciruna, B. ptk7 mutant zebrafish models of congenital and idiopathic scoliosis implicate dysregulated Wnt signalling in disease. *Nat. Commun.* **2014**, *5*, 4777. [CrossRef]

12. Zhao, S.; Zhang, Y.; Chen, W.; Li, W.; Wang, S.; Wang, L.; Zhao, Y.; Lin, M.; Ye, Y.; Lin, J.; et al. Diagnostic yield and clinical impact of exome sequencing in early-onset scoliosis (EOS). *J. Med. Genet.* **2021**, *58*, 41–47. [CrossRef]

13. Chen, N.; Zhao, S.; Jolly, A.; Wang, L.; Pan, H.; Yuan, J.; Chen, S.; Koch, A.; Ma, C.; Tian, W.; et al. Perturbations of genes essential for Mullerian duct and Wolffian duct development in Mayer-Rokitansky-Kuster-Hauser syndrome. *Am. J. Hum. Genet.* **2021**, *108*, 337–345. [CrossRef]

14. Kircher, M.; Witten, D.M.; Jain, P.; O'Roak, B.J.; Cooper, G.M.; Shendure, J. A general framework for estimating the relative pathogenicity of human genetic variants. *Nat. Genet.* **2014**, *46*, 310–315. [CrossRef] [PubMed]

15. Adzhubei, I.A.; Schmidt, S.; Peshkin, L.; Ramensky, V.E.; Gerasimova, A.; Bork, P.; Kondrashov, A.S.; Sunyaev, S.R. A method and server for predicting damaging missense mutations. *Nat. Methods* **2010**, *7*, 248–249. [CrossRef] [PubMed]

16. Fu, W.; O'Connor, T.D.; Jun, G.; Kang, H.M.; Abecasis, G.; Leal, S.M.; Gabriel, S.; Rieder, M.J.; Altshuler, D.; Shendure, J.; et al. Analysis of 6,515 exomes reveals the recent origin of most human protein-coding variants. *Nature* **2013**, *493*, 216–220. [CrossRef]

17. Karczewski, K.J.; Francioli, L.C.; Tiao, G.; Cummings, B.B.; Alföldi, J.; Wang, Q.; Collins, R.L.; Laricchia, K.M.; Ganna, A.; Birnbaum, D.P.; et al. The mutational constraint spectrum quantified from variation in 141,456 humans. *Nature* **2020**, *581*, 434–443. [CrossRef]

18. Jaganathan, K.; Kyriazopoulou Panagiotopoulou, S.; McRae, J.F.; Darbandi, S.F.; Knowles, D.; Li, Y.I.; Kosmicki, J.A.; Arbelaez, J.; Cui, W.; Schwartz, G.B.; et al. Predicting Splicing from Primary Sequence with Deep Learning. *Cell* **2019**, *176*, 535–548. [CrossRef] [PubMed]

19. Lei, Y.; Kim, S.E.; Chen, Z.; Cao, X.; Zhu, H.; Yang, W.; Shaw, G.M.; Zheng, Y.; Zhang, T.; Wang, H.Y.; et al. Variants identified in PTK7 associated with neural tube defects. *Mol. Genet. Genom. Med.* **2019**, *7*, e00584. [CrossRef] [PubMed]

20. Grimes, D.T.; Boswell, C.W.; Morante, N.F.; Henkelman, R.M.; Burdine, R.D.; Ciruna, B. Zebrafish models of idiopathic scoliosis link cerebrospinal fluid flow defects to spine curvature. *Science* **2016**, *352*, 1341–1344. [CrossRef] [PubMed]

21. Paudyal, A.; Damrau, C.; Patterson, V.L.; Ermakov, A.; Formstone, C.; Lalanne, Z.; Wells, S.; Lu, X.; Norris, D.P.; Dean, C.H.; et al. The novel mouse mutant, chuzhoi, has disruption of Ptk7 protein and exhibits defects in neural tube, heart and lung development and abnormal planar cell polarity in the ear. *BMC Dev. Biol.* **2010**, *10*, 87. [CrossRef]

22. Berger, H.; Wodarz, A.; Borchers, A. PTK7 Faces the Wnt in Development and Disease. *Front. Cell Dev. Biol.* **2017**, *5*, 31. [CrossRef] [PubMed]

23. Gong, Y.; Mo, C.; Fraser, S.E. Planar cell polarity signalling controls cell division orientation during zebrafish gastrulation. *Nature* **2004**, *430*, 689–693. [CrossRef]

24. Roszko, I.; Sawada, A.; Solnica-Krezel, L. Regulation of convergence and extension movements during vertebrate gastrulation by the Wnt/PCP pathway. *Semin. Cell Dev. Biol.* **2009**, *20*, 986–997. [CrossRef] [PubMed]

25. Happe, H.; de Heer, E.; Peters, D.J. Polycystic kidney disease: The complexity of planar cell polarity and signaling during tissue regeneration and cyst formation. *Biochim. Biophys. Acta* **2011**, *1812*, 1249–1255. [CrossRef]

26. Jessen, J.R.; Topczewski, J.; Bingham, S.; Sepich, D.S.; Marlow, F.; Chandrasekhar, A.; Solnica-Krezel, L. Zebrafish trilobite identifies new roles for Strabismus in gastrulation and neuronal movements. *Nat. Cell Biol.* **2002**, *4*, 610–615. [CrossRef] [PubMed]

27. Golubkov, V.S.; Strongin, A.Y. Downstream signaling and genome-wide regulatory effects of PTK7 pseudokinase and its proteolytic fragments in cancer cells. *Cell Commun. Signal.* **2014**, *12*, 15. [CrossRef]

28. Koehler, A.; Schlupf, J.; Schneider, M.; Kraft, B.; Winter, C.; Kashef, J. Loss of Xenopus cadherin-11 leads to increased Wnt/beta-catenin signaling and up-regulation of target genes c-myc and cyclin D1 in neural crest. *Dev. Biol.* **2013**, *383*, 132–145. [CrossRef]

29. Martinez, S.; Scerbo, P.; Giordano, M.; Daulat, A.M.; Lhoumeau, A.C.; Thome, V.; Kodjabachian, L.; Borg, J.P. The PTK7 and ROR2 Protein Receptors Interact in the Vertebrate WNT/Planar Cell Polarity (PCP) Pathway. *J. Biol. Chem.* **2015**, *290*, 30562–30572. [CrossRef] [PubMed]

MDPI
St. Alban-Anlage 66
4052 Basel
Switzerland
Tel. +41 61 683 77 34
Fax +41 61 302 89 18
www.mdpi.com

Genes Editorial Office
E-mail: genes@mdpi.com
www.mdpi.com/journal/genes